Fluid Power Systems

FLUID POWER SYSTEMS

Yanjun Yan
Nikiforos Stamatiadis

Kruger Brentt
Publishers
2025

Kruger Brentt Publishers UK. LTD.
Company Number 9728962

Regd. Office: 68 St Margarets Road, Edgware, Middlesex HA8 9UU

ISBN: 978-1-78715-315-8

For information on all our publications visit our website at http://krugerbrentt.com/

PREFACE

"Fluid Power Systems" immerses readers into the dynamic world of hydraulic and pneumatic technologies, offering a comprehensive exploration of the principles, components, applications, and advancements that define fluid power systems. This book serves as an essential resource for students, engineers, technicians, and professionals involved in the design, operation, and maintenance of fluid power systems across various industries.

Fluid power systems, including hydraulic and pneumatic systems, play a crucial role in powering machinery, equipment, and vehicles in industries ranging from manufacturing and construction to aerospace and transportation. By harnessing the power of pressurized fluids, these systems enable precise and efficient control of motion, force, and energy transfer in a wide range of applications.

The primary objective of this book is to provide readers with a comprehensive overview of fluid power systems, covering key topics such as fluid properties, hydraulic and pneumatic components, system design considerations, control strategies, maintenance practices, and safety standards. Each chapter offers insights, practical examples, and real-world applications to help readers understand the fundamental principles and practical aspects of fluid power technology.

Contributions from leading experts in the field delve into various aspects of fluid power systems, including hydraulic actuators, pumps, valves, cylinders, filters, pneumatic compressors, actuators, valves, and control systems. Special emphasis is placed on emerging trends and technologies in fluid power, such as electro-hydraulic and electro-pneumatic systems, digital control, energy efficiency, and sustainability.

As industries strive for greater efficiency, flexibility, and automation, fluid power systems continue to evolve and innovate to meet the demands of modern applications. By mastering the principles and practices of fluid power technology, engineers and technicians can optimize system performance, reduce downtime, and enhance productivity in their respective fields.

We extend our sincere gratitude to the contributors whose expertise and dedication have enriched this book. It is our hope that "Fluid Power Systems" will serve as a valuable resource and companion for students, professionals, and enthusiasts passionate about the transformative potential of fluid power technology in shaping the future of engineering and industry.

Yanjun Yan
Nikiforos Stamatiadis

CONTENTS

01 ENERGY SAVINGS IN HYDRAULIC HYBRID TRANSMISSIONS THROUGH DIGITAL HYDRAULICS TECHNOLOGY

1. INTRODUCTION: AN OVERVIEW

Fluid power is a broad field of engineering technology that utilizes pressurized fluids to create, regulate, and transmit power [1]. Hydraulic oil is commonly utilized for its low freezing point, high boiling temperature, significant bulk modulus, and self-lubricating properties[2]. Fluid power encompasses a distinct study field and academic pursuits that cater to many industries such as agriculture, construction, transportation, aerospace, marine, manufacturing, and others that necessitate high power-to-weight ratios[3]. Fluid power is known for its capacity to achieve high power densities, flexibility in transmission, ease of operation, low cost, controllability, and manageability[4]. Fluid power technology offers great power density and reliability, making it suitable for heavy-duty applications such as construction trucks, material handling equipment, and military activities [5].

The fluid power sector is a crucial component of the United States economy [6]. Thousands of companies in the United States use hydraulic power systems, as reported by the National Fluid Power Association (NFPA). These corporations are well-known as the leading industrial firms in the United States, employing over 845,000 workers with an annual payroll above $60 billion [7]. The fluid power industry has various application areas. Mobile hydraulics, industrial hydraulics, pneumatics, and aerospace are the four primary application areas [8].

Hydraulic hybrid technology, a branch of mobile hydraulics, is recognized as a pioneering system in the automotive industry. Scholars extensively study this research issue to enhance the efficiency of power transfer between a prime mover and the wheels. This technology enables significant fuel efficiency, energy savings, and several environmental advantages, making it a popular choice over traditional drivetrain setups [9]. Hydraulic hybrid drivetrains have numerous advantages compared to traditional manual/automatic gearboxes and continuously variable transmissions (CVTs). Hydraulic hybrid drivetrains differ from traditional drivetrains and CVTs by including two power sources: the prime mover (an internal combustion engine or electric motor) and a positive variable displacement pump/motor paired with a high-pressure storage device (accumulator). The variable displacement pump enables different speeds by controlling the flow rate, hence avoiding gearshift interruptions in traditional drivetrains (manual/automatic) when adjusting speed. CVTs do not encounter significant gearshift interruptions when adjusting speed. CVTs have a restricted number of belts, referred to as chain strengths, that lead to torsional moments, resulting in high frictional forces and torsional moments [11]. In addition to this benefit, incorporating a hydraulic accumulator in hydraulic hybrid drivetrains enhances power management [12]. The accumulator allows for the storage and release of energy, leading to increased efficiency, energy savings, reduced carbon emissions, and improved fuel economy [13].

Hydraulic hybrid drivetrains offer numerous advantages over electric hybrid transmissions, including their dominance over manual/automatic drivetrains and CVTs. Electric hybrid systems, like other systems, need a transmission source and an energy storage source (batteries) [14,15]. However, the batteries in electric systems can only utilize about 20% of the available energy during braking. The hydraulic hybrid transmissions use accumulators that can store and transmit up to 80% of the braking energy, depending on the duty cycle [16]. Performing this action while the vehicle is accelerating or decelerating is recommended to enhance fuel efficiency and lower emissions according to sources (18, 19). It allows for the quick transmission of a significant amount of energy, resulting in increased power densities and power-to-weight ratios [20].

These advantages make the hydraulic hybrid technology more convenient for heavy-duty operations [**21**], like heavy equipment manufacturing, road appliances, heavy load material handling equipment, and many other types of equipment that require high power densities and energy capture. For instance, commercial vehicles, such as heavy load vehicles undergoing high dynamic-drive-cycle operations, have started to rely on this transmission technology. They carry bulky loads while experiencing continual stops, requiring more braking and high energy capture [**15**]. Besides heavy-duty vehicles, some utility vehicles designed to carry light weights also rely on hydraulic hybrid technology. For example, a sport utility vehicle was developed in a research laboratory to demonstrate the concept of a hydraulic hybrid transmission in a passenger vehicle. The vehicle was designed to operate the system in four different modes: hydrostatic driving mode, hybrid driving mode, a combination of hydrostatic and hybrid driving modes, and braking mode [**22**].

This chapter is structured as follows: Section 2 presents an overview of the currently available hydraulic hybrid transmissions. The text presents various hydraulic hybrid powertrain configurations: series, parallel, and series-parallel, and explains the operation of each. Section 3 discusses various significant uses of hydraulic hybrid drivetrains. Section 4 of this paper presents the problem statement, highlighting the shortcomings of the current hydraulic hybrid drivetrains. It also emphasizes and presents the approach used to address these shortcomings. Section 5 provides a summary of digital hydraulics technology in fluid power. It discusses the architecture of a modern digital pump, its fundamental functioning, digital motoring/pumping data, and the potential to attain high efficiencies at various pump displacements. Section 6 shows the enhanced efficiency of the hydraulic hybrid transmission when digital hydraulics are used. This is achieved by simulating a physical series hydraulic hybrid system using the specified digital motoring/pumping data. The document includes the circuit schematic, simulation model, and outcomes. Section 7 ends with a concise overview of the key concepts acquired from the study.

2. EXISTING HYDRAULIC HYBRID TRANSMISSION CONFIGURATIONS

Hydraulic hybrid transmission drivetrains can be classified into three main topologies: parallel, series, and series-parallel arrangements. Hydraulic hybrid transmission systems consist of a variable displacement pump/motor, a high-pressure reservoir for energy storage (often a nitrogen accumulator), and a low-pressure reservoir (fluid tank) [23]. The variable displacement pump/motor functions as a motor by turning hydraulic energy into mechanical energy to propel the vehicle. It functions as a pump by turning mechanical energy into hydraulic energy, which is stored in the accumulator and released when required. The next sections provide a concise overview of current hydraulic hybrid drivetrains (parallel, series, and series-parallel) and explain how they work.

2.1. Parallel Hydraulic Hybrid Configuration

A parallel hydraulic hybrid has a straightforward structure, design, and execution in comparison to other setups [24]. The system consists of two sub-systems: a traditional mechanical transmission linked to the output shaft by a differential gearbox and a hydraulic transmission. The sub-systems are connected in parallel utilizing a mechanical power transmitter such as a gearbox, chain, or belt. Figure 1 displays a parallel hydraulic hybrid system that incorporates a belt mechanical power transmitter. The drivetrain comprises a mechanical gearbox (7), a clutch (2), and the prime mover (1), usually an internal combustion engine, as depicted in Figure 1. The mechanical transmission is linked to the engine through the clutch component and to the differential gearbox on the drive shaft to distribute power to each wheel. The hydraulic system includes a variable displacement pump/motor (4), an accumulator (5), a fluid tank (6), and many other hydraulic components such as valves and fittings. The hydraulic pump/motor (4) is coupled to the mechanical transmission (7) through the belt linkage (3) in a parallel configuration. The clutch enables the engine shaft to be connected or disconnected from the system by utilizing the energy stored in the accumulator, hence regulating the engine's on/off signal. Decoupling the engine from the system happens when there is sufficient energy stored in the accumulator to power the wheels during acceleration. Engaging the engine shaft with the transmission shaft is necessary to propel the wheels when the accumulator storage levels are low.

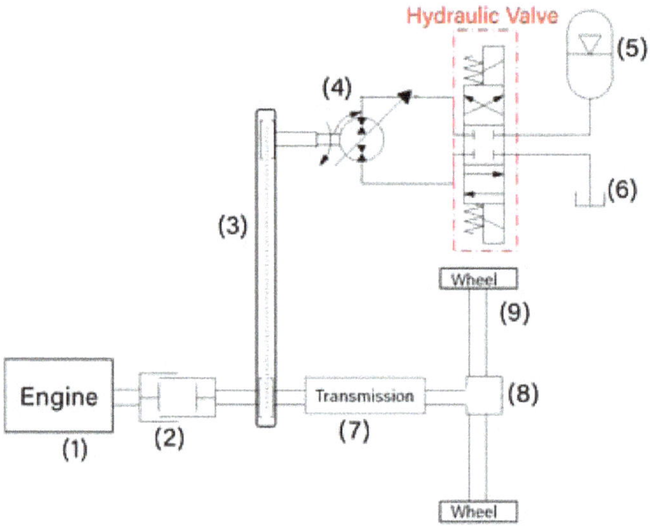

Item	Description
1	Engine
2	Mechanical Clutch
3	Mechanical Transmission Belt
4	Variable Displacement Motor/Pump
5	Gas Accumulator
6	Reservoir
7	Mechanical Transmission
8	Differential Gear Box
9	Output Shaft

Figure 1. Schematic for the parallel hydraulic hybrid configuration.

We will now briefly examine the general operation of the parallel hybrid setup after describing its basic principle. The transmission duty cycle consists of three primary stages: accelerating, cruising, and braking. The accumulator functions as a hydraulic launch assist during acceleration. The pressurized fluid is released to operate the hydraulic pump/motor (4) as an energy source, converting high-pressure energy into mechanical energy at the motor's shaft. The mechanical belt (3) transfers the mechanical power produced to the drive shaft (9), increasing the movement of the wheels. During the cruising phase, when the accumulator's pressure drops below its pre-charged pressure, it triggers the clutch to connect the engine shaft to the system, powering the drive shaft (9) and propelling the wheels. During the last stage, the clutch (2) disengages the engine shaft from the driving shaft (9) to slow down the vehicle. The rotational movement at the wheels is transferred through the belt (3) to power the hydraulic pump/motor (4), which then functions as a pump to fill the accumulator. This cycle recurs with each braking operation.

2.2. Series Hydraulic Hybrid Configuration

The series hydraulic hybrid configuration does not use a mechanical transmission, unlike the parallel hydraulic hybrid configuration. The system uses a variable displacement pump that is directly linked to the main engine through a clutch [27]. Figure 2 illustrates the replacement of the mechanical transmission with a hydraulic variable displacement pump (3) in series with a hydraulic drivetrain that includes a variable displacement pump/motor (6), accumulator (5), fluid tank (4), and other hydraulic components.

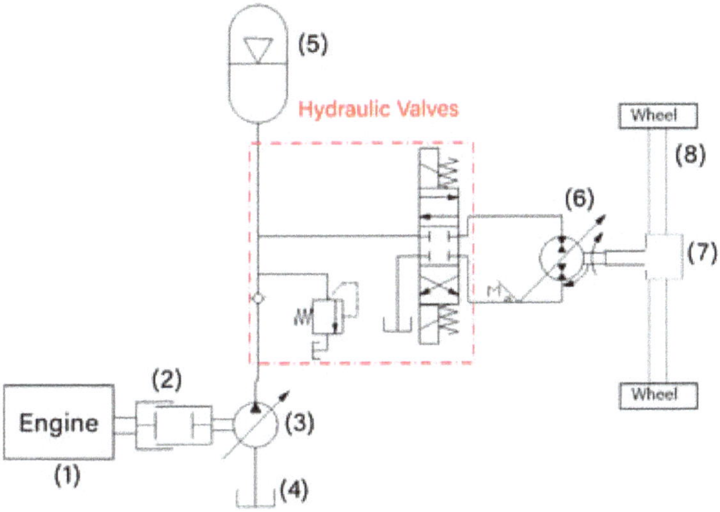

Item	Description
1	Engine
2	Mechanical Clutch
3	Variable Displacement Pump
4	Reservoir
5	Gas Accumulator
6	Variable Displacement Motor/Pump
7	Differential Gear Box
8	Output Shaft

Figure 2. Schematic for the series hydraulic hybrid configuration.

The operation consists of three primary phases: accelerating, cruising, and braking, similar to the parallel architecture. During the acceleration phase, the clutch disengages the engine shaft from the pump shaft, and pressurized hydraulic fluid from the accumulator is directed into the hydraulic pump/motor, enabling it to function as a motor. This process transforms the hydraulic energy stored in the accumulator into mechanical energy at the output shaft (8). Following acceleration, cruising commences, causing the fluid in the accumulator to decrease to its minimal capacity. When the pressure drops below the pre-charge pressure of nitrogen, it causes the engine to start operating and the clutch to engage. The engine's power is transferred to the hydraulic pump (3), enabling it to move fluid from the low-pressure reservoir (4) to the hydraulic pump/motor (6) to power the wheels. The surplus high-pressure fluid generated by the hydraulic pump is stored in the accumulator for future use. When braking, the engine stops, the clutch disconnects, and the hydraulic pump/motor (6) switches to pump mode to move fluid from the low-pressure reservoir to the accumulator. The kinetic energy generated by the wheels is transformed into hydraulic energy and stored in the accumulator. This energy will be utilized to propel the vehicle forward as required.

2.3. Series-Parallel Hydraulic Hybrid Configuration

The power-split hydraulic hybrid system, also called series-parallel, consists of the series and parallel hydraulic hybrid layouts [28]. The concept combines both hydraulic hybrid types into one. The power-split design, shown in Figure 3, involves two sub-systems connected by a mechanical transmitter, such as a belt: a traditional mechanical powertrain and a hydraulic drivetrain. The mechanical powertrain consists of a mechanical transmission (8) linked directly to the output shaft (10) through a differential gearbox (9) and connected to the engine (1) by a mechanical clutch (2). The hydraulic drivetrain consists of two variable displacement pump/motors (4 & 5), accumulator (6), low-pressure reservoir (7), and other hydraulic components. The hydraulic transmission is immediately linked to the output shaft (10) in a manner akin to the series architecture.

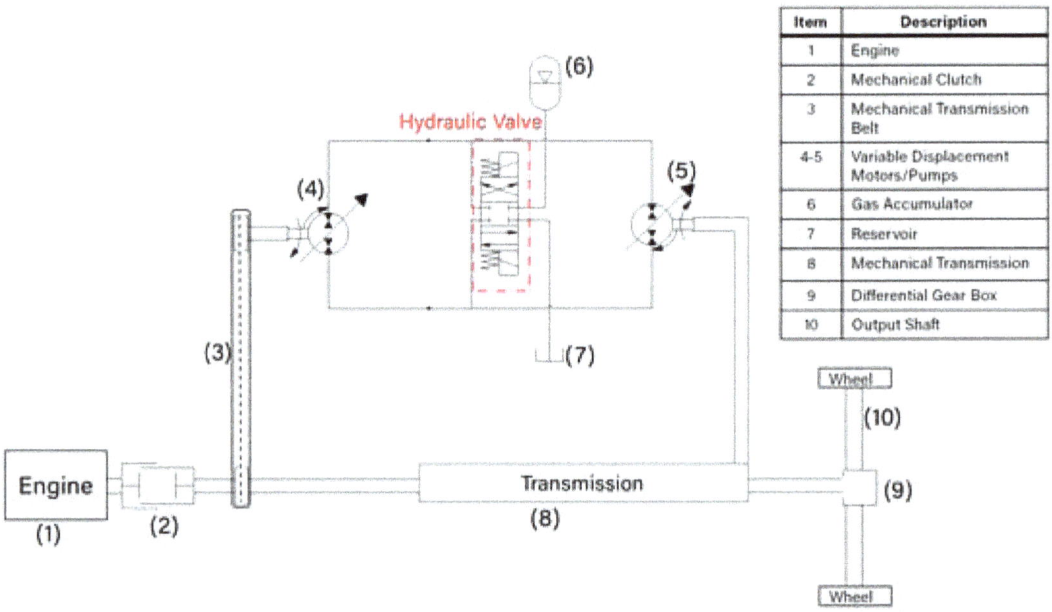

Figure 3. Schematic for the series-parallel hydraulic hybrid configuration.

This setup allows the device to operate in two modes: pure hydraulic mode and hybrid mode. The pure hydraulic mode enables the variable displacement pump/motors (4 & 5) to operate simultaneously. The hybrid mode enables the smooth functioning of the mechanical transmission (8) and the hydraulic drivetrain [30]. The smooth connection between the mechanical powertrain and the hydraulic system enables the transfer of power from the engine to the wheels while ensuring the engine operates at its best. This mechanism allows for obtaining superior mechanical transmission performance, along with efficient energy storage and regeneration, resulting in improved overall efficiency and reduced fuel usage. The design results in higher system complexity, weight, and cost.

3. SUMMARY OF NOTABLE MOBILE APPLICATIONS USING HYDRAULIC HYBRID DRIVETRAINS

This section provides a concise overview of significant applications and suggested frameworks for hydraulic hybrid drivetrains. Stecki et al. [31] provided an extensive study of hydraulic hybrid computer simulation models up to 2005 and outlined the simulation results. The report covers the design, execution, and simulation of over eight prototype automobiles utilizing CVT transmissions with hydrostatic regeneration. The majority of the vehicles analyzed were on-road automobiles. The authors referenced ADVISOR, an Advanced Vehicle Simulator developed by the National Renewable Energy Laboratory located in Golden, CO, USA, which is a simulation tool used for creating hybrid drivetrain models. The discussion included a tactical vehicle designed in MATLAB/Simulink by MathWorks, an American business based in California, and implemented in ADVISOR [32]. A Permo-Drive regenerative energy management system for an army tactical vehicle was developed and simulated using MATLAB/Simulink in this study. Subsequently, the model was included into the vehicle simulator known as ADVISOR, which is a simulation tool used for creating hybrid drivetrain models. The authors ended their review by indicating that additional fuel trials are necessary to verify the accuracy of the chosen fuel map and to further validate the model. Thy determined that the accumulators could provide a safety risk and suggested that their use should involve more advanced accumulator technologies.

In addition to the research conducted by Stecki et al., Stelson et al. [10] conducted a simulation study evaluating three hydraulic hybrid drivetrain configurations: series, parallel, and power-split. The prime mover is replicated as a combustion engine in all their simulation investigations. The approach utilized dynamic programming, using fuel usage as the optimization variable. The created models show that the three configurations of the on-road cars (series, parallel, and power-split) enhance fuel efficiency for both urban and highway travel cycles. During the urban drive cycle, the series, parallel, and power-split hydraulic hybrid designs demonstrate enhancements of 14.7%, 53.2%, and 10.0%, respectively.

The series, parallel, and power-split architectures for the highway driving cycle were improved by 8.9%, 11.9%, and 8.5%, respectively. Additionally, during the simulation, the pump/motors used were larger than necessary. As a result, the units had to function with incomplete displacements, resulting in a reduced hydraulic efficiency overall.

In 2004, Wu et al. [33] used dynamic programming optimization to enhance the accumulator's charging/discharging process for a Class VI parallel hydraulic hybrid delivery truck. The researchers utilized a 99.9-liter accumulator running at a maximum pressure of 36 megapascals during their study. The control method under consideration includes two optimization parameters: the fuel consumption variable as the main control parameter and the state of charge of the accumulator (SOC) as the secondary optimization parameter. The researchers calculated that the fuel efficiency could vary between 28% and 47% based on the effectiveness of the hydraulic components. In 2009, Van Batavia introduced a novel control method to oversee and manage the hydraulic launch assist function. The method involves running the engine at its maximum torque-RPM performance. The Diesel engine, rated at 110 kW, operates within a range of approximately 1000 to 5500 RPM and 125 to 200 N.m. The study asserts that the suggested control method enhanced fuel efficiency by 50% and decreased emissions by 40%. The suggested setup utilized a variable displacement bent axis pump and a distinct variable displacement bent axis motor. Bent axis units are highly efficient, reaching up to 98% efficiency at 100% displacements. However, this high efficiency is only attainable at high displacements and decreases significantly at low displacements.

Buchwald et al. [36] presented a Hydraulic—Internal combustion hybrid idea and four configurations for a passenger bus in Denmark. In addition to their previous efforts, Eaton Hydraulics introduced their hydraulic launch aid system for parallel drivetrain layouts in 2008. The goal was to enhance energy efficiency by adding more engine power while starting the vehicle from a standstill. The system was designed for vehicles that operate with frequent stop-and-go cycles, such as waste trucks, delivery vehicles, and buses. The system was created in partnership with the U.S. Environmental Protection Agency and relied on a cutting-edge pump/motor unit that was custom-made for a parallel drivetrain setup with an axial piston design.

Tollefson et al. [38] investigated a novel internal combustion engine/hydro-pneumatic accumulator energy storage setup for cars. The suggested design necessitates a single hydrostatic pump-motor (P/M) unit. Computer simulations were performed with a diverse set of parameter changes utilizing an appropriate control approach. The simulation results showed that the new idea resulted to substantial fuel savings. The suggested hydraulic accumulator passenger vehicle might reach a maximum of 65 mpg in urban driving situations, in contrast to current cars that often manage just 25–30 mpg in similar conditions.

Kepner [39] demonstrates a Hydraulic Power Assist (HPA) system on a full-size sport utility vehicle. The document includes details on the vehicle, fuel economy data, as well as performance and noise results. The car is a result of a collaboration between Ford Motor Company, the United States Environmental Protection Agency (EPA), and the Advanced Technology Division under a Cooperative Research and Development Agreement (CRAD). The system incorporates a valve block from Ifield Technologies, a pump/motor connected in parallel to the driveshaft of the traditional engine, and carbon fiber accumulators functioning as HPA. The HPA increased the efficiency of the Environmental Protection Agency (EPA) and the Heavier Acceleration city cycles by 23.6% and 35.5%, respectively, as indicated by the results. Bozic [40] introduced a hybrid system consisting of a combustion engine as the main power source, hydrostatic transmission, and a concurrent electric generation system for the vehicle. The literature shows that using hybrid powertrains can lead to energy savings of up to 50% in some circumstances [33,34]. The system's efficiency may be affected by the efficiency of the core components of the hybrid transmission, such as the pump, motor, or pump/motor units. These units can achieve up to 98% efficiency while operating at 100% displacement, but this high efficiency may only occur during a certain portion of the duty cycle and may not be consistent across all working situations.

4. INADEQUACIES OF THE EXISTING HYDRAULIC HYBRID DRIVETRAINS

Although fluid power transmission is widely used, current fluid power drivetrains are inefficient according to recent research [3]. A research released by the United States Department of Energy indicates that fluid power drives in various industries have efficiencies ranging from 9% to 60%, with an average efficiency of no more than 22%. The inefficiencies in fluid power transmissions are mostly linked to the performance of the hydraulic devices being used. The variable displacement pump/motor is a crucial hydraulic component used in all hydraulic applications; it serves as the core of fluid power systems. The efficiency has a major impact on the performance and overall efficiency of fluid power transmission [42,43].

Traditional variable displacement pump/motors, which are current top-of-the-line hydraulic pumps, are most effective when running at full capacity. However, their efficiency drops significantly when working at partial capacity, reaching as low as 30%. Conventional pumps are not effective at partial or low displacements, which impacts the system's efficiency because of the high pressure that can be left in the displacement chamber, as well as leakage and other mechanical losses. The hydraulic hybrid drivetrains have a relatively low overall efficiency, ranging from 64% to 81%, due to the restricted efficiency of traditional variable displacement pump/motors. [47]. Many researchers in various fluid power applications have been driven by this significant issue to suggest, create, and execute new approaches and strategies to enhance the efficiency of hydraulic hybrid transmissions by improving the maximum efficiency of variable displacement vane and piston pumps [48,49,50,51,52,53,54,55]. The study in reference [49] introduces a dynamic seven-stage model to forecast the thickness of the fluid film between the valve plate and cylinder block in axial piston pumps for precise lubrication information. The authors enhanced the efficiency map of axial piston pumps by reducing friction losses through the application of nanocoatings on the pump's slippers using a coating process in reference [51]. In addition to these publications, the study in reference [54] presents research findings from a prior research project. The findings can improve the effectiveness of several technological systems in power engineering and transportation that utilize hydraulic pumps and motors. The study in reference [55] aims to enhance the numerical technique employed to solve the optimal control problem related to the three-dimensional heat equation, with potential benefits for various technological systems such as hydraulic units. The techniques suggested in the previously mentioned studies nevertheless lead to low overall efficiency when working with partial volumetric displacements. The pumps' lowest efficiency at low volumetric displacements does not show any significant increase [56]. The study described in reference [42] suggested incorporating digital technology into fluid power systems by developing digital variable displacement pump/motors to maintain high efficiency across a broad range of displacements.

This study seeks to improve the performance and efficiency of a hydraulic hybrid transmission by using a digitally validated pump/motor based on previous research findings. The digital pump/motors aim to improve the efficiency of hydraulic hybrid drivetrains by raising the maximum total efficiency of the transmission, as demonstrated in a physical-based simulation.

5. DIGITAL HYDRAULICS

Digital Hydraulics is an innovative sub-discipline of fluid power that utilizes sophisticated on/off control technology to improve the performance of hydraulic systems, drive progress in fluid power applications, and reduce energy usage. This technology can enhance efficiency by substituting traditional fluid power components like directional control valves with numerous switching on/off valves arranged in a parallel manner [41]. The valves' configurations enhance the design in this application by utilizing a poppet-style actuation that prevents leaking, a major issue with spool valves. This configuration decreases the system's losses, enhances control, boosts performance and efficiency, and lowers the cost of maintaining components [60,61,62]. This technology has been used in several ways to enhance the effectiveness of standard systems, such as substituting typical check valves on fixed displacement pumps. The technology also allows for controlling and redirecting flow in traditional fixed displacement pumps to permit variable displacements [44].

Conventional piston pump/motors, such as reciprocating piston pumps, use mechanical check valves on the input and exit ports to separate the high-pressure and low-pressure sides of the system. While pumping, this setup permits fluid to flow in through the low-pressure check valve when the pump is in its intake stroke.

As the pump empties its cylinder, pressure causes the intake check valve to close and the exhaust check valve to open, moving fluid to the high-pressure part of the system. This setup restricts the pump to fixed displacements as the pressure differential in the chamber is the sole means of regulating the intake and exhaust valves' actuation. The work described in reference [42] introduces an electrically controlled digital inline 3-piston pump that substitutes traditional valves with digital on/off switching valves. The digital pump/motor being suggested is a 28cc/rev unit that consists of three displacement chambers, each equipped with two high-speed switching on/off valves. Figure 4 displays the pump/motor's setup for a single displacement chamber. Valve 1 regulates flow on the low-pressure side, whereas Valve 2 regulates flow on the high-pressure side. The suggested setup enables the valves to be opened or closed at any point throughout the pumping or driving cycle. This setup enhances the system's controllability and flexibility by allowing the implementation of different digital pumping and motoring approaches known as flow diverting and restricting operating strategies [44,63].

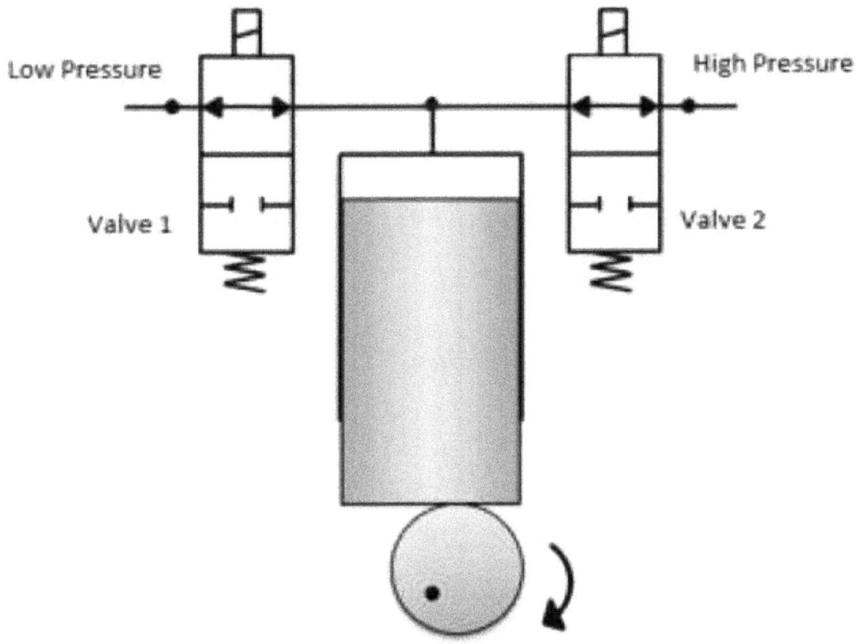

Figure 4. Single-piston digital pump/motor configuration [64].

In addition to the electrically operated digital pump, our team created and put into operation a mechanically operated digital pump that can likewise achieve high efficiency across a broad range of displacements. The subsequent sections provide an overview of the internal mechanical structure of the mechanically actuated digital pump, provide the most efficient operating strategy for the electrically actuated pump, and describe the experimental testing and validation of the electrically actuated pump.

5.1. Internal Mechanical Structure of the Mechanically Actuated Digital Pump

This section provides a concise overview of the mechanical configuration of the mechanically operated digital pump, capable of achieving efficiencies above those documented in this study. Figure 5 displays the internal mechanical construction of the digital pump, consisting of three primary sub-systems: gear-cam unit, cam-follower unit, and valve manifold. The gear-cam unit employs three primary half-masking cams, each including two sub-cams that are phased in relation to each other to enable the creation of various cam profiles. This unit controls the phasing mechanism of the camshafts, enabling the adjustment of the cams' position in relation to the shaft as it rotates. The cam-follower unit comprises three followers, their mounting bracket, and three valves to be operated by the cam-follower mechanism. This device transforms the rotational movement from the cam into linear motion for the follower, which regulates the opening and closing of the valves. Varying the cam profile through the phasing mechanism allows for adjusting the opening and closing of the valves, resulting in the ability to provide varied pump displacement. A comprehensive explanation of the mechanical framework and the pumping mechanisms can be found in reference [65].

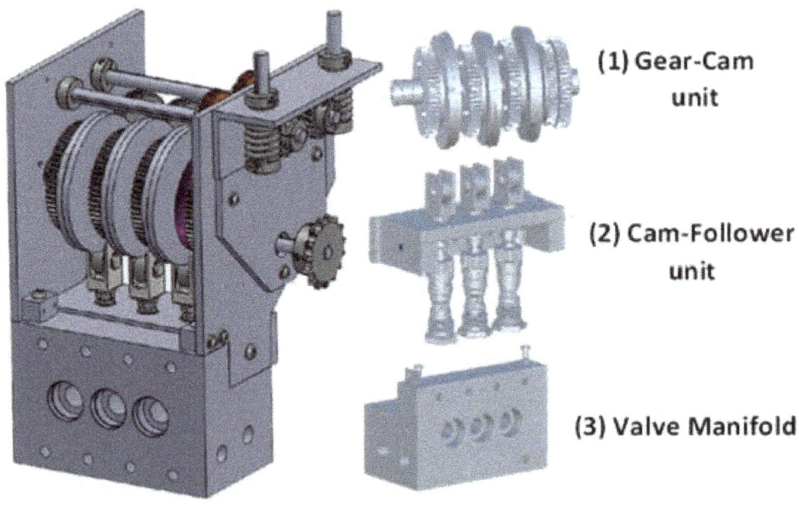

Figure 5. CAD drawing showing the basic units of the mechanically actuated digital pump.

5.2. Operating Strategies of the State-of-the-Art Electrically Actuated Digital Pump

The electrically operated digital variable displacement pump/motor [42] operates based on the control flow that enters and exits the displacement chambers. Controlling the actuation of two on/off valves on the intake and exhaust ports of the pump can vary the pump/motor displacement. Controlling when the valves open and close allows for different pumping and driving techniques to be used. Four distinct operational approaches have been examined and evaluated using the cutting-edge digital pump. The four operating modes are Partial Flow-Diverting (PFD), Partial Flow-Limiting, Sequential Flow-Diverting (SFD), and Sequential Flow-Limiting (SFL). The most effective operational approach identified is the SFL [63]. SFL is a method that allows for the activation or deactivation of the states of each cylinder, resulting in either 100% or 0% displacement being sent to the output valve, respectively. This approach is applicable for both pumping and driving cycles. Figure 6 displays an illustration of a pumping cycle with 100% displacement. This cycle accomplishes comparable functionality to the mechanical check valve. The intake valve remains open during the full stroke (A) and closes at the start of the exhaust stroke (B) during pumping. Subsequently, the exhaust valve is opened and maintained until the exhaust stroke is finished. This process transfers all the fluid that enters the piston chamber during the intake stroke, without considering the dead volume in the cylinder.

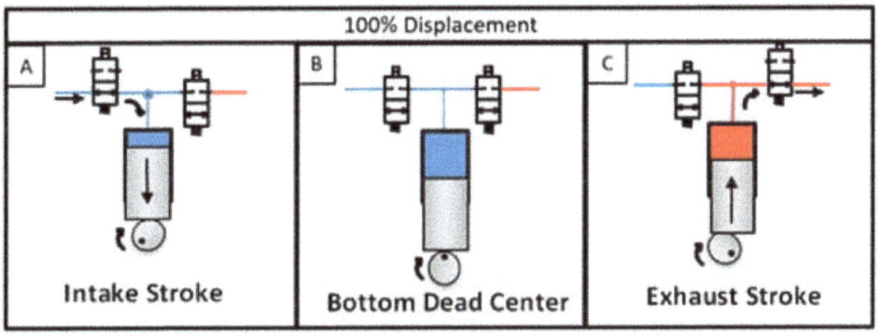

Figure 6. Enabled chamber operating at 100% displacement during SFL mode.

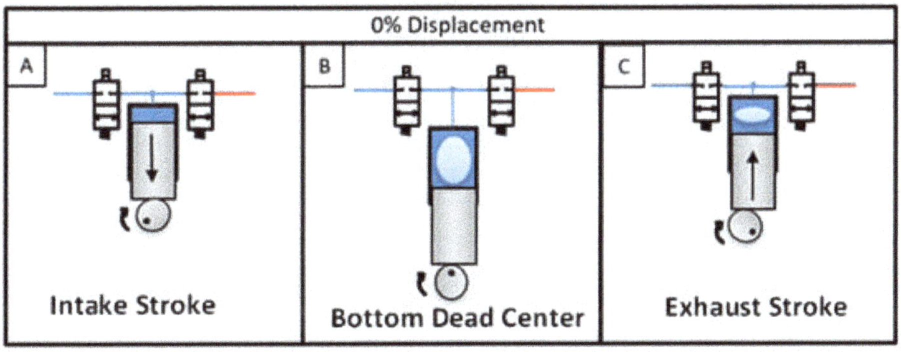

Figure 7. Disabled chamber operating at 0% displacement during SFL mode.

Figure 7 illustrates an instance where there is 0% displacement during pumping. Throughout this cycle, both intake and exhaust valves are sealed shut, enabling the fluid to undergo compression and decompression. During the intake stroke, the piston decompresses the fluid in the dead volume of the cylinder until it reaches the Bottom Dead Center, creating a partial vacuum in the chamber. During the exhaust stroke (C), the fluid is recompressed to its initial state once the piston starts moving. As the intake and exhaust valves stayed shut throughout this cycle, the cylinder provided 0% displacement. This cycle enables the depressurization and pressurization of the chamber's dead volume fluid, with no delivery of fluid to the system.

The partial vacuum formed during the intake stroke is referred to as chamber voiding. [66] experimentally examined the chamber voiding phenomenon. An experiment demonstrated that the gas produced during the chamber voiding phase, referred to as (B), would completely dissolve when the fluid is recompressed. This strategy's fundamental premise involves changing the state of the chambers sequentially to alter the displacement of the pump. For example, Figure 8 displays a three-piston pump arrangement. For a total pumping displacement of 66%, chambers 1 and 2 would need to operate at 100% displacement, while chamber 3 would run at 0% displacement.

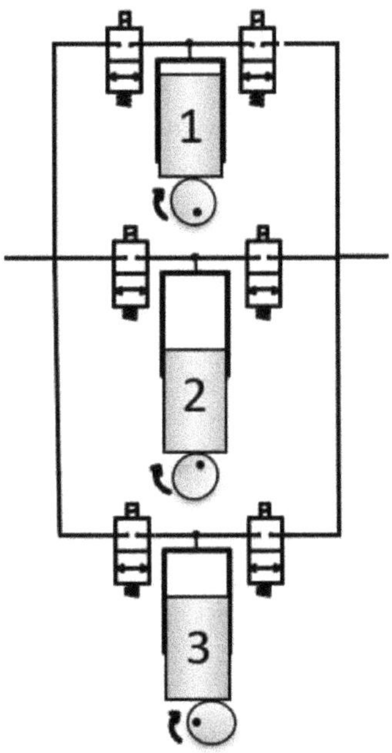

Figure 8. Three-piston digital inline piston pump configuration.

Sequencing is averaged over various pump revolutions to achieve a wide range of displacements between 0, 33%, 66%, and 100% [**42**]. As a case in point, to achieve 11.11% displacement, the first cycle of the pump would turn on one cylinder (1) and turn off two cylinders (2 & 3). The following two pump revolutions would sequence all cylinders (1, 2 & 3) off. This operation achieves 1/9th, or 11%, displacement over the three cycles. A ~1% displacement resolution can be achieved by averaging the sequences over 32 shaft revolutions.

5.3. Experimental Validation of the Electrically Actuated Digital Pump/ Motor

Figure 9 displays the experimental test stand utilized to test the SFL operating strategy. The test stand includes a 28 cc/rev digital pump, two accumulators (one on the suction side and one on the discharge side), a manifold housing with six on/off valves (two for each piston chamber), and other necessary hardware for the hydraulic circuit, data acquisition, and control. The two accumulators were incorporated into the system to mitigate pressure and flow fluctuations and prevent cavitation issues during the suction phase. If the hydraulic system encounters abrupt halts during suction, rapid fluctuations in system pressure will create pressure ripples that lead to shock waves. In this scenario, the system's pressure will exceed the pre-charged pressure of the accumulator. The accumulator at the discharge port will serve as a shock absorber. The surplus flow will enter, compressing the nitrogen bladder to cushion and reduce shock. The second accumulator at the suction port is necessary due to the pump's bi-directional nature, allowing it to discharge fluid from either port. A comprehensive account of the hardware utilized is provided in reference [66].

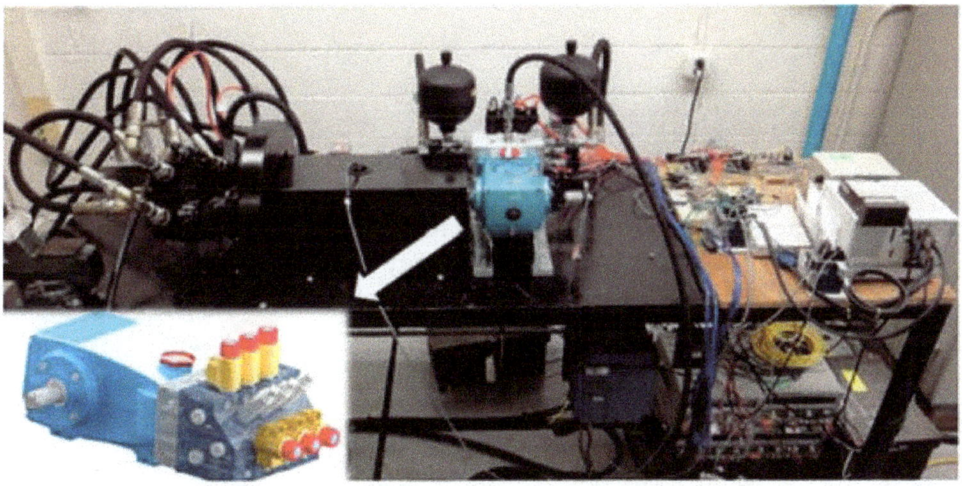

Figure 9. The experimental test stand for testing different operating strategies of the electrically actuated digital 3-piston pump [42].

The SFL pumping/motoring technique was experimentally tested under steady-state conditions at a temperature of 50 °C using pump displacements of 25%, 50%, 75%, and 100%. The SFL pumping and motoring cycles were analyzed at various system pressures between 4 MPa and 18 MPa, together with shaft speeds reaching up to 700 RPM. To operate the digital pump at shaft speeds above 700 RPM, high-speed switching valves are necessary. However, this can impact the valves' flexibility and lead to non-linearity and inaccuracies in their timing (opening and shutting time). An algorithm for correcting timing valve errors was created in reference [67] and used during experimental testing to address and compensate for the valve timing fault. The algorithm calculates the delay in valve timing by analyzing pressure measurements from low- and high-pressure ports to control the opening and closing times of the valves. The algorithm allowed for rapid switching, however the valves' opening and shutting speeds are insufficient. The maximum working shaft speed was restricted to 700 RPM during the whole research period. The experimental testing gathered data on torque, pressure, and flow at different pump displacements (25%, 50%, 75%, and 100%), system pressures (4 MPa, 10 MPa, and 18 MPa), and shaft speeds (300 RPM, 500 RPM, and 700 RPM). The efficiencies were computed based on the data collected.

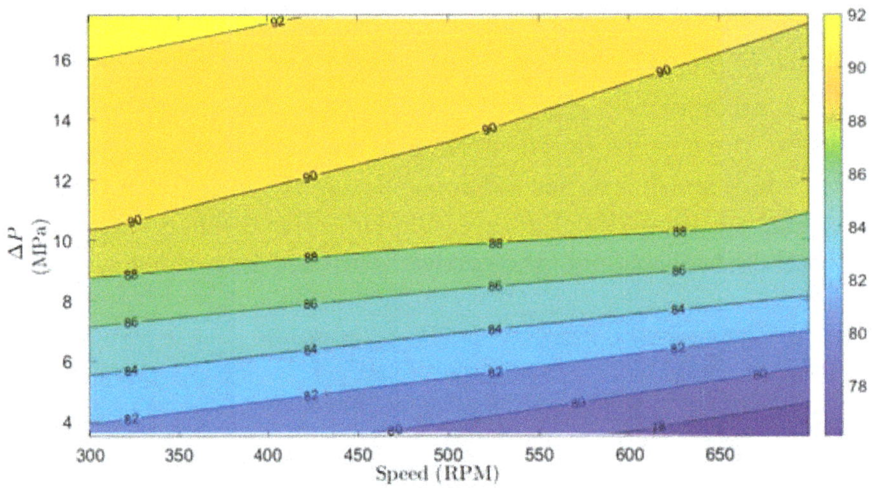

Figure 10. Efficiency map of the electrically actuated digital pump/motor operated in the SFL pumping strategy at 100% displacement.

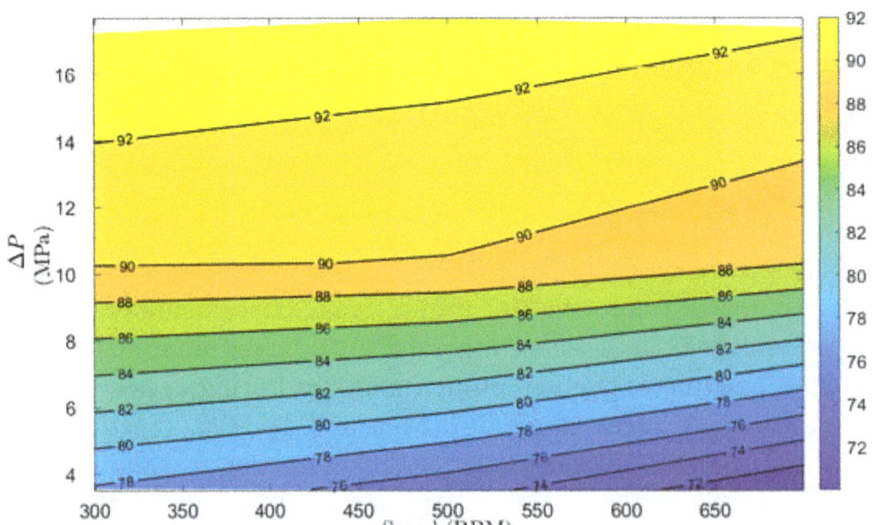

Figure 11. Efficiency map of the electrically actuated digital pump/motor operated in the SFL motoring strategy at 100% displacement.

The efficiency map for operating the pump using the SFL method at 100% displacement during the motoring mode is depicted in Figure 11. The motoring mode depicted in the image allows for efficiency ranging from 72% to 92%. In driving mode, the efficiency curves are nearly parallel to the x-axis, indicating that the shaft's speed has minimal effect on the pump's efficiency, unlike in pumping mode. We can obtain efficiencies ranging from 72% to 92% at any RPM, depending on the system's pressure.

6. SIMULATION MODEL: HYDRAULIC HYBRID DRIVETRAINS UTILIZING DIGITAL PUMP/MOTORS

This section describes a physics-based simulation of a hydraulic hybrid drivetrain using MATLAB Simscape, a graphical programming product developed by MathWorks. The MATLAB Simscape model simulates a series hydraulic hybrid transmission used in normal terrain hydraulic vehicles. The model aims to analyze the enhancement of hydraulic hybrid transmission performance by substituting conventional pump/motors with digital pump/motor components. The cutting-edge digital units are simulated by using the digital motoring/pumping data collected at various displacements (25%, 50%, 75%, and 100%), pressures (4 Mpa, 10 MPa, and 18 MPa), and shaft speeds (300 RPM, 500 RPM, and 700 RPM) on the physical simulation components (pump and motor). The digital input is converted into lookup tables and applied to the physical components of the simulation model, specifically the actual pump and motor, for use throughout the simulation. The subsequent parts will showcase the circuit schematic, simulation model, design, and operation over the three transmission stages (acceleration, cruising, and stopping), followed by an analysis of the consequences.

6.1. Circuit Schematics

Figure 12 shows the adopted series hydraulic transmission system, which includes a variable displacement pressure compensated pump (5) powered by an electric motor (2), a mechanical clutch (3) for connecting and disconnecting the motor's shaft, a pressurized gas accumulator (1) for energy storage and boost, a low-pressure reservoir (8), and a variable displacement pump/motor (11) for power transmission. The pump/motor (11) functions as a motor during acceleration and cruising (motoring mode) and as a pump during braking (pumping mode). A check valve (6) is installed to prevent backflow (flow in both directions between the accumulator and the pump) from entering the discharge port of the pump. Two 4/3-way (9 & 10) and one 3/2-way (7) directional flow control valves facilitate smooth operation in driving and pumping modes. The 4/3 way valves (9 & 10) are closed-center valves with spring return and two directly actuated solenoid coils, while the 3/2 way valve (7) is a spring return valve with one actuated solenoid coil. Valve (9) is utilized in the driving mode for acceleration and cruising in the simulation model. The user adjusts the valve according to the direction of travel, whether it is forward or backward drive. Valve (10) is utilized during regenerative braking in the pumping mode and is regulated according to the rotation direction of the unit's shaft (11). Valve (7) permits the switching between valves (9) and (10) depending on the system's operational state. Valve (7) remains in its un-actuated state during motoring mode, which enables fluid to be pumped into the motor (11) through valve (9) during acceleration and cruising. During the pumping mode, when braking, the mechanical braking signal is transformed into an electrical signal at valve (7), causing it to switch positions and enabling motor (11) to pump flow from the tank into the accumulator through valve (10).

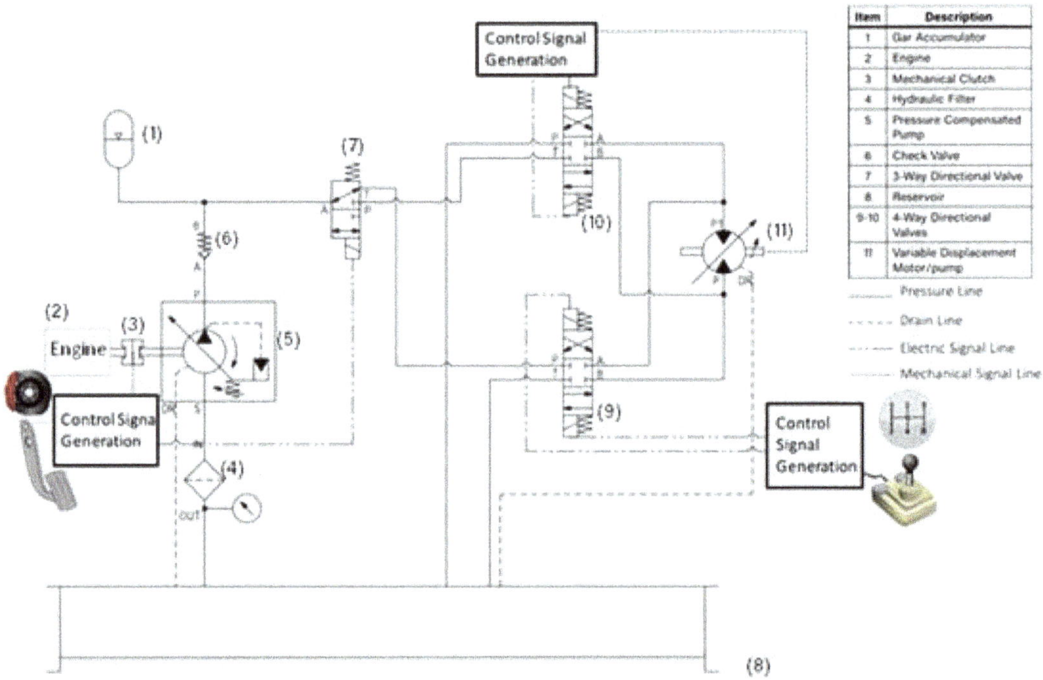

Figure 12. A circuit illustrating the simulated series hydraulic hybrid transmission.

6.2. Simulation Model, Design, and Operation

Figure 13 shows the physical model of the series hydraulic hybrid transmission depicted in Figure 12. The system is designed and simulated using Simscape Fluids and Drivetrains libraries in MATLAB-Simulink. The simulation model is created by combining the necessary numerical block diagrams with the physical paradigms. The hydraulic physical components, such as the pump, motor, valves, and other hydraulic devices, are assembled via interfaceable physical connections. The diagram illustrates a model comprising a pressure compensated pump and a variable displacement motor. The motor functions as a pump when receiving a mechanical signal at its output shaft and as a motor while pumping high-pressure fluid into one of its ports. The model utilizes physical hydraulic sensors to detect the flow and pressure at specific locations within the system. For instance, pressure-flow sensors are installed at the pump's discharge port, accumulator's port, and motor's inlet and outlet ports. The flow sensor of the accumulator is mounted inside the accumulator. A positive flow rate at the accumulator's port indicates flow into the accumulator during charging. Conversely, a negative flow indicates that the flow is exiting the accumulator while discharging. In addition to pressure-flow sensors, mechanical sensors such as rotational motion and torque sensors are installed on the pump and motor shafts to monitor the speed and torque of the shaft. Obtaining this data (torque and speed) enabled us to calculate the input and output mechanical powers. The power data was exported to the MATLAB workspace to calculate the average input and output powers. The average input and output powers were computed for the entire cycle using the mean function in MATLAB.

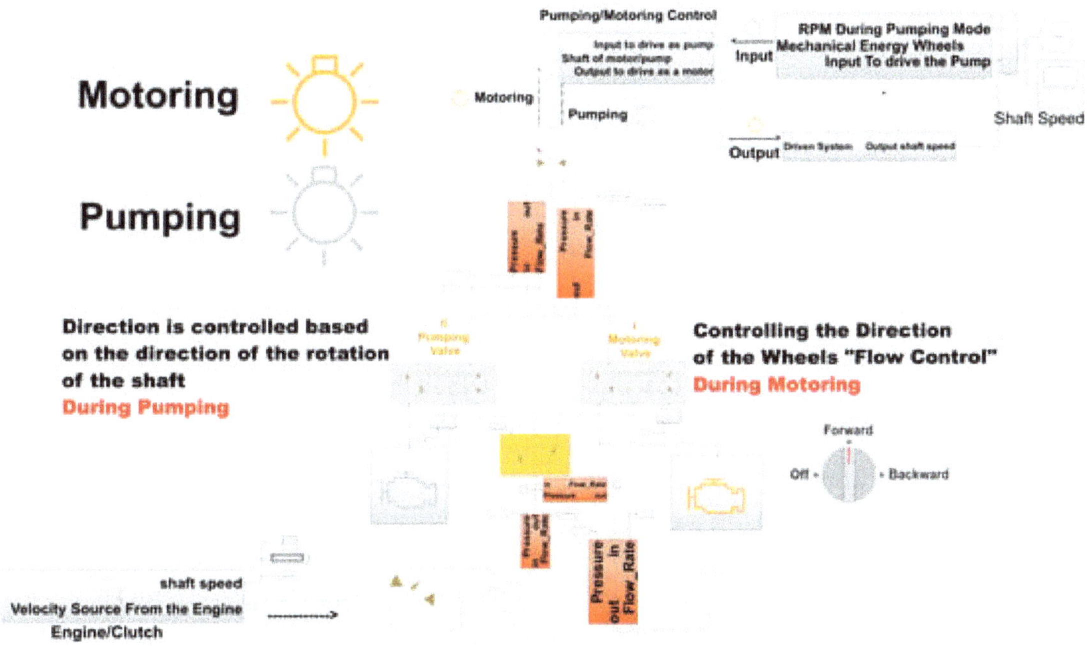

Figure 13. A MATLAB Simscape simulation model for the series hydraulic hybrid transmission.

The digital data from Section 5.2 is transformed into lookup tables and then transferred to the physical pumping and motoring components (the pressure-compensated pump and variable displacement motor) to integrate the previously introduced digital pump and motor. The pumping data is loaded into the physical motor (11) as a lookup table during braking. The fluid is transferred into the physical pump (5) when cruising, considering that the pump will not be operational during acceleration. The motoring data is used as a lookup table on the physical motor for acceleration and cruising. The physical pump and motor (pressure-compensated pump and variable displacement motor) are sized according to imported lookup tables to establish the maximum and minimum pressures in the system as well as the desired flow rate. The accumulator and other hydraulic components such as valves, fittings, hoses, etc., are chosen according to the pressure and flow rate of the system. The hydraulic units (5) and (11) will function within a speed range of 300 to 700 RPM based on the digital pumping/motoring data, with a maximum displacement of 28 cc/rev. The hydraulic units will operate with a volumetric efficiency of up to 94% and an overall efficiency ranging from 78% to 90% during pumping. Unit (11) will function with volumetric and total efficiencies of up to 99% and 92%, respectively, when motoring. The simulation model also considers the system's internal losses, such as losses in hydraulic pipelines, fittings, and valves. The internal losses are approximated in the distributed parameter model by incorporating hydraulic resistance, capacitance, and inductance using simscape block models designed for hydraulic pipelines, fittings, and hoses. Within the hydraulic hybrid transmission series, the gas accumulator functions as a hydraulic lunch assist device to supply more flow during high demand situations, particularly in the acceleration phase, causing the gas in the accumulator to expand and compress under adiabatic conditions.

Based on the intended application, the accumulator is sized using the following equations [**69**]:

$$P_1 V_1^k = P_2 V_2^k \tag{1}$$

$$V_1 = \frac{V_x (P_3/P_1)^{1/k}}{1 - (P_3/P_2)^{1/k}} \tag{2}$$

where $P1$, $P2$, and $P3$ in Pa denote the accumulator's pre-charged pressure, system maximum operating pressure, and system minimum operating pressure, respectively, set based on the experimentally validated data of the digital pump. $V1$ and $V2$ in L designate the nitrogen's volume at $P1$ and $P2$, respectively, Vx in L denotes the required fluid volume, and k=1.4 is the nitrogen constant during adiabatic operation. To prevent the accumulator's poppet (poppet that opens and closes, allowing fluid to enter and leave) from fully closing during the expansion of the nitrogen bladder, the accumulator's pre-charged pressure $P1$ has to be set between 60% to 80% of the minimum system pressure [**70,71**]. Thus, for $P2$ = 1.9e + 7 Pa, $P3$ = 5.3e + 6 Pa and $P1$ = 0.75 $P3$, we get $V1$ = 36.3 L and Vx = 18 L.

The digital data from Section 5.2 is converted into lookup tables and then sent to the physical pumping and motoring components (the pressure-compensated pump and variable displacement motor) to incorporate the previously introduced digital pump and motor. The pumping data is stored in the physical motor (11) as a lookup table during braking. Fluid is transported to the physical pump (5) when cruising, as the pump will not be in operation during acceleration. The driving data serves as a reference table for the physical motor's acceleration and cruising. The pressure-compensated pump and variable displacement motor are sized based on lookup tables to determine the system's maximum and minimum pressures and the desired flow rate. The selection of the accumulator and other hydraulic components, such as valves, fittings, hoses, etc., is based on the pressure and flow rate requirements of the system. Hydraulic units (5) and (11) will operate at speeds ranging from 300 to 700 RPM according to the digital pumping/motoring data, with a maximum displacement of 28 cc/rev. The hydraulic units will function with a volumetric efficiency of up to 94% and an overall efficiency between 78% and 90% while pumping. Unit (11) will operate with volumetric and total efficiencies of 99% and 92%, respectively, during motoring. The simulation model accounts for internal losses inside the system, including losses in hydraulic pipelines, fittings, and valves. The distributed parameter model approximates internal losses by integrating hydraulic resistance, capacitance, and inductance with simscape block models tailored for hydraulic pipelines, fittings, and hoses.

The gas accumulator in the hydraulic hybrid transmission series acts as a hydraulic launch aid device to provide additional flow during high-demand situations, especially in the acceleration phase. This causes the gas in the accumulator to expand and compress under adiabatic conditions.

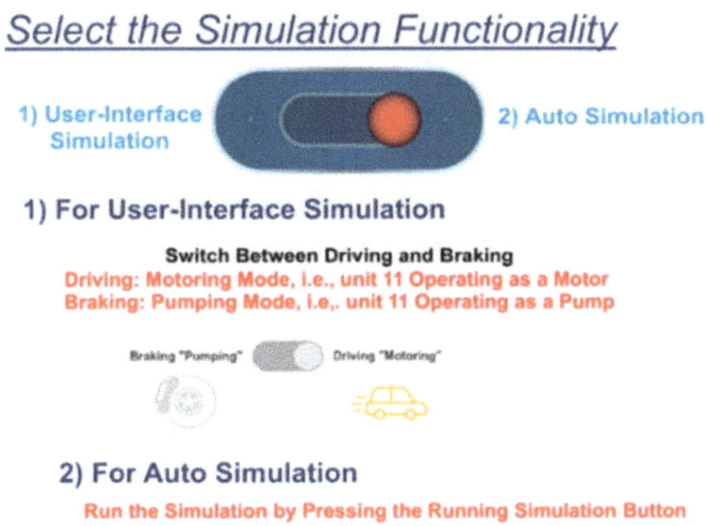

Figure 14. User interface illustrating the functionality of the simulation model.

Relying on option two, several drive cycles were adopted to test the simulation model's capability over the three transmission stages. The total time frame of the tested drive cycles ranged from 200 s/cycle to 300 s/cycle, where the acceleration and braking phases required long times up to 60 s and 40 s, respectively. The acceleration and braking timing is due to the accumulator's charging and discharging time, which was selected based on the experimentally acquired pump data that was limited with a maximum and minimum pressure around 19.1 MPa and 5.3 MPa, respectively.

One of the tested duty cycles is shown in **Figure 15**. The figure shows that the physical system is simulated for 400 s for two consecutive cycles. Each cycle comprises the three stages of the series hydraulic hybrid transmission (acceleration, cruising, and braking). During the first cycle, the acceleration phase occurs between 0 and 60 s, the cruising starts at time = 60 s and continues until time = 140 s, and the re-regenerative braking takes place within 140 and 180 s. The second cycle starts 20 s after the first cycle, which accounts for the complete stopping of the vehicle. Similarly, the three transmission stages (acceleration, cruising, and braking) occur for 60, 80, and 40 s, respectively, during the second cycle.

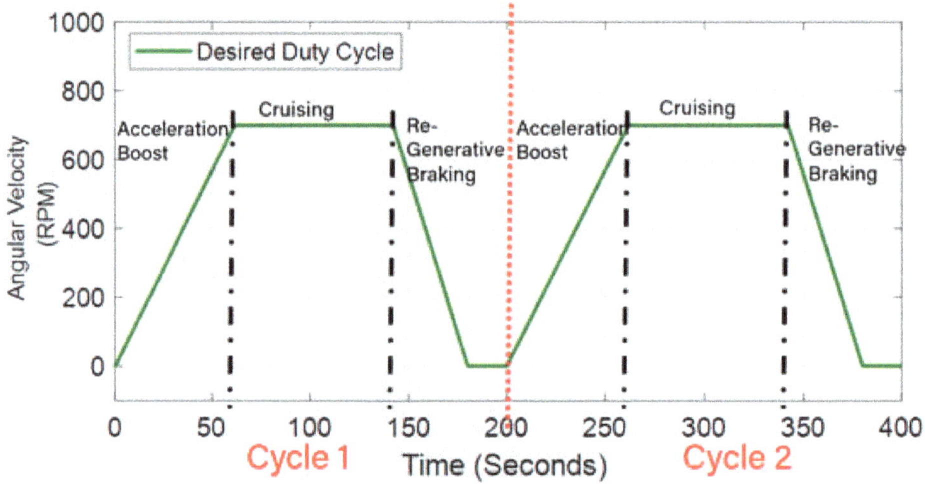

Figure 15. The desired duty cycle.

To accurately replicate the hydraulic hybrid series transmission system and achieve the desired duty cycle shown in Figure 15, it is necessary to set the desired operating angular velocity at the system's input port and define its load. The system's operating RPM is simulated by utilizing an input signal builder with a physical velocity source. The entire weight of the transmission was accounted for in the hydraulic units and components used in the Simscape environment. Four Simscape Tire blocks are used to simulate the transmission load by modeling the tires based on their radius and ideal inertia. Valeski estimated the average tire's weight to be approximately 10 kg and its diameter to be around 40 cm. The moment of inertia calculated for each tire was around 0.37 kg·m2. A modest physical torque was given to the motor's shaft, which has an ideal moment of inertia of 1.5 kg·m2, based on the computed moment of inertia on each tire. We will now elucidate the functioning of the simulation model at each level.

6.2.1. Acceleration

While accelerating, the accumulator is filled with pressured fluid, and its initial pressure exceeds the system's normal operating pressure. Currently, the pump is not operating, thus the accumulator functions as a supplementary power source. The accumulator releases its stored energy into the motor (11) while it is in its driving phase to accelerate the vehicle. The valves are operated based on the pre-charged pressure of the accumulator. During this phase, when the pre-charged pressure of the nitrogen in the accumulator exceeds the system's pressure, no signal is produced at valve (7). Valve (7) is maintained in its un-actuated state to allow the flow through valve (9) and obstruct it at valve (10).

6.2.2. Cruising

During the acceleration phase, the fluid in the accumulator decreases, causing the pre-charged pressure to drop below the system's pressure. The accumulator has depleted most of its stored energy, necessitating an extra hydraulic power source to operate the wheels. During the cruising stage, a decrease in the pre-charged pressure causes an electric signal to activate at the engine, prompting the clutch to connect the engine to the rest of the system. The pump transfers fluid from the low-pressure reservoir (hydraulic tank) to the hydraulic motor (11), which functions as a motor to drive the motion of the wheels. The fluid is supplied to the motor through valve (9) during the acceleration phase, while valve (10) remains closed.

6.2.3. Braking

When braking, the motor (11) shifts to pumping mode, and the clutch disengages the engine from the system, causing the pump (5) to stop. The pre-charged nitrogen gas pressure in the accumulator falls below the system's pressure. An electric signal activates valve (7), causing it to move to the second position, which then pumps the fluid through valve 10 and blocks it at valve 9. The motor uses the kinetic energy from braking to transfer fluid from the tank to the accumulator, transforming mechanical energy into hydraulic energy stored in the accumulator for acceleration.

7. RESULTING OUTCOMES

After introducing the simulation model and discussing its fundamental operation, throughout the following sections: **Section 7.1** and **Section 7.2**, we present the simulation outcomes resulting from applying the digital motoring/pumping data into units (5) and (11) at moderate load-speed conditions (1.5 kg·m^2 and 700 RPM).

7.1. Pressure-Flow Outcomes

Figures 16 and 17 depict the fluctuations in the system's pressure and flow during the suggested duty cycle. The pressure-flow results at various ports in the system are displayed during the three transmission phases: accelerating, cruising, and braking. The figures demonstrate that the pump's shaft angular velocity remains at zero during the acceleration period (0 to 60 s in cycle one and 200 to 260 s in cycle two), and is initiated after 40 s. The pump began functioning after 40 seconds due to the accumulator having sufficient stored energy at the beginning of the acceleration period (from 0 to 40 seconds). The clutch disengages the engine's shaft from the pump's shaft, resulting in a zero RPM at the pump's input shaft. The accumulator functions as a hydraulic launch assist by releasing its stored energy into the motor. The flow rate at the accumulator's port moves in the negative direction, from the accumulator into the system, as depicted in Figure 16.

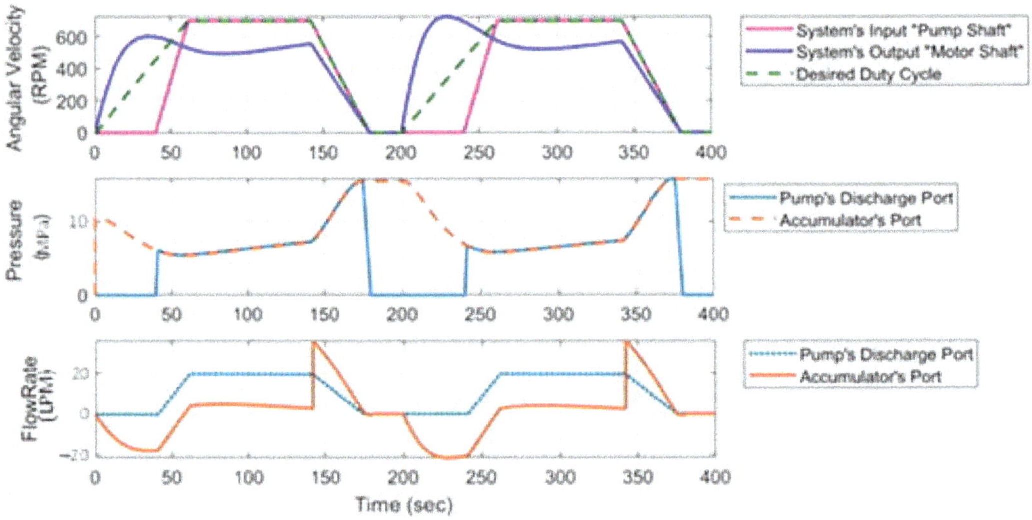

Figure 16. The resulting pressure and flow simulation outcomes at the pump's discharge and accumulator's ports throughout the desired duty cycle.

A pressure difference of approximately 10 MPa is established between the input and outlet ports of the motor, depicted in Figure 17. This enables the conversion of hydraulic energy into mechanical energy, facilitating the acceleration of the wheels until the output shaft speed reaches 600 RPM. The accumulator supplies auxiliary power, causing a decrease in fluid pressure at the port indicated in Figure 16, as the accumulator's poppet remains open to provide fluid to the system. At 40 seconds, when the accumulator has released approximately 80% to 85% of its stored energy, the pump begins functioning through the clutch-engine mechanism. This action keeps the system's output angular velocity stable at around 600 RPM, causing the discharge pressure, as shown in Figure 16, to rise to 6–7.5 MPa. After 20 seconds, at time 60 seconds, the vehicle enters the cruising stage. The pump runs at 700 RPM and operates at an operational pressure of around 7.5 to 8 MPa.

The motor's output shaft operates at a speed of 500 RPM, which is considered modest. At this time, the pressure at the accumulator's port, as depicted in Figure

16, equals the system's pressure due to the closed poppet of the accumulator. The accumulator is currently in a state of neither charging nor discharging. After 80 seconds, the vehicle enters the braking phase, during which the clutch initiates the disengagement of the engine shaft from the pump shaft, leading to a reduction in its RPM. During this phase, the motor (11) transitions to its pumping mode, moving the fluid from the tank into the accumulator. The pressure at the motor's outlet port, seen in Figure 17, rises to 17.5 MPa, resulting in high pressure at the accumulator's port and the pump's discharge port, as illustrated in Figure 16. The accumulator begins to charge as the flow rate, as depicted in Figure 16, moves in the positive direction, from the system into the accumulator.

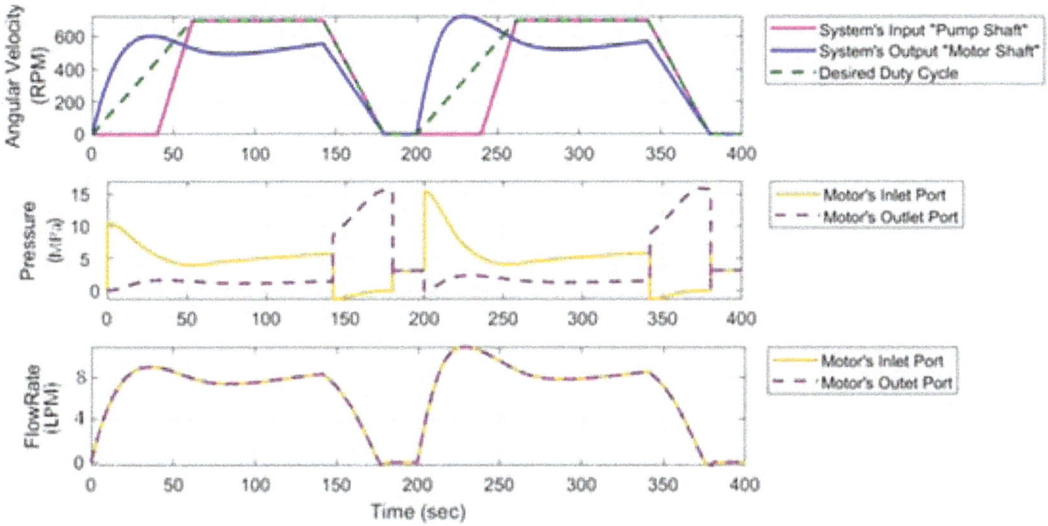

Figure 17. The resulting pressure and flow simulation outcomes at the Motor's inlet and outlet ports throughout the desired duty cycle.

Figure 18 displays the changes in the accumulator's fluid volume, while Figure 19 illustrates the motor's pressure decrease over the specified duty cycle across the three transmission stages: accelerating, cruising, and braking. During the acceleration phase, the accumulator releases its stored energy into the motor's inlet port to speed up the motion of the wheels. The motor is operating in its driving mode. The discharge of the accumulator results in a reduction in fluid volume, as depicted in Figure 18, and a positive pressure decrease of approximately 10 MPa at the motor, as illustrated in Figure 19. The 10 MPa pressure drop across the motor lowers significantly until reaching 5–6.5 MPa, as depicted in Figure 19. During the cruising phase, the accumulator maintains a nearly constant fluid level as it is neither charging nor discharging, as depicted in Figure 18. The pressure drop over the motor, as depicted in Figure 19, is consistently kept at around 7.5 MPa in this period. When braking, the fluid volume in the accumulator, as depicted in Figure 18, increases due to the charging of the accumulator. A negative pressure differential occurs across the motor while it is in pumping mode, as depicted in Figure 19.

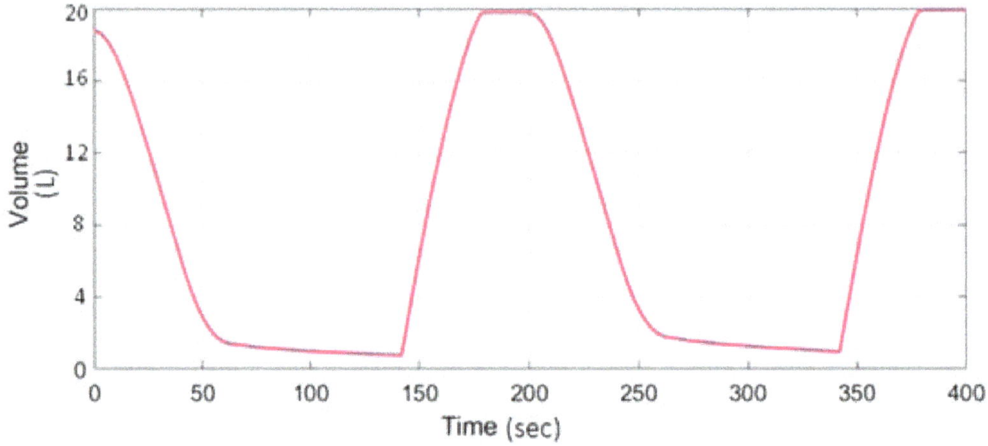

Figure 18. Simulation outcomes for the fluid volume inside the accumulator.

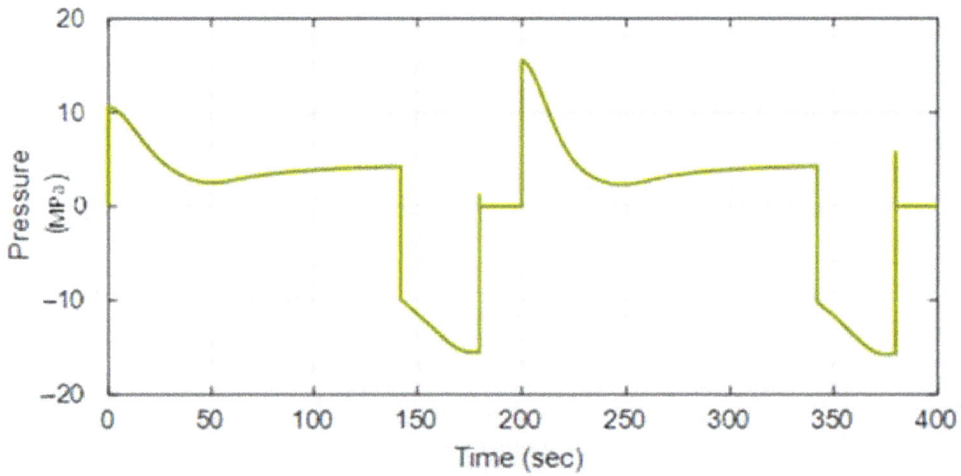

Figure 19. Simulation results for the pressure drop across the motor, i.e., the pressure difference between the inlet and outlet motor's pressure.

7.2. Power-Efficiency Outcomes

Figure 20 illustrates the load-speed results and the mechanical power at the pump and motor shafts, which correspond to the input and output shafts of the system, over the three transmission stages: acceleration, cruising, and braking. During acceleration, the pump is not functioning, which means the accumulator is the only power source in the system. The accumulator discharges its accumulated energy into the motor. The motor converts hydraulic energy into mechanical energy, generating an output mechanical power of about 1300 W in the first cycle and roughly 2000 W in the second cycle. The difference in mechanical powers between cycles one and two is due to the initial stored energy of the accumulator. During cruising, the pump starts operating at about 700 RPM, producing an input mechanical power of approximately 2100 watts. The addition of 1.5 kg·m^2 inertia at the wheels leads to an output mechanical power of around 1000 W at the motor's shaft. The motor transitions to its pumping mode upon braking. The system harnesses the braking energy and transforms it into hydraulic energy, which is then stored in the accumulator.

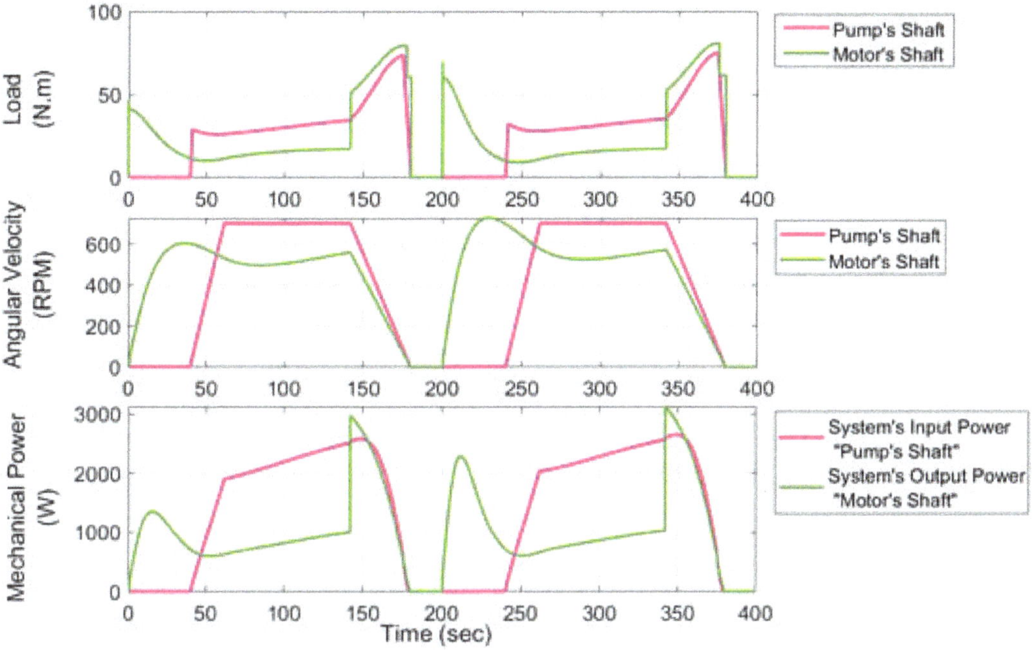

Figure 20. Load-Speed and mechanical power simulation results.

When simulating the system to reach the target duty cycle under moderate load-speed settings (1.5 kg·m2 and 700 RPM), the average input power at the system's input shaft is 876 W. The system's output shaft produces an average power of 780 W, resulting in an overall efficiency of approximately 89%. This outcome indicates a notable enhancement in the effectiveness of the series hydraulic hybrid drivetrains. The total average efficiency of the series hydraulic hybrid transmission ranges from 64% to 81% due to the rated efficiency of conventional hydraulic pump/motors, which is considered rather low[47]. The use of the advanced multi-piston digital pump improved the system's performance by boosting overall efficiency by a minimum of 8% and up to 25%.

8. CONCISE SUMMARY

Series hydraulic hybrid drivetrains typically have an average efficiency ranging from 64% to 81%, mostly because they use traditional variable displacement pumps that are inefficient while operating at partial displacements. This study evaluates the enhancement in efficiency of a sequential hydraulic hybrid transmission through the integration of hydraulic hybrid technology with digital hydraulics. The proposal suggests replacing the traditional piston pump/motor used in series hydraulic hybrid drivetrains with the multi-piston digital pump created and utilized in previous studies. The suggested approach is evaluated using a physical-based simulation of a hydraulic hybrid drivetrain series using MATLAB Simscape. The advanced multi-piston digital pump is replicated by inputting its digital driving and pumping data into the power supply units, specifically the pump and motor used in the simulation model. The simulation model that was selected enabled the assessment of the series hydraulic hybrid drivetrain's performance and facilitated the analysis of the system's enhanced overall efficiency. The final simulation results showed an overall efficiency of 89%, surpassing the standard series hydraulic hybrid drivetrain efficiency range of 64% to 81%. The use of advanced digital pumps can increase the overall efficiency of hydraulic hybrid drivetrains by a minimum of 8% and up to 25%. The achieved 89% efficiency can be further improved by utilizing a more sophisticated digital unit. The digital pump used in the simulation model was a prototype designed as a proof of concept, with a maximum shaft speed of 700 RPM. A revised model would operate across a wider variety of inputs and produce superior outcomes. Our future goal is to upgrade the proof-of-concept digital pump into a faster and more advanced unit for experimental testing of the planned transmission.

9. AT A GLANCE

Hydraulic hybrid drivetrains, fluid power technologies used in cars, offer a popular alternative to traditional drivetrain designs because of their significant energy efficiency, versatile power transmission, and user-friendly operation. Hydraulic hybrid drivetrains have numerous environmental advantages in comparison to alternative power transmission technologies. They offer robust vehicles such as commercial transportation, construction equipment, wagon handling, drilling machines, and military trucks that have the capability to improve fuel efficiency and reduce carbon emissions. Although hydraulic hybrid transmissions are common, modern hydraulic hybrid drivetrains have very low efficiency, ranging from 64% to 81%. The inefficiency is caused by using traditional variable displacement pumps and motors, which result in significant power losses during operation and struggle to maintain high efficiency at lower volumetric displacements. The study introduces and confirms a novel approach to enhance the overall effectiveness of hydraulic hybrid drivetrains by substituting traditional pump/motor units with digital versions. The digital pump/motor can achieve higher overall efficiency across a broad spectrum of operating conditions compared to traditional pump/motors. A prototype of a digital pump/motor was constructed and evaluated as a proof of concept. The experimental data were incorporated into a complex physics-based simulation model of a hydraulic hybrid transmission with many domains. The proposed methodology improves the overall efficiency of a hydraulic hybrid transmission series, leading to energy savings. Operating the system under moderate load-speed conditions resulted in a total efficiency of around 89%. Our simulation results show that using the state-of-the-art digital pump can increase the total efficiency of the series hydraulic hybrid powertrain by up to 25% compared to the average efficiency.

REFERENCES

1. McCloy, D.; Martin, H.R. Control of Fluid Power: Analysis and Design, 2nd ed.; Ehlh: Chichester, UK, 1980; Available online: **https://ui.adsabs.harvard.edu/abs/1980ehlh.book.....M/abstract** (accessed on 10 October 2021).
2. Durfee, W.; Sun, Z. Fluid Power System; A National Science Foundation Engineering Research Center: Minneapolis, MN, USA, 2009; pp. 1–48.
3. Vacca, A. Energy Efficiency and Controllability of Fluid Power Systems. Energies **2018**, 11, 1169. **[Google Scholar] [CrossRef] [Green Version]**
4. Konami, S.; Nishiumi, T. Hydraulic Control Systems; John Wiley & Sons: Singapore, 2016; pp. 187–299. **[Google Scholar] [CrossRef]**
5. Fabis-Domagala, J.; Domagala, M.; Momeni, H. A Concept of Risk Prioritization in FMEA Analysis for Fluid Power Systems. Energies **2021**, 14, 6482. **[Google Scholar] [CrossRef]**
6. What is Fluid Power? National Fluid Power Association. 2021. Available online: **https://www.nfpa.com/home/AboutNFPA/What-is-Fluid-Power.htm** (accessed on 1 December 2021).
7. U.S. Fluid Power Industry Brief. 2021. Available online: **https://www.nfpa.com/home/industry-stats/Industry-Brief.htm** (accessed on 1 December 2021).
8. Love, L.; Lanke, E.; Alles, P. Estimating the Impact (Energy, Emissions and Economics) of the US Fluid Power Industry. In Oak Ridge National Laboratory (Issue December). 2012. Available online: **http://www.osti.gov/servlets/purl/1061537/** (accessed on 1 December 2021).
9. Kwon, H.; Sprengel, M.; Ivantysynova, M. Thermal modeling of a hydraulic hybrid vehicle transmission based on thermodynamic analysis. Energy **2016**, 116, 650–660. **[Google Scholar] [CrossRef]**
10. Stelson, K.A.; Meyer, J.J.; Alleyne, A.G.; Hencey, B. Optimization of a passenger hydraulic hybrid vehicle to improve fuel economy. Proc. JFPS Int. Symp. Fluid Power **2008**, 2008, 143–148. **[Google Scholar] [CrossRef] [Green Version]**

11. Chen, J.-S. Energy Efficiency Comparison between Hydraulic Hybrid and Hybrid Electric Vehicles. Energies **2015**, 8, 4697–4723. **[Google Scholar]** **[CrossRef]** **[Green Version]**
12. Gary, J. Hydrid Transmission. Acad. Manag. Rev. **2006**, 31, 386–408. **[Google Scholar]**
13. Zhang, Z.; Chen, J.; Wu, B. The control strategy of optimal brake energy recovery for a parallel hydraulic hybrid vehicle. Proc. Inst. Mech. Eng. Part D J. Automob. Eng. **2012**, 226, 1445–1453. **[Google Scholar]** **[CrossRef]**
14. Rydberg, K.-E. Energy Efficient Hydraulic Hybrid Drives. In Proceedings of the 11th Scandinavian International Conference on Fluid Power, SICFP'09, Linköping, Sweden, 2–4 June 2009. **[Google Scholar]**
15. Tvrdić, V.; Podrug, S.; Šuljić, I.; Matić, B. Hydraulic hybrid vehicle configurations and comparison with hybrid electric vehicle. In Proceedings of the Comtemporary Issues in Economy & Technology—CIET 2018, Split, Croatia, 1–2 June 2018; pp. 548–556. **[Google Scholar]**
16. Florida, N.; López, C.; Pocomucha, V. CORE View metadata, citation and similar papers at core. ac. uk. J. Dedik. Pendidik. **2012**, 2, 35–43. **[Google Scholar]**
17. Shan, M. Modeling and Control Strategy for Series Hydraulic Hybrid Vehicles. Ph.D. Thesis, The University of Toledo, Toledo, OH, USA, December 2009. **[Google Scholar]**
18. Tavares, F.; Johri, R.; Salvi, A.; Baseley, S.; Filipi, Z. Hydraulic Hybrid Powertrain-In-the-Loop Integration for Analyzing Real-World Fuel Economy and Emissions Improvements. SAE Technical Paper. 2011. Available online: **https://doi.org/10.4271/2011-01-2275** (accessed on 20 October 2021).
19. Midgley, W.; Cebon, D. Comparison of regenerative braking technologies for heavy goods vehicles in urban environments. Proc. Inst. Mech. Eng. Part D J. Automob. Eng. **2012**, 226, 957–970. **[Google Scholar]** **[CrossRef]**
20. Li, C.-T.; Peng, H. Optimal configuration design for hydraulic split hybrid vehicles. In Proceedings of the 2010 American Control Conference, Baltimore, MD, USA, 30 June–2 July 2010; pp. 5812–5817. **[Google Scholar]** **[CrossRef]**
21. Kim, N.; Rousseau, A. A Comparative Study of Hydraulic Hybrid Systems for Class 6 Trucks. SAE Tech. Pap. **2013**, 2. **[Google Scholar]** **[CrossRef]**
22. Sprengel, M.; Bleazard, T.; Haria, H.; Ivantysynova, M. Implementation of a Novel Hydraulic Hybrid Powertrain in a Sports Utility Vehicle. IFAC-PapersOnLine **2015**, 48, 187–194. **[Google Scholar]** **[CrossRef]**
23. Bottiglione, F.; Mantriota, G.; Valle, M. Power-Split Hydrostatic Transmissions for Wind Energy Systems. Energies **2018**, 11, 3369. **[Google Scholar]** **[CrossRef]** **[Green Version]**

24. Jae, Y.; Filipi, Z.; Kim, Y.J.; Filipi, Z. Simulation Study of a Series Hydraulic Hybrid Prop System for a Light Truck. SAE Trans. **2021**, 116, 147–161. [**Google Scholar**]

25. Hui, S.; Junqing, J. Research on the system configuration and energy control strategy for parallel hydraulic hybrid loader. Autom. Constr. **2010**, 19, 213–220. [**Google Scholar**] [**CrossRef**]

26. Zhou, S.; Walker, P.; Tian, Y.; Zhang, N. Mode switching analysis and control for a parallel hydraulic hybrid vehicle. Veh. Syst. Dyn. **2020**, 59, 928–948. [**Google Scholar**] [**CrossRef**]

27. Sim, T.P.; Li, P.Y. Analysis and Control Design of a Hydro-Mechanical Hydraulic Hybrid Passenger Vehicle. In Proceedings of the Dynamic Systems and Control Conference, Hollywood, CA, USA, 12–14 October 2009; pp. 667–674. [**Google Scholar**] [**CrossRef**] [**Green Version**]

28. Cheong, K.L. Design and Analysis of Hydraulic Hybrid Passenger Vehicles. Ph.D. Thesis, University of Minnesota, Minneapolis, MN, USA, 2015. [**Google Scholar**]

29. Ramdan, M.I.; A Stelson, K. Optimal design of a power-split hybrid hydraulic bus. Proc. Inst. Mech. Eng. Part D J. Automob. Eng. **2016**, 230, 1699–1718. [**Google Scholar**] [**CrossRef**]

30. Sprengel, M.; Ivantysynova, M. Recent Developments in a Novel Blended Hydraulic Hybrid Transmission. SAE Tech. Pap. **2014**, 2014. [**Google Scholar**] [**CrossRef**]

31. Stecki, J.; Matheson, P. Advances In Automotive Hydraulic Hybrid Drives. Proc. JFPS Int. Symp. Fluid Power **2005**, 2005, 664–669. [**Google Scholar**] [**CrossRef**] [**Green Version**]

32. Matheson, P.; Stecki, J. Development and simulation of a hydraulic-hybrid powertrain for use in commercial heavy vehicles. SAE Tech. Pap. **2003**, 112, 114–123. [**Google Scholar**] [**CrossRef**]

33. Wu, B.; Lin, C.-C.; Filipi, Z.; Peng, H.; Assanis, D. Optimal Power Management for a Hydraulic Hybrid Delivery Truck. Veh. Syst. Dyn. **2004**, 42, 23–40. [**Google Scholar**] [**CrossRef**]

34. Van Batavia, B.L. Hydraulic Hybrid Vehicle Energy Management System; SAE Technical Paper; SAE International: Warrendale, PA, USA, 2009. [**Google Scholar**] [**CrossRef**]

35. Achten, P.A.J.; Schellekens, M.P.A.; Murrenhoff, H.; Deeken, M. Efficiency and Low Speed Behavior of the Floating Cup Pump. J. Commer. Veh. **2004**, 113, 366–376. [**Google Scholar**] [**CrossRef**]

36. Buchwald, P.; Christensen, G.; Larsen, H.; Sunn Pedersen, P. Improvement of citybus fuel economy using a hydraulic hybrid propulsion system—A theoretical and experimental study. SAE Tech. Pap. **1979**, 88, 1042–1056. [**Google Scholar**] [**CrossRef**]

37. Hydraulic Launch Assist. 2013. Available online: **https://www.eaton.com/SEAsia/ProductsSolutions/Hydraulics/ProductsServices/HydraulicLaunchAssist/index.htm** (accessed on 1 October 2021).

38. Tollefson, S.; Beachley, N.H.; Fronczak, F.J. Studies of an Accumulator Energy-Storage Automobile Design with a Single Pump/Motor Unit. SAE Trans. **1985**. [**Google Scholar**] [**CrossRef**]

39. Kepner, R.P. Hydraulic power assist—A demonstration of Hydraulic Hybrid Vehicle regenerative braking in a road vehicle application. SAE Tech. Pap. **2002**, 111, 826–833. [**Google Scholar**] [**CrossRef**]

40. Bozic, A. Introducing Hydraulic-Electric Synergy into Hybrid Transmission Using the Free-piston Engine Technology. In Proceedings of the Commercial Vehicle Engineering Congress and Exhibition and Powertrain & Fluid Systems Conference, Rosemont, IL, USA, 29 October–1 November 2007. [**Google Scholar**] [**CrossRef**]

41. Breidi, F.; Garrity, J.; Lumkes, J. Design and Testing of Novel Hydraulic Pump/Motors to Improve the Efficiency of Agricultural Equipment. Trans. ASABE **2017**, 60, 1809–1817. [**Google Scholar**] [**CrossRef**]

42. El-Breidi, F. Investigation of digital pump/motor control strategies. Ph.D. Thesis, Purdue University, West Lafayette, IN, USA, 2016. [**Google Scholar**]

43. Yang, H.-Y.; Pan, M. Engineering research in fluid power: A review. J. Zhejiang Univ. A **2015**, 16, 427–442. [**Google Scholar**] [**CrossRef**] [**Green Version**]

44. Merrill, K.J.; Breidi, F.Y.; Lumkes, J. Simulation based design and optimization of digital pump/motors. In Fluid Power Systems Technology; American Society of Mechanical Engineers: New York, NY, USA, 2013. [**Google Scholar**]

45. Li, M.; Foss, R.; Stelson, K.A.; Van De Ven, J.D.; Barth, E.J. Design, Dynamic Modeling, and Experimental Validation of A Novel Alternating Flow Variable Displacement Hydraulic Pump. IEEE/ASME Trans. Mechatron. **2019**, 24, 1294–1305. [**Google Scholar**] [**CrossRef**]

46. Wilhelm, S.R.; Van De Ven, J.D. Synthesis of a Variable Displacement Linkage for a Hydraulic Transformer. In Proceedings of the International Design Engineering Technical Conferences and Computers and Information in Engineering Conference, Washington, DC, USA, 28–31 August 2011. [**Google Scholar**] [**CrossRef**] [**Green Version**]

47. Chen, Y.L.; Liu, S.A.; Jiang, J.H.; Shang, T.; Zhang, Y.K.; Wei, W. Dynamic analysis of energy storage unit of the hydraulic hybrid vehicle. Int. J. Automot. Technol. **2013**, 14, 101–112. **[Google Scholar] [CrossRef]**

48. Ivantysynova, M. Innovations in pump design-what are future directions? Proc. JFPS Int. Symp. Fluid Power **2008**, 2008, 59–64. **[Google Scholar] [CrossRef] [Green Version]**

49. Zhang, C.; Huang, S.; Du, J.; Wang, X.; Wang, S.; Zhang, H. A new dynamic seven-stage model for thickness prediction of the film between valve plate and cylinder block in axial piston pumps. Adv. Mech. Eng. **2016**, 8. **[Google Scholar] [CrossRef] [Green Version]**

50. Shang, L.; Ivantysynova, M. A temperature adaptive piston design for swash plate type axial piston machines. Int. J. Fluid Power **2016**, 18, 38–48. **[Google Scholar] [CrossRef]**

51. Rizzo, G.; Massarotti, G.; Bonanno, A.; Paoluzzi, R.; Raimondo, M.; Blosi, M.; Veronesi, F.; Caldarelli, A.; Guarini, G. Axial piston pumps slippers with nanocoated surfaces to reduce friction. Int. J. Fluid Power **2015**, 16, 1–10. **[Google Scholar] [CrossRef]**

52. Inaguma, Y.; Hibi, A. Reduction of friction torque in vane pump by smoothing cam ring surface. Proc. Inst. Mech. Eng. Part C J. Mech. Eng. Sci. **2007**, 221, 527–534. **[Google Scholar] [CrossRef]**

53. Wieczorek, U.; Ivantysynova, M. Computer Aided Optimization of Bearing and Sealing Gaps in Hydrostatic Machines—The Simulation Tool Caspar. Int. J. Fluid Power **2002**, 3, 7–20. **[Google Scholar] [CrossRef]**

54. Sazonov, Y.A.; Mokhov, M.A.; Gryaznova, I.V.; Voronova, V.V.; Tumanyan, K.A.; Frankov, M.A.; Balaka, N.N. Development and Prototyping of Jet Systems for Advanced Turbomachinery with Mesh Rotor. Emerg. Sci. J. **2021**, 5, 775–801. **[Google Scholar] [CrossRef]**

55. Kostikov, Y.A.; Romanenkov, A.M. Approximation of the Multidimensional Optimal Control Problem for the Heat Equation (Applicable to Computational Fluid Dynamics (CFD)). Civ. Eng. J. **2020**, 6, 743–768. **[Google Scholar] [CrossRef]**

56. Manring, N.D. Valve-Plate Design for an Axial Piston Pump Operating at Low Displacements. J. Mech. Des. **2003**, 125, 200–205. **[Google Scholar] [CrossRef]**

57. Helmus, T.; Breidi, F.; Lumkes, J. Simulation of a variable displacement mechanically actuated digital pump unit. In Proceedings of the Eight Workshop on Digital Fluid Power, Tampere, Finland, 24–25 May 2016; pp. 95–106. **[Google Scholar]**

58. Chehade, A.; Breidi, F.; Pate, K.S.; Lumkes, J. Data-driven Adaptive Thresholding Model for Real-time Valve Delay Estimation in Digital Pump/Motors. Int. J. Fluid Power **2019**, 20, 271–294. **[Google Scholar]** **[CrossRef]**

59. Winkler, B. Recent Advances in Digital Hydraulic Components and Applications. In Proceedings of the The Ninth Workshop on Digital Fluid Power, Aalborg, Denmark, 7–8 September 2017; Available online: **https://www.et.aau.dk/digitalAssets/377/377525_id117-recent-advances-in-digital-hydraulic-components-and-applications.pdf** (accessed on 1 December 2021).

60. Brandstetter, R.; Deubel, T.; Scheidl, R.; Winkler, B.; Zeman, K. Digital hydraulics and "Industrie 4.0". Proc. Inst. Mech. Eng. Part I J. Syst. Control. Eng. **2016**, 231, 82–93. **[Google Scholar]** **[CrossRef]**

61. Peng, S.; Branson Iii, D.T.; Guglielmino, E.; Boaventura, T.; Caldwell, D.G. Performance Assessment of Digital Hydraulics in a Quadruped Robot Leg. 2012. Available online: **http://asmedigitalcollection.asme.org/ESDA/proceedings-pdf/ESDA2012/44861/227/4451304/227_1.pdf?casa_token=fWcoONtKGfUAAAAA:9c3YrOxP0rRp-PmWLcv8DUFtas4Qe6HnyCq02vGig-lSARq94FAcQnGOpL9W-XpcM-sdwGk** (accessed on 5 November 2021).

62. Pinto, L.P.G.; Belan, H.C.; Locateli, C.C.; Krus, P.; De Negri, V.J.; Lantto, B. New perspectives on digital hydraulics for aerospace applications. In Proceedings of the Aerospace Technology Congress, Solna, Stockholm, 11–12 October 2016. **[Google Scholar]**

63. Breidi, F.; Helmus, T.; Lumkes, J. The impact of peak-and-hold and reverse current solenoid driving strategies on the dynamic performance of commercial cartridge valves in a digital pump/motor. Int. J. Fluid Power **2015**, 17, 37–47. **[Google Scholar]** **[CrossRef]**

64. Breidi, F.; Chehade, A.; Lumkes, J. Monitoring Digital Technologies in Hydraulic Systems Using CUSUM Control Charts. In Fluid Power Systems Technology; American Society of Mechanical Engineers: New York, NY, USA, 2019. **[Google Scholar]**

65. Helmus, T. Investigation and Implementation of Mechanically Actuated Valves for Digital Hydraulic Units. Ph.D. Thesis, Purdue University, West Lafayette, IN, USA, 2017. **[Google Scholar]**

66. Holland, M.A. Design of Digital Pump/Motors and Experimental Validation of Operating Strategies Doctor of Philosophy. 2012. Available online: **http://www.purdue.edu/policies/pages/teach_res_outreach/c_22.html** (accessed on 1 December 2021).

67. Breidi, F.; Garrity, J.; Lumkes, J., Jr. Investigation of a real-time pressure based valve timing correction algorithm. In Fluid Power Systems Technology; American Society of Mechanical Engineers: New York, NY, USA, 2017. **[Google Scholar]**

68. Simscape—MATLAB & Simulink. 2021. Available online: **https://www.mathworks.com/products/simscape.html** (accessed on 6 October 2021).

69. Sizing Tool—Reasontek Corp. 2021. Available online: **https://www.reasontek.com/sizing-tool/** (accessed on 23 November 2021).

70. Cheema, J. Air and Oil Do Mix: Selecting the Right Accumulator Is Critical for Wind Farm Applications. Hydraulics and Pneumatic. 2013. Available online: **http://ezproxy.derby.ac.uk/login?url=http://search.ebscohost.com/login.aspx?direct=true&db=edsbl&AN=RN332050940&site=eds-live** (accessed on 28 November 2021).

71. Xu, Z.; Li, W.; Liu, X.; Chen, Z. Dynamic characteristics of coupling model of valve-controlled cylinder parallel accumulator. Mech. Ind. **2019**, 20, 306. **[Google Scholar] [CrossRef]**

72. Valeski, B. Average Tire Weight (With 10 Examples)—Survival Tech Shop. 2021. Available online: **https://www.survivaltechshop.com/tire-weight/** (accessed on 1 December 2021).

02 COMPARING THE ENERGY EFFICIENCY OF LOAD SENSING SYSTEM WITH ELECTRO-HYDRAULIC SOLUTIONS ON A 9-TON EXCAVATOR

1. INTRODUCTION: AN OVERVIEW

Electro-hydraulic systems can be created with electro-hydrostatic actuators (EHA), which have small size, excellent energy efficiency, and the benefit of facilitating plug-and-play setups. Several research articles contrast EHA systems with conventional valve-controlled hydraulic structures. Schmidt and colleagues [10] suggest multiple options that offer pump control along with load holding features. The energy distribution and potential for energy recovery are outlined for all analyzed options, with a comparison to a conventional hydraulic system controlled by valves. Padovani et al. propose an alternative electro-hydraulic drive method for single-rod cylinders that can hold a load passively. The authors demonstrate that when implementing this design on a single-arm crane, the positional error of the arm is limited to ±2 mm, and the energy efficiency during handling is approximately 60%. Takahashi et al. [12] have suggested an electro-hydraulic system for a single-rod double-acting actuator used in aircraft landing gear, among other aeronautical applications. Ketelsen et al. [13] conducted an in-depth analysis of the scientific literature, examining electro-actuated systems with variable displacement hydraulic pumps and solutions with variable speed electric motors to compare the pros and cons of each design. The authors document a growing shift towards electric motor-based alternatives, distancing from diesel engines. The EHA architecture allows for energy recovery and spreading it among several actuators.

According to Abekawa et al. [14], excavators account for approximately 60% of the CO_2 emissions generated by construction machines. Implementing novel ways to reduce greenhouse gas emissions from excavators can effectively cut world CO_2 emissions. Niraula et al. and Zhang et al. demonstrated that utilizing an electro-hydraulic actuation system in 1-ton mini excavators might reduce energy consumption by up to 50% compared to the typical load sensing circuit architecture. Budden et al. [17] upgraded a 20-ton excavator by installing a digital displacement pump in place of the swashplate machine. The results indicate a notable rise in productivity linked to fuel conservation.

This article discusses additional findings from a research study conducted on a 9-ton excavator, focusing on the analysis of several electro-hydraulic solutions. The results regarding mechanical energy and fuel usage were compared to those achieved using the typical load sensing system. The results were derived from simulations conducted with a mathematical model of the excavator developed in the Simcenter Amesim© software. The machine›s standard load sensing model, developed using a lumped parameter method, includes simulating the diesel engine, the variable displacement main pump, the post-compensated flow sharing distributors, and the kinematics of the front equipment. The model has been confirmed through rigorous experimental testing, allowing for the quantification of fuel consumption during work cycles [18,19,20,21,22,23,24,25]. The excavator›s mathematical model has been altered to analyze novel EHA circuit structures. Specifically, an open-circuit and two closed-circuit solutions have been examined. The energy efficiency of different setups was compared to a load sensing system. The findings indicate that electro-hydraulic circuit solutions provide notably better energy efficiency than a conventional valve-controlled system, resulting in reduced fuel consumption. The authors in a previous study explored an EHA architecture using an open-circuit solution. In this setup, the accumulator was linked to the actuators' side with lower average pressure. However, this approach had a drawback as the side with lower pressure could vary with the duty cycle. To address this limitation, a shuttle valve has been incorporated to link the accumulator to the lower pressure side of the actuator [13,27,28,29]. Additionally, this work explores two closed-circuit solutions that were not examined in reference [26]. The initial closed-circuit shown is akin to the open-circuit to facilitate a dependable comparison. The second closed-circuit solution [13] introduces a new architecture that requires additional control of a solenoid valve. The methodology involves comparing all researched options with the typical load sensing circuit in terms of mechanical energy and fuel usage savings. This paper's contribution lies not in developing new circuit architectures, which are already documented in the cited papers, but in the potential of mathematical modeling to provide accurate information on the implementation of various solutions in a mobile machine such as the excavator under study. The document is organized as follows. Section 2 outlines the mathematical basis of the conventional load sensing hydraulic excavator. Section 3 outlines various circuit solutions utilizing EHA. The simulation results for all architectures are shown in Section 4. Section 5 compares the conventional model LS and EHA solutions. Section 6 will present the conclusions and future developments.

2. MATHEMATICAL MODEL OF THE STANDARD HYDRAULIC EXCAVATOR

The machine being discussed is a 9-ton hydraulic excavator. The standard version has a 46-kW diesel engine that powers an axial piston pump to supply a hydraulic system controlled by load sensing logic (LS) with flow sharing distributors. Figure 1 displays the ISO diagram of the machine's circuit configuration, highlighting the sensors utilized for testing. A mathematical model utilizing a lumped parameter method has been created in the Simcenter Amesim® environment. The comprehensive and intricate model description is available in prior works [18,20,22,23].

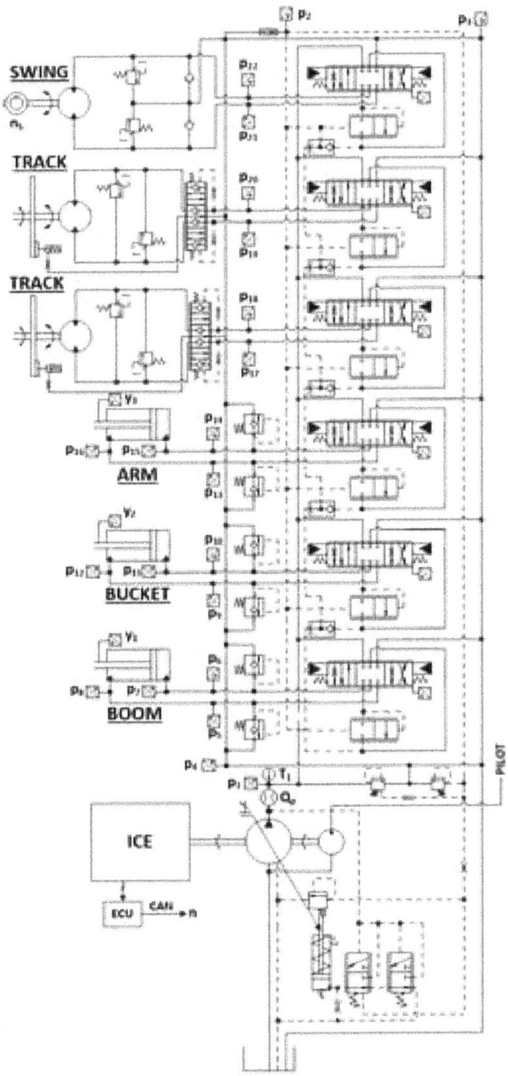

Figure 1. ISO scheme of the standard excavator hydraulic circuit. The sensors used in the test phase are also indicated [23].

The standard hydraulic circuit, as depicted in the ISO figure, has two pumps. The machine is a fixed displacement device with external gears that supplies the circuit specifically designed for pilots. The primary pump that provides for the consumers is a variable displacement axial piston pump, which includes a pressure compensator, flow compensator, and torque limiter. The pump has been represented using a gray box method, with white box models for the regulators and a black box approach for the pump flow characteristic, utilizing hydromechanical and volumetric efficiency maps derived from experimental work.

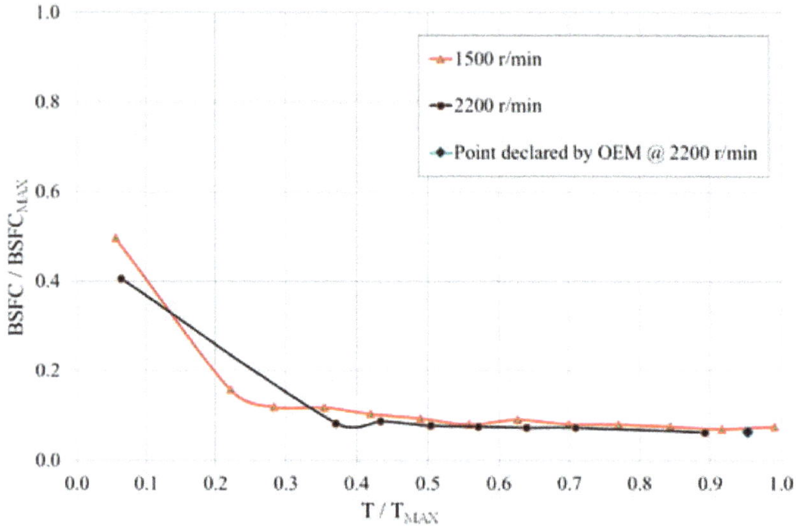

Figure 2. Diesel engine—Brake specific fuel consumption map.

The directional valves are post-compensated LS flow-sharing valves that enable the extraction of load sensing pressure to be directed to the pump regulators. This sort of valve's primary function is to ensure a consistent pressure decrease in each segment, even when the main pump hits saturation levels. The distributor model has been confirmed by experimental results from tests conducted in the laboratory of the Department of Engineering and Architecture at the University of Parma, Italy, as outlined in references [19,24]. The hydraulic cylinders are single-rod, double-acting actuators. Each cylinder model accounts for linear friction and leakage between the two chambers. The Coulomb and viscous friction coefficients were determined by experimental testing. The hydraulic cylinders were simulated using the continuity equation and the fluid state equation for two control volumes: one on the piston side and one on the rod side. The actuators' model has not changed in the additional circuit designs suggested in this article. A kinematic model has been created for the front attachment and swing to accurately determine the forces and torques affecting the hydraulic actuators. The front attachment comprises a boom, arm, and bucket, modeled as rigid structures connected by rotary joints and linear actuators. The mass, moment of inertia, and center of gravity parameters have been determined based on the CAD geometry of the components. The swing model assumes constant inertia and incorporates both Coulomb friction and viscous friction factors based on experimental experiments conducted with rapid and slow rotation cycles. The kinematic model has stayed consistent in all the circuit solutions discussed in this paper. The internal combustion engine model, incorporating manufacturer-provided data and experimentally collected data, enables the calculation of instantaneous fuel consumption based on torque and engine speed, as shown in Figure 2 [20,22].

3. ELECTRO-HYDRAULICS SOLUTIONS

The validated excavator model serves as the foundation for developing EHA solutions to conduct energy comparisons. This paper reintroduces an open-circuit configuration named "Open Circuit—Layout 1" from a previous article [26], with modifications in the accumulator's position. Additionally, two new closed-circuit configurations, labeled "Layout 2" and "Layout 3," have been created.

3.1. Open Circuit—Layout 1

The circuit configuration depicted in the ISO diagram in Figure 3 consists of a pair of hydraulic machines with fixed displacement for each user. These machines are coupled to an electric unit that can function as either a pump or a motor based on the pressure differential applied. The swing was intended to be operated only by an electric motor. The internal combustion engine in the typical excavator powers an electric generator, which then delivers energy to the electric motor used by each user. The hydraulic machines are external gear units with models that show volumetric and hydromechanical efficiency based on speed and delivery pressure.

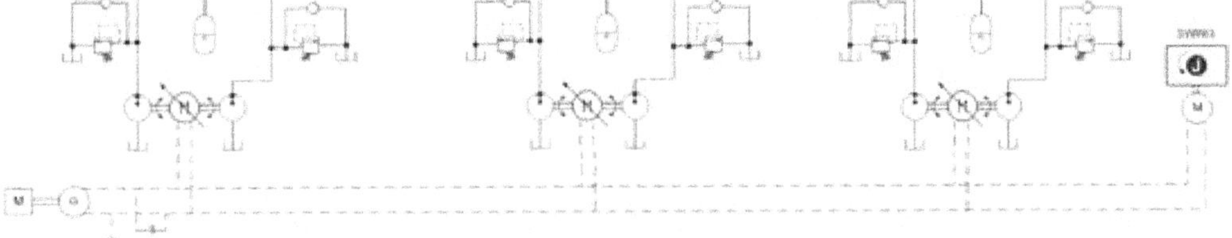

Figure 3. ISO scheme of Open Circuit—Layout 1.

The hydraulic machine sizing is determined based on the actuator displacement. The ratio between the hydraulic machine's displacement on the piston side and the rod side should be similar to the ratio between the active surfaces of the actuator. It is recommended to maintain the difference between these ratios below 2%. A hydraulic accumulator is used to compensate for discrepancies in flow rate caused by changes in displacements of hydraulic machines and internal losses of components. The accumulator is linked to the side with the lowest pressure using a shuttle valve. If there is excess load, the hydraulic motor can recover energy by functioning as a generator through the electric unit. Therefore, it is essential to incorporate a battery to store the generated energy.

Each side of the actuator is equipped with electrically controlled on/off valves to block the load, which includes a pressure drop. They are activated solely when required to move the cylinder. The valves are equipped with a spring mechanism to prevent the load from moving, even if there is a failure in the electrical power supply system for safety purposes.

3.2. Closed Circuits

Two distinct circuit solutions utilizing EHA architecture were suggested and their outcomes were contrasted with those of the conventional LS circuit and the open-circuit. The suggested setups are closed-circuit layouts. All layouts include a bidirectional fixed displacement hydraulic pump powered by an electric motor. The pump's delivery port is connected to either the rod side or the piston side of the actuator based on the electric motor's rotation direction for each cylinder. Electric motors are powered by a generator that is operated by the internal combustion engine, which is identical to the one used in a regular excavator. An electric motor has replaced the hydraulic motor for moving the swing, without taking the tracks into account. All recommended options allow for the removal of the variable displacement main pump and the directional valves commonly seen in the standard LS system.

The hydraulic machines are external gear pumps that operate in two directions and have been designed based on efficiency maps that evaluate volumetric and hydromechanical efficiency in relation to rotation speed and delivery pressure. The pumps and their electric motors are identical in all suggested setups.

It is crucial to balance the differential flow in actuators due to the presence of distinct zones between the rod side and the piston side. The two solutions vary in their approach to achieving flow compensation.

3.3. Closed Circuit—Layout 2

Figure 4 displays the ISO method for Layout 1. The solution involves a flushing valve that links the low-pressure chamber of the cylinder with the accumulator to offset the uneven flow rate caused by the differential actuator [13]. When the piston side is supplied, the pump requires a higher flow rate, causing depressurization in the rod side chamber. The valve then switches to link the accumulator with the rod side, supplying the additional flow rate needed by the pump. When the rod side is activated, the surplus flow from the piston side is directed into the accumulator.

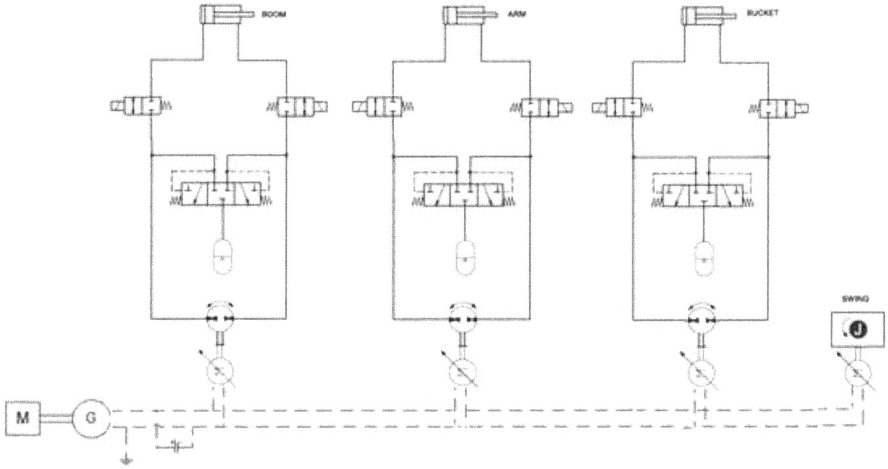

Figure 4. ISO scheme of Closed Circuit—Layout 2.

In the case of an overrunning load, the hydraulic machine works as a motor and drives the electric machine, which, acting as a generator, allows energy to be recovered, storing it in the battery.

Two block valves have been inserted, similarly to the open-circuit layout, to ensure the position of the actuator even in the event of power failure. These on/off valves are opened only when it is necessary to move the cylinder.

3.4. Closed Circuit—Layout 3

Figure 5 shows the circuit diagram of Layout 3. This solution has been reported by Padovani et al. **[13]**. Each user consists of an electric motor and a fixed displacement hydraulic machine driving the double-acting single-rod cylinder, in a closed-circuit layout.

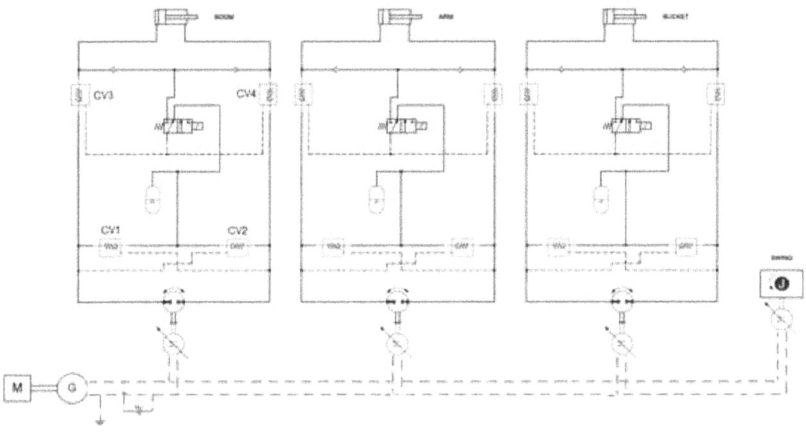

Figure 5. ISO scheme of Layout 3.

The differential flow managed by the low-pressure accumulator is controlled by two piloted check valves (CV1 and CV2). There are also two pilot operated check valves (CV3 and CV4) for load holding, which, unlike Layout 1, are fully hydraulic valves. The actuation of the solenoid valve transmits the signal of the maximum pressure of the actuator to the pilot line to allow for the opening of the two load holding valves to permit the movement of the actuator. When the solenoid valve is not powered, the piloting line of the load holding valves is connected to the low-pressure line and does not allow for their opening. In the event of a power failure, the solenoid valve assumes this latter position by means of a spring, preventing the movement of the load.

The next sections describe the features common to all of the proposed EHA configurations.

3.5. Model of the Operator

The digging duty cycle defined by the JCMAS standard is replicated by the operator model. In the standard model of the excavator, the position of the actuator is compared with the desired position and, consequently, the opening of the directional valve is defined.

In all of the proposed EHA solutions, the position of the actuator depends on the flow rate delivered, given by the rotation speed of the hydraulic machine: the error between the desired position and the actual one controls the speed of the electric motor by means of a PI type controller, as indicated in **Figure 6.**

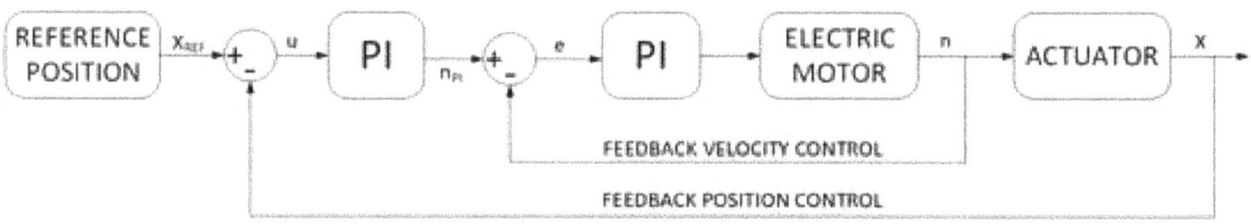

Figure 6. Actuator control scheme.

3.6. Electric Components

The electrical circuit is a common feature in all analyzed EHA setups. The system has three electric devices, a battery, and a generator linked to the diesel engine. Introducing a battery is necessary for energy recovery during overrunning load phases. The battery is simulated as an ideal storage device without accounting for its dynamic behavior or thermal impacts. The information presented in Table 1 pertains to a lithium-ion battery with properties documented in scientific literature [32].

The electric machine model in Amesim© can operate in all four quadrants of the torque-speed map, functioning as a generator to produce electricity stored in the battery. Electric machines are believed to operate within a consistent range where a steady average efficiency is maintained.

Table 1. Main features of the electric components.

Component	Parameter	Value
Electric motor	Voltage (V]	400
	Max speed (r/min]	4000
	Moment of inertia (kg/m^2]	0.049
Generator	Min speed (r/min]	1000
	Rated speed (r/min]	2200
	Voltage (V]	400
Battery	Specific Energy (Wh/kg]	130
	Nominal voltage (V]	400
	Specific power (W/kg]	2000
	Power (kW]	50

Table 1 displays the common features of the electrical components in all investigated combinations.

4. RESULTS

The simulations were utilized to compare the energy requirements of the various systems under investigation. Table 2 displays the mechanical energy provided by the diesel engine and needed by the pumps for each configuration. In the LS solution, the value indicates the energy needed by the main pump, whereas in the EHA solutions, the numbers represent the energy required by the alternator. The results indicate that EHA solutions significantly decrease the mechanical energy needed, resulting in savings of more than 60% for all configurations. This is primarily because EHA systems do not require pressure compensators and directional valves, which are common in load sensing systems.

Closed-circuit solutions yield greater improvements compared to open-circuit methods. Layout 2 allows for a 2% greater mechanical energy saving compared to Layout 3. These patterns are further supported by examining the fuel savings. Table 3 displays the fuel consumption of the diesel engine for each configuration, taking into account the engine efficiency map. The fuel savings are notable but less significant compared to the mechanical energy savings due to the thermal engine's efficiency being closely linked to the absorbed torque value. Throughout the cycle, the diesel engine requires less power for EHA solutions compared to the conventional LS circuit, as depicted in Figure 9. Typically, the internal combustion engine experiences reduced efficiency when operating under lighter loads. The reduced power needed by the examined solutions makes it feasible to decrease the size of the diesel engine.

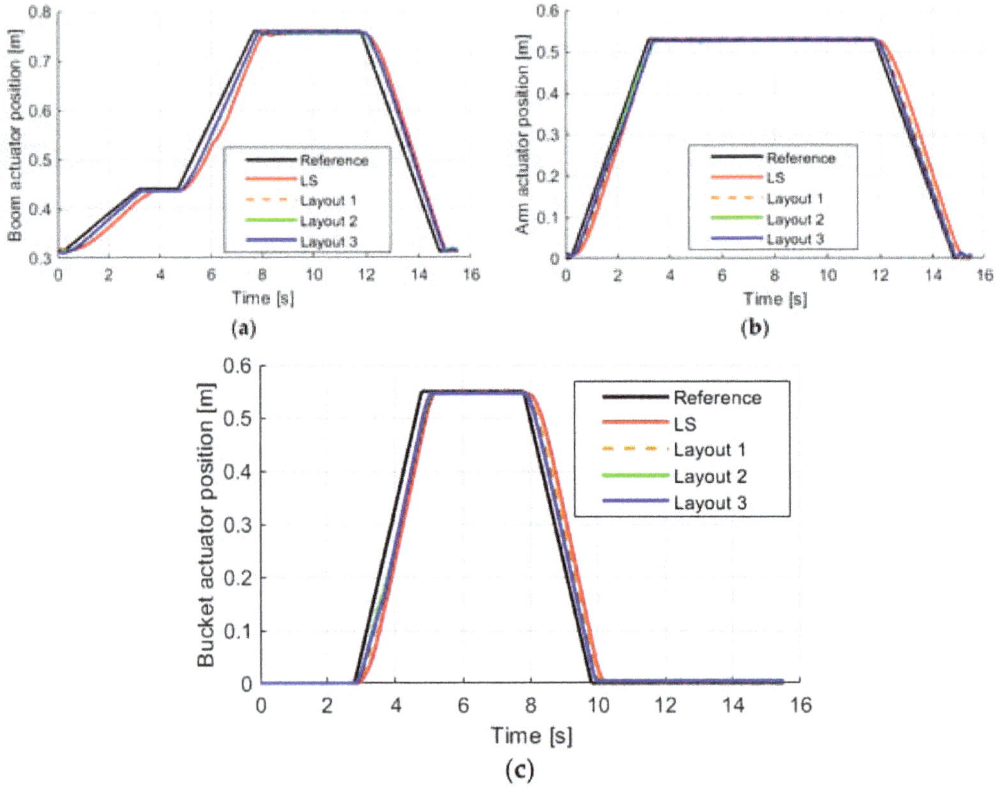

Figure 7. Actuator response to reference position of the digging cycle: simulations results: (a) boom, (b) arm, and (c) bucket.

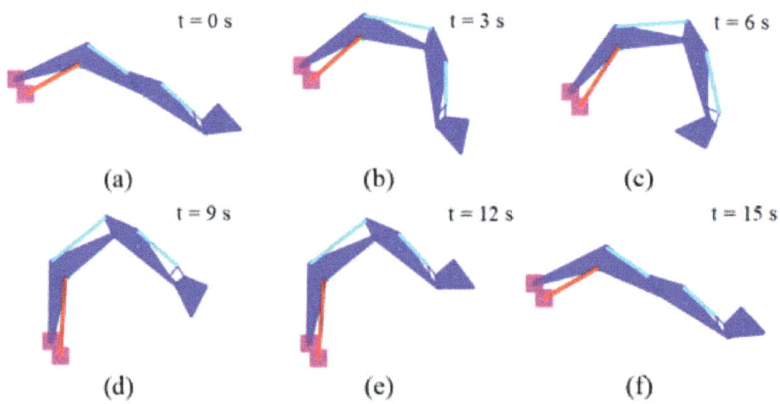

Figure 8. Sequence of front implement movements during the digging cycle.

This approach could lead to increased fuel efficiency by enabling the engine to operate at average loads that are closer to the maximum, where efficiency is better. This work did not explore the potential for reducing the size of the diesel engine, but it references fascinating findings in [26] related to an open-circuit EHA solution.

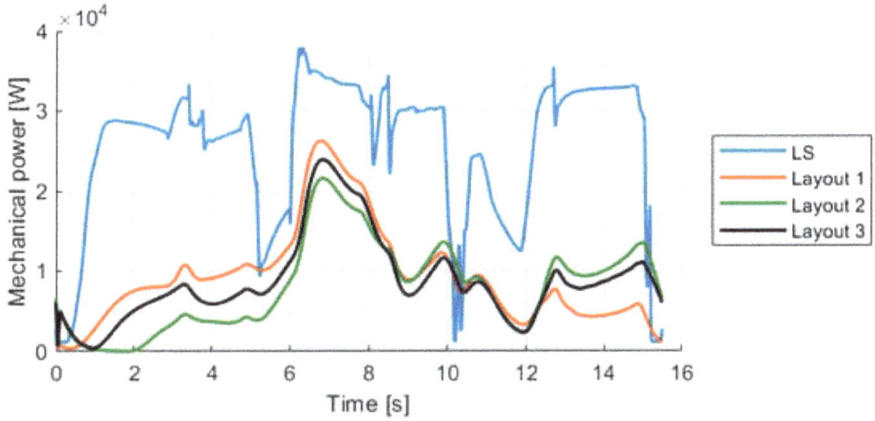

Figure 9. Profile of mechanical power during the digging cycle.

Table 2. Mechanical energy.

Solutions	Mechanical Energy Supplied by Diesel Engine (kJ/Cycle)	Mechanical Energy Saving (%)	Mechanical Energy Supplied to the Pumps (kJ/cycle)	Mechanical Energy Saving (%)
LS	393.6	/	393.6	/
Open Circuit—Layout 1	143.6	63.5%	91.6	76.7%
Closed Circuit—Layout 2	112.2	71.5%	84.7	78.5%
Closed Circuit—Layout 3	136.2	65.4%	109.2	72.3%

Table 3. Diesel engine fuel consumption.

Solutions	Fuel Consumption (g/cycle)	Fuel Saving (%)
LS	34.49	/
Open Circuit—Layout 1	23.92	30.6%
Closed Circuit—Layout 2	22.50	33.1%
Closed Circuit—Layout 3	23.60	31.6%

Figures 10, 11, 12, and 13 display Sankey diagrams illustrating the distribution of mechanical input energy among various components in the circuit, enabling the quantification of energy losses. The valve block in the LS standard arrangement, as shown in Figure 10, is responsible for over 57% of the energy loss. In Figure 11, the open-circuit EHA solution shows that the majority of energy losses are focused on the electric and hydraulic machinery group. In contrast, the closed-circuit versions exhibit comparable energy distributions, as illustrated in Figure 12 and Figure 13, with reduced losses compared to the open-circuit option. Layout 3 experiences lower energy losses on hydraulic machines compared to Layout 2. However, it incurs significantly higher losses on load holding valves due to the piloted check valves creating greater pressure drops than the on/off valves in Layout 2.

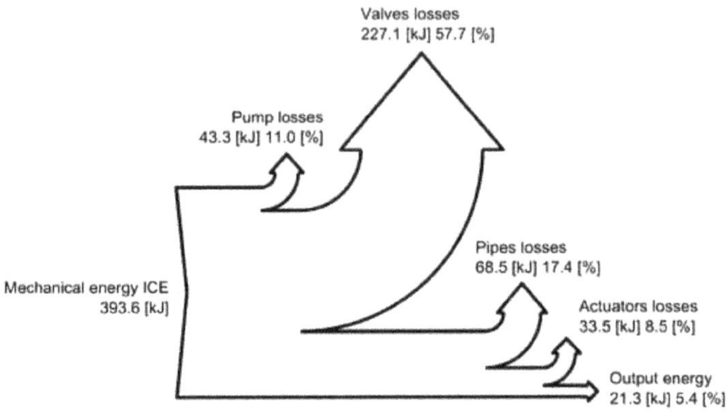

Figure 10. Energy distribution for the LS standard configuration during the digging cycle.

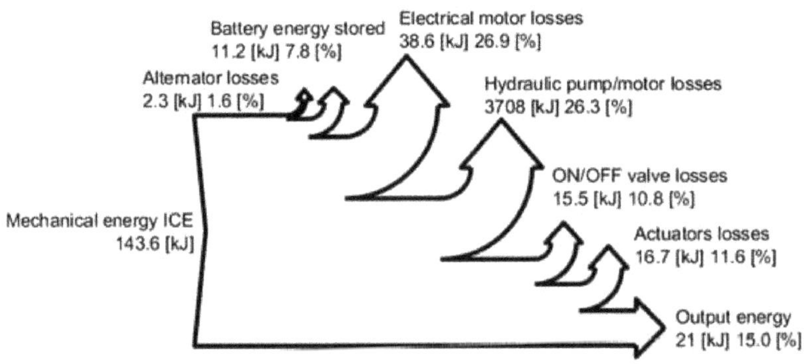

Figure 11. Energy distribution for configuration during the digging cycle for Open Circuit—Layout 2.

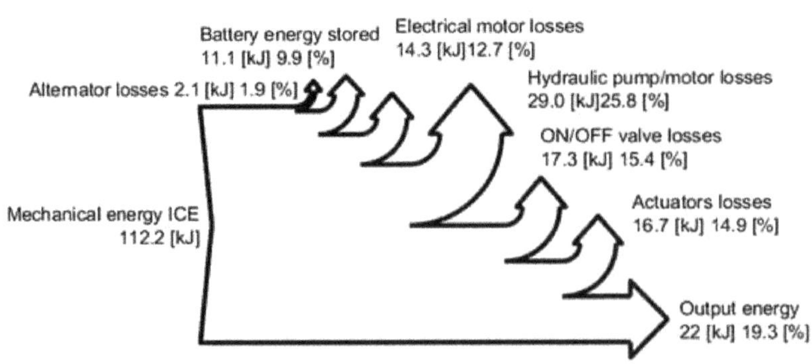

Figure 12. Energy distribution configuration during the digging cycle for Closed Circuit—Layout 2.

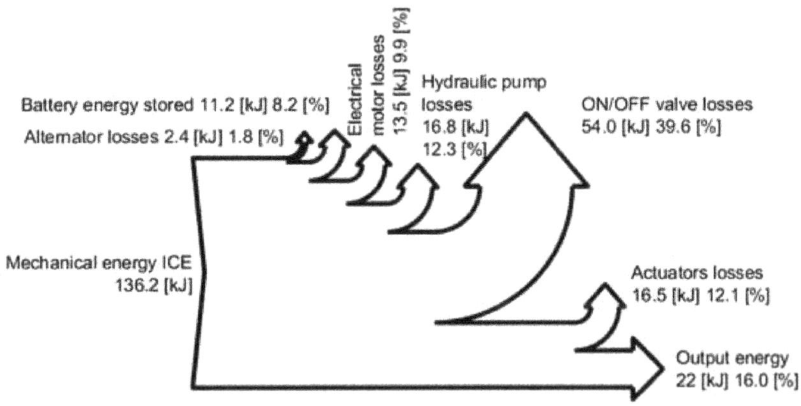

Figure 13. Energy distribution configuration during the digging cycle for Closed Circuit—Layout 3.

5. DISCUSSION

This project focused on implementing electro-hydraulic systems on a 9-ton excavator to enhance energy efficiency.

The results are derived from simulations conducted in the Amesim© environment, utilizing a verified mathematical model of an excavator in LS logic as a baseline. Three EHA setups were evaluated: one in an open-circuit configuration and two in closed-circuit configurations.

The energy distributions for the classic LS system and the EHA configurations (open- and closed-circuit) have been estimated using Sankey diagrams for a 9-ton excavator. EHA solutions have demonstrated the ability to decrease the mechanical energy needed by more than 60% by eliminating the directional valves used in typical load sensing circuits. Closed-circuit systems offer more advantages compared to open-circuit options. The circuit layout significantly impacts the efficiency and total consumption of the system and requires meticulous design. Closed-circuit technologies offer benefits beyond only energy savings. The open-circuit method requires two hydraulic machines for each actuator, resulting in higher costs, larger dimensions, and increased weight compared to closed-circuit solutions. The Layout 2 approach enables a reduction in the number of components while ensuring optimal energy performance.

Despite having fewer components, closed-circuit solutions still face the common problem of all EHA solutions, which is the greater cost. Electric motors, alternators, batteries, and electro-systems are necessary components for managing electric drives, which now make this technology less competitive than entirely hydraulic solutions. However, the benefits outlined in this article may assist to mitigate these drawbacks. The results assess the energy efficiency of various circuit solutions by using specific work cycles. The simulations utilized the digging cycle outlined in the JCMAS standard, which occurs without any payload. The simulation demonstrated the potential of electro-hydraulic systems when applied to a medium-sized excavator. Future advancements will focus on exploring other techniques to compensate for the varying flow rates by utilizing a differential cylinder. The diesel engine consistently requires less mechanical power than the typical LS circuit. Future research is focusing on the potential to decrease the size of diesel engines to ensure they consistently run at their peak efficiency, leading to decreased fuel consumption.

6. CONCISE SUMMARY

The article discusses the outcomes of implementing electro-hydraulic actuation (EHA) on a 9-ton excavator. A verified mathematical model of the conventional excavator with a load sensing circuit is being used as a basis for further energy evaluations. Three EHA setups were considered: one in an open-circuit layout and the other two in closed-circuit layouts. The simulations allow for quantifying the potential energy savings provided by each analyzed solution and for detailing the sources of energy losses.

The typical load sensing circuit includes directional valves with pressure compensators, which result in significant energy wastage. EHA systems enable the removal of these components, leading to a notable decrease in hydraulic losses and enabling substantial fuel savings.

Among all the EHA solutions analyzed, the diesel engine requires less mechanical power than the LS standard circuit. This finding suggests potential for future advancements by decreasing the diesel engine's size to enhance efficiency and reduce fuel consumption.

7. AT A GLANCE

Researchers are focusing on exploring new methods to enhance efficiency and save energy in off-road mobile devices due to strict air quality rules and emission restrictions for internal combustion engines. Pump-controlled or displacement-controlled systems are being considered for use in off-road working machines to accomplish this purpose. Typically, these systems are created by combining a hydraulic machine with an electric one to form compact components that may be placed close to the actuator. This study focuses on a 9-ton excavator with a hydraulic circuit based on load sensing logic. The validated mathematical model, created previously in the Simcenter Amesim© environment, represents the starting point for developing electro-hydraulic solutions. Electric components have been added to form various structures, including open and closed-circuit designs, to evaluate the energy efficiency of these configurations compared to the standard load sensing system. The simulations demonstrate the energy advantages of electro-hydraulic solutions, which significantly decrease the mechanical energy used by the diesel engine and, as a result, reduce fuel consumption. This is mostly achievable due to the removal of directional valves and pressure compensators, which are essential in a load sensing circuit but also contribute significantly to energy dissipation. Closed-circuit solutions demonstrate superior energy efficiency compared to open-circuit solutions, as indicated by the results. Closed-circuit topologies need fewer components, resulting in more compact, lighter, and cost-effective solutions.

REFERENCES

1. Vukovic, M.; Leifeld, R.; Murrenhoff, H. Reducing Fuel Consumption in Hydraulic Excavators—A Comprehensive Analysis. Energies **2017**, 10, 687. **[Google Scholar] [CrossRef] [Green Version]**
2. Inderelst, M.; Weidner, F.D.; Stammen, C. Quantification of Energy Saving Influencers. 21t Excavator Hydraulic System—A Holistic Investigation? In Proceedings of the 11th International Fluid Power Conference, Aachen, Germany, 19–21 March 2018. **[Google Scholar]**
3. Joo, C.; Stangl, M. Application of Power Regenerative Boom system to excavator. In Proceedings of the 10. IFK: International Fluid Power Conference, Dresden, Germany, 8–10 March 2016; Volume 3, pp. 175–184. **[Google Scholar]**
4. Li, J.; Zhao, J.; Zhang, X. A Novel Energy Recovery System Integrating Flywheel and Flow Regeneration for a Hydraulic Excavator Boom System. Energies **2020**, 13, 315. **[Google Scholar] [CrossRef] [Green Version]**
5. Guan, C.; Xu, X.; Lin, X.; Wang, S.-H. Recovering system of swing braking energy in hydraulic excavator. Zhejiang Daxue Xuebao (Gongxue Ban)/J. Zhejiang Univ. (Eng. Sci.) **2012**, 46, 142–149. **[Google Scholar] [CrossRef]**
6. Padovani, D.; Rundo, M.; Altare, G. The Working Hydraulics of Valve-Controlled Mobile Machines: Classification and Review. J. Dyn. Syst. Meas. Control **2020**, 142, 070801. **[Google Scholar] [CrossRef]**
7. Rosero, J.A.; Ortega, J.A.; Aldabas, E.; Romeral, L. Moving towards a more electric aircraft. IEEE Aerosp. Electron. Syst. Mag. **2007**, 22, 3–9. **[Google Scholar] [CrossRef] [Green Version]**
8. Smith, S.; Irving, J.; Irving, J. Electro Hydrostatic Actuators for Control of Undersea Vehicles. In Proceedings of the Joint Undersea Warfare Technology Fall Conference, Groton, CT, USA, 11–14 August 2016. **[Google Scholar]**
9. Schneider, M.; Koch, O.; Weber, J. Green Wheel Loader—Improving fuel economy through energy efficient drive and control concepts. In Proceedings of the 10th International Fluid Power Conference, Dresden, Germany; 2016. **[Google Scholar]**

10. Schmidt, L.; Ketelsen, S.; Brask, M.H.; Mortensen, K.A. A Class of Energy Efficient Self-Contained Electro-Hydraulic Drives with Self-Locking Capability. Energies **2019**, 12, 1866. **[Google Scholar] [CrossRef] [Green Version]**

11. Padovani, D.; Ketelsen, S.; Hagen, D.; Schmidt, L. A Self-Contained Electro-Hydraulic Cylinder with Passive Load-Holding Capability. Energies **2019**, 12, 292. **[Google Scholar] [CrossRef] [Green Version]**

12. Takahashi, N.; Kondo, T.; Takada, M.; Masutani, K.; Okano, S.; Tsujita, M. Development of Prototype Electro-Hydrostatic Actuator for Landing Gear Extension and Retraction. In Proceedings of the 7th JFPS International Symposium on Fluid Power, Toyama, Japan, 15–18 September 2008. **[Google Scholar]**

13. Ketelsen, S.; Padovani, D.; Andersen, T.O.; Ebbesen, M.K.; Schmidt, L. Classification and Review of Pump-Controlled Differential Cylinder Drives. Energies **2019**, 12, 1293. **[Google Scholar] [CrossRef] [Green Version]**

14. Abekawa, T.; Tanikawa, Y.; Hirosawa, A. Introduction of Komatsu Genuine Hydraulic Oil KOMHYDRO HE; Komatsu Technical Report; Yumpu: Komatsu, Japan, 2010; Volume 56, No. 163. **[Google Scholar]**

15. Niraula, A.; Zhang, S.; Minav, T.; Pietola, M. Effect of Zonal Hydraulics on Energy Consumption and Boom Structure of a Micro-Excavator. Energies **2018**, 11, 2088. **[Google Scholar] [CrossRef] [Green Version]**

16. Zhang, S.; Minav, T.; Pietola, M. Improving efficiency of micro excavator with decentralized hydraulics. In Proceedings of the 2017 Bath/ASME Symposium on Fluid Power and Motion Control, FPMC, Sarasota, FL, USA, 16–19 October 2017. **[Google Scholar]**

17. Budden, J.J.; Williamson, C. Danfoss Digital Displacement® Excavator: Test Results and Analysis. In Proceedings of the ASME/BATH 2019 Symposium on Fluid Power and Motion Control, FPMC2019, Longboat Key, FL, USA, 7–9 October 2019. **[Google Scholar] [CrossRef]**

18. Casoli, P.; Gambarotta, A.; Pompini, N.; Riccò, L. Development and Application of Co-simulation and "Control-oriented" Modeling in the Improvement of Performance and Energy Saving of Mobile Machinery. Energy Procedia **2014**, 45, 849–858. **[Google Scholar] [CrossRef] [Green Version]**

19. Casoli, P.; Anthony, A.; Riccò, L. Modeling simulation and experimental verification of an Excavator hydraulic System–Load sensing flow sharing valve model. In Proceedings of the SAE 2012 Commercial Vehicle Engineering Congress, Rosemont, IL, USA, 13–14 September 2012. **[Google Scholar] [CrossRef]**

20. Casoli, P.; Gambarotta, A.; Pompini, N.; Riccò, L. Coupling excavator hydraulic system and internal combustion engine models for the real-time simulation. Control Eng. Pract. **2015**, 41, 26–37. **[Google Scholar] [CrossRef]**

21. Casoli, P.; Pompini, N.; Riccò, L. Simulation of an Excavator Hydraulic System Using Nonlinear Mathematical Models. Stroj. Vestn.-J. Mech. Eng. **2015**, 61, 583–593. **[Google Scholar] [CrossRef] [Green Version]**

22. Casoli, P.; Riccò, L.; Campanini, F.; Lettini, A.; Dolcin, C. Mathematical model of a hydraulic excavator for fuel consumption predictions. In Proceedings of the ASME/BATH Symposium on Fluid Power & Motion Control, Chicago, IL, USA, 12–14 October 2015; Paper No. FPMC2015-9566. ISBN 978-0-7918-5723-6. **[Google Scholar] [CrossRef]**

23. Bedotti, A.; Pastori, M.; Casoli, P. Modelling and energy comparison of system layouts for a hydraulic excavator. Energy Procedia **2018**, 148, 26–33. **[Google Scholar] [CrossRef]**

24. Casoli, P.; Riccò, L.; Campanini, F.; Bedotti, A. Hydraulic Hybrid Excavator—Mathematical Model Validation and Energy Analysis. Energies **2016**, 9, 1002. **[Google Scholar] [CrossRef] [Green Version]**

25. Casoli, P.; Anthony, A.; Rigosi, M. Modeling of an Excavator System—Semi Empirical Hydraulic Pump Model. SAE Int. J. Commer. Veh. **2011**, 4, 242–255. **[Google Scholar] [CrossRef]**

26. Casoli, P.; Scolari, F.; Minav, T.; Rundo, M. Comparative Energy Analysis of a Load Sensing System and a Zonal Hydraulics for a 9-Tonne Excavator. Actuators **2020**, 9, 39. **[Google Scholar] [CrossRef]**

27. Weber, J.; Beck, B.; Fischer, E.; Ivantysyn, R.; Kolks, G.; Kunkis, M.; Lohse, H.; Lübbert, J.; Michel, S.; Schneider, M.; et al. Novel System Architectures by Individual Drives. In Proceedings of the 10th International Fluid Power Conference, Dresden, Germany, 8–10 March 2016. **[Google Scholar]**

28. Michel, S.; Weber, J. Energy-efficient electrohydraulic compact drives for low power applications. In Proceedings of the ASME/BATH 2012 Fluid Power and Motion Control, Bath, UK, 12–14 September 2012; pp. 93–107. **[Google Scholar]**

29. Costa, G.K.; Sepehri, N. Four-Quadrant Analysis and System Design for Single-Rod Hydrostatic Actuators. J. Dyn. Syst. Meas. Control **2018**, 141, 021011. **[Google Scholar] [CrossRef]**

30. Minav, T.; Zhang, S.; Pietola, M. Eliminating sizing error in direct-driven hydraulics. In Proceedings of the 10th JFPS International Symposium on Fluid Power, Fukuoka, Japan, 24–27 October 2017. **[Google Scholar]**

31. Järf, A.; Minav, T.; Pietola, M. Nonsymmetrical flow compensation using hydraulic accumulator in direct driven differential cylinder. In Proceedings of the ASME 2016 9th FPNI Ph.D. Symposium on Fluid Power, FPNI2016, Florianópolis, Brazil, 26–28 October 2016. [**Google Scholar**]
32. Rahn, C.D.; Wang, C.-Y. Battery Systems Engineering; John Wiley& Sons: New York, NY, USA, 2013; ISBN 9781119979500. [**Google Scholar**]
33. Japanese Technical Standard JCMAS H020: 2007; Earth-Moving Machinery—Fuel Consumption on Hydraulic Excavator—Test Procedure. JCMAS: Tokyo, Japan, 2007.

03 INFLUENCE OF OPERATING PARAMETERS ON EFFICIENCY OF SWASH-PLATE TYPE AXIAL PISTON PUMP

1. INTRODUCTION: AN OVERVIEW

Hydraulic systems are commonly utilized in both stationary and mobile devices due to their high power density and ability to incorporate many functionalities. These systems usually exhibit subpar energy efficiency, particularly in situations where the system's operational conditions fluctuate during the work cycle and prevent it from consistently operating at or near its optimal performance level. Low energy efficiency results in high energy consumption compared to a system's energy output and performance. Subsequently, this necessitates the implementation of a cooling system that further elevates the energy consumption of the system.

The energy efficiency of a pump relies on factors such as the operation point, pressure difference between inlet and outlet ports, rotational speed, displacement setting value (for variable displacement pumps), fluid temperature, viscosity, and density. Pumps have specific operation points or ranges where their efficiency is maximized. Operating the pump in these places requires specialized information that is not readily accessible, as pump manufacturers often do not provide detailed performance data of their pumps. Efficiency data for certain pumps can be found in research publications, such as the one released by Innas [1]. Pump manufacturers typically provide 2D efficiency curves that show efficiency as a function of pressure at one or a few specific rotating speeds, rather than just theoretical or ideal values. This data does not meet the requirements of a system designer aiming to create a system, such as an electrical motor-driven system, that utilizes a motor and pump combination to regulate the system's actuators.

To build an energy-efficient hydraulic system, one needs a thorough understanding of the pump's characteristics and performance throughout all operational parameters, especially at the minimum pressure, rotational speed, and displacement settings for variable displacement pumps. The only method for pump users to accomplish this is by measuring the pump properties. However, this necessitates a substantial investment in research equipment and facilities, making it a feature that is only used by a small number of machine manufacturers.

This study aimed to investigate how the performance of a swash-plate type axial piston pump is affected by the system factors indicated above.

2. MEASUREMENT SYSTEMS AND DETERMINING PUMP'S STEADY-STATE PERFORMANCE

The methods for measuring positive displacement pumps' and motors' characteristics and performance are determined in two ISO standards. ISO 8426:2008 [**2**] describes the methods for determining the derived capacity of these machines, and ISO 4409:2019 [**3**] describes the methods for determining their steady state performance. These standards determine the basic structure of the measurement systems (e.g., **Figure 1** open circuit for unidirectional pumps), measured quantities, positions of the transducers in relation to the measured machine and each other, system fluid with its filtration and temperature, magnitudes of systematic errors permitted in each measurement accuracy class, and the form of measurement reporting.

When conducting pump measurements, two other ISO standards should also be taken into account. ISO 9110-1:2020 [**4**] and ISO 9110-2:2020 [**5**] discuss general measurement related issues as measurement uncertainty, factors affecting it, determination of uncertainty, and calibration of measurement equipment.

Figure 1 presents the basic open circuit measurement system for unidirectional pumps as described in [**2,3**]. Positions of pressure and temperature measuring devices (hereafter referred as transducers) are determined by the measure of inner diameter of the fluid conduits, but the positions of flow transducers are not so rigorously governed. The recommendation is to place it close to the pump outlet port, and when not possible, as shown in the Figure with the caption "alternative position", the inlet port of the flow transducer has to be equipped with pressure and temperature transducers, and the produced actual pump flow rate is to be calculated with the methods announced in the standards. This calculation is required because the fluid temperature and pressure at the far-placed flow transducer may deviate significantly from the values prevailing at the pump's outlet port. Based on pressure and flow rate measurement at the outlet of the pump, the produced hydraulic power, i.e., the produced output power, can be determined. Determination of the pump's power losses, and thus also its efficiencies, also requires the measurement of the pump's input power. This is a product of the driving rotational frequency and torque at the pump's axle.

When a pump is operated in a real application, its characteristics are affected by the pressure difference prevailing between outlet and inlet ports, the rotational speed of the pump, the pump displacement setting in variable displacement pumps, the characteristics of the fluid used, the impurity level, both the free and dissolved air content of the fluid, and the operational temperatures, i.e., fluid temperature and ambient temperature. In pump measurements, the emphasis is in solving the effects of pressure, rotational speed and displacement on pump characteristics, and, therefore, the effects of all other factors should be eliminated. This calls for adequate filtering and temperature management of the fluid, as well as ambient temperature management. In addition, to be able to produce comparable results between various pumps and measurers, ISO 4409 recommends using fluids of certain viscosity grades and indexes, classes, and densities. Naturally, the fluid used in the measurements has to meet the demands of the pump manufacturer.

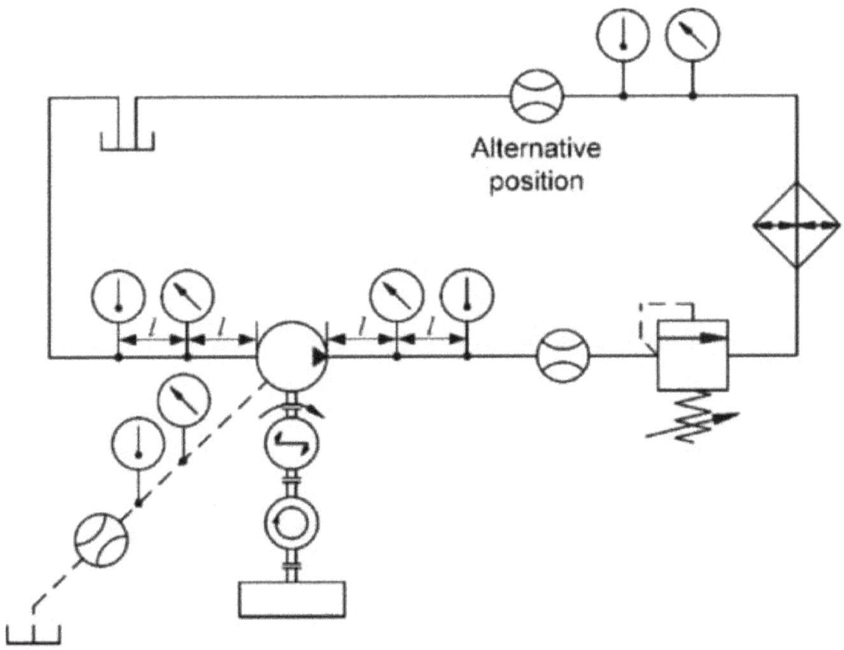

Figure 1. Basic measurement system for determining derived capacity and steady-state performance of a unidirectional positive displacement pump according to standards ISO 8426 and ISO 4409 [2,3].

Control of pressure difference across the pump ports requires some sort of loading device. In the system of **Figure 1** this is a simple manually operated pressure relief valve, but when conducting a wide or comprehensive series of measurements with multiple pressure levels, a more user-friendly and accurate device, such as an electrically controlled proportional pressure relief valve or an electrically controlled flow control valve, should be used instead. If the effect of the rotational speed on the pump characteristics lies in the field of interest, it calls for control of the pump's drive motor's rotational speed. This is typically implemented by a frequency converter. In the case of variable displacement pumps, the applied displacement control should be such that it prevents unintentional alterations of setting value during measurements, and should also be equipped with a displacement indicating transducer to verify the value of the displacement setting. The more accurate and easily operated the controls are for all of these available, the shorter time it takes to shift from one measurement point (operating point) to another and conduct a series of measurements.

Determination of the pump's power losses and efficiency by calculation from the measurement data necessitates determination of the pump's derived capacity first. A basic method for this is presented in ISO 8426:2008 [**2**], but also more advanced methods have been proposed [**6**]. The ISO 8426 method is based on measurements on one rotational speed and several outlet pressures, whose number is dictated by the measurement class striven for. Measurement results are placed in a q_V/p set of coordinates and the pump output flow at zero pressure difference between the pump's inlet and outlet ports is determined by the least squares method and extrapolation. The value of derived capacity is then determined by dividing this flow rate by the used rotational speed. In more advanced methods, such as those of Toet and Wilson-II [**6,7**], determining the derived capacity is based on measurements on several pressure and rotational speed levels. In these methods the determination is two-phased, e.g., in the Wilson-II method the first phase is practically similar to ISO 8426, but the q_V/p dependence is determined at several levels of constant rotational speeds, after which the calculated speed-dependent angular coefficients V_i are placed in a V_i/n set of coordinates, and the derived capacity is determined with the least squares method and extrapolated to speed value zero. In the Toet method, the procedure is reversed to that of the Wilson-II method, i.e., first is the determination of the flow rate dependence on rotational speed at several pressures, then placing the solved pressure dependent angular coefficients V_i in a V_i/p set of coordinates, and the derived capacity is determined with the least squares method and extrapolated to pressure value zero.

Figure 2 presents an example of a comparison of derived capacity values calculated with four methods using the same measurement data, and using the value given by the Wilson-II method as a reference. The now replaced ISO 8426 method from 1988 was based on measurements at one low-pressure level and with several rotational speeds, which resulted in a lower than real derived capacity. When used in calculating efficiencies, this would result in higher than real volumetric efficiency values and lower than real hydromechanical efficiency values. The current ISO 8426 method from 2008 results in derived capacity, which is very close to the values given by the Toet and Wilson-II methods even though it is a single-phased determination method. The reason for this is that the volumetric losses of pumps are mainly pressure-related and the effect of rotational speed, which the new ISO 8426 method omits, is minor on the flow losses. Regardless of which one of the three latter methods is used in determining the derived capacity, the data used for it should be selected carefully and any data affected by obvious measurement error should be discarded from calculation, since it might have a strong distorting effect on the resulting value of the derived capacity.

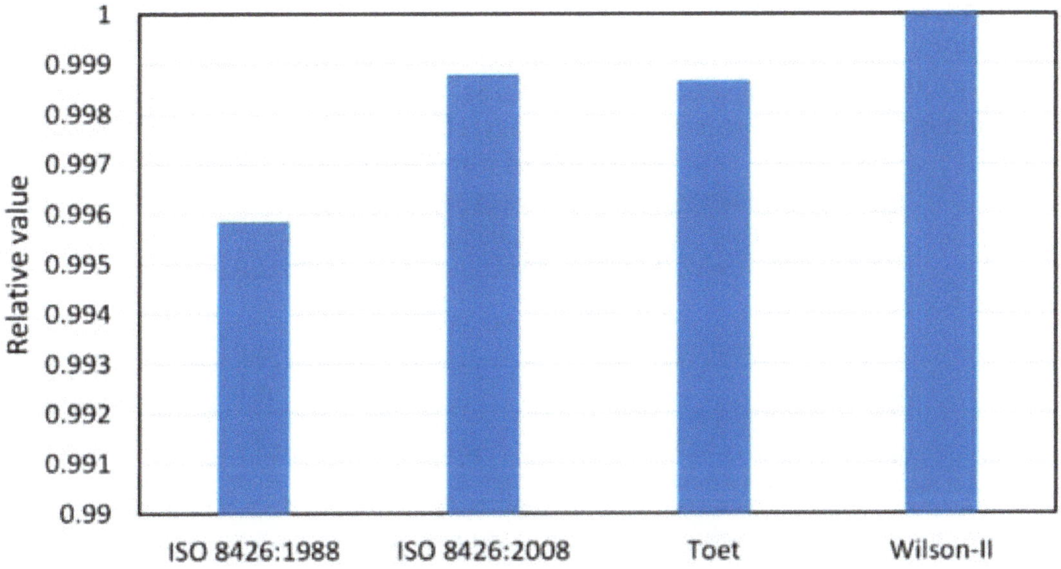

Figure 2. Derived capacities determined with different methods. Values relative to the result given by the Wilson-II method.

The minor deviation between the results of the Toet and Wilson-II methods illustrates the importance of keeping the values of parametrizing quantities accurately at selected levels, since even minor deviations from these will induce errors in calculated angular coefficients in the first calculation phase and therefore also in the second calculation phase. If the deviations in the selected measurement quantity levels were zero, the Toet and Wilson-II methods would give the same result. In the measurements on which the results of **Figure 2** are based, the value of pump outlet pressure between separate measurement points has varied relatively more than the rotational speed, thus the minor value of the derived capacity of the Toet method.

Determining the value of derived capacity is of great importance as it has a major impact on the calculated values of power, torque and flow losses, and therefore also on the calculated efficiencies. An incorrect value will lead to either over- or underestimating the losses and efficiencies depending on the direction of the error in derived capacity. Therefore, the measurements should be conducted as accurately as possible.

ISO 4409 classifies the measurements in to three accuracy classes: A, B and C. A certain measurement point in steady-state measurements is acceptable for calculation only when the indicated value of the measurement parameter is within the limits presented in **Table 1.** In ISO 8426, which determines the conditions for derived capacity measurements, the differences to the values presented in the table are that the pressure boundary value is at 0.15 bar and the allowed temperature variation is half of the value presented in **Table 1.** In addition, ISO 8426 also determines a permissible variation region for ambient temperature.

Table 1. Maximum permissible variations in the values of measurement parameters in each measurement accuracy class when measuring steady state performance of pumps [3].

Measurement Parameter and Unit	Limits of Permissible Variations in Measurement Accuracy Class		
	A	B	C
Rotational frequency, %	±0.5	±1.0	±2.0
Torque, %	±0.5	±1.0	±2.0
Flow rate, %	±0.5	±1.5	±2.5
Pressure ($p < 2$ bar), bar	±0.01	±0.03	±0.05
Pressure ($p > 2$ bar), %	±0.5	±1.5	±2.5
Temperature, °C	±1.0	±2.0	±4.0

Equations for calculating a pump's performance are presented in standard ISO 4391 [**8**]. These equations are simple to use and require no knowledge of the pump's internal structure. The methods and equations presented in this standard and older version of ISO 4409 are, however, not fully consistent in a methodological sense. Some criticism has been presented towards the equations presented in the standards, e.g., the comparison of pressurized flow rate with zero pressure flow rate, i.e., compressed fluid volume to uncompressed fluid volume, to produce volumetric efficiency.

Achten et al. [**9**] pointed out two main deficiencies in the calculation methods: omitting the effect of fluid compressibility and omitting the effect of the pump's dead volume in displacement chambers. Instead of announcing the pump performance in terms of efficiency and dividing the total efficiency into volumetric and hydromechanical efficiency as in ISO standards, Achten et al. proposed presenting the pump performance in terms of power losses and dividing these into overall, volumetric, and hydromechanical power losses. By taking into account the two deficiencies mentioned, they derived new equations for losses using thermodynamical analysis of pump power as a starting point. The heat losses were neglected and the kinetic energies of inlet and outlet flows were assumed to be equal. As a result, Achten et al. proposed two power loss factors, which take into account the fluid's compressibility and the dead volume in the displacement chamber, to be added to the standards' efficiency equations and the loss equations derived from them.

Later, both Li and Barkei [**10**], and Schänzle and Pelz [**11**] continued the work on developing more accurate definitions for the efficiencies of hydrostatic machines. Similar to Achten et al., both of them used the thermodynamical analysis of pumping processes as a starting point, but after pointing out some inconsistencies in the work of Achten et al., they resulted in different efficiency definitions, and unlike Achten et al., they were able to present a definition for the volumetric efficiency as well. The efficiency definitions of both Li and Barkei, and Schänzle and Pelz are basically the same, but since the latter researchers chose to make some simplifying assumptions concerning the compressibility of the fluid and the relationship between the density of the fluid and the prevailing pressure, which the former did not make, the work of Schänzle and Pelz resulted in somewhat simpler efficiency definitions compared to Li and Barkei. In this way, Schänzle and Pelz, however, managed to achieve one of their main goals, which was to create practical and easily applicable efficiency definitions. The more complete and accurate Li and Barkei definitions for hydromechanical and total efficiencies include a quantity called volume specific enthalpy, which requires more measurement effort from the user when determining its value for the efficiency equations.

Figure 3 presents factors affecting effective outlet flow in a variable displacement pump. In fixed displacement pump, the control flow can be omitted, and depending on the pump's structure type and implementation, the external leakage may also be omitted.

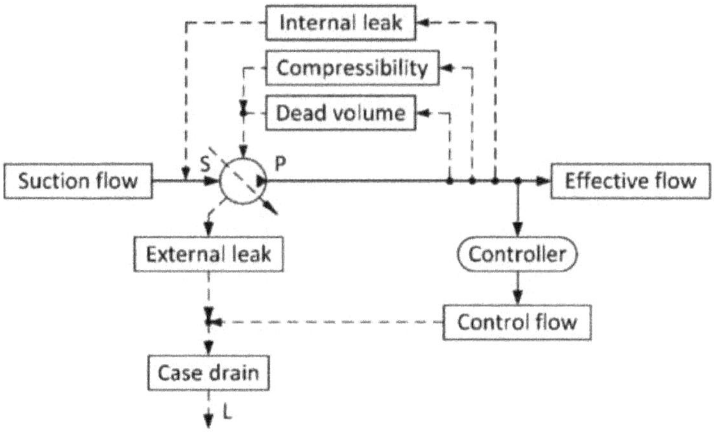

Figure 3. Flow rates of a variable displacement pump.

The displacement parts of the pump convert the low-pressure flow at the inlet port into high-pressure flow at the exit port in Figure 3. The effective output flow is reduced compared to the inlet flow due to the compressibility of the fluid, notwithstanding the possibility of equal mass flow if there were no flow losses within the pump. These losses do exist. Some of the fluid volume is utilized for lubricating the moving parts of the pump. This fluid flow moves from high-pressure areas to low-pressure areas within the pump structure, flowing from the pump outlet back to the pump inlet or external drain port through the pump casing. Some of the input flow is lost due to fluid compression, as discussed by the researchers mentioned above. In variable displacement pumps, a part of the flow is also lost to the displacement control system if it is hydraulically operated. The effective flow rate of the pump is the flow rate that remains after accounting for these losses.

Calculating pump performance using the Schänzle and Pelz efficiency determinations involves considering the averaged isentropic compressibility, which relies on the fluid's bulk modulus and its variations with pressure, temperature, and dissolved air content. These conclusions require knowledge of the pump's dead volume, which in turn necessitates understanding the pump's structure and measurements. These details are often difficult to discover and typically outside the grasp of the individual conducting the performance measurements. To utilize these efficiency calculations, an experimental technique must be created to determine the dead volume.

The discrepancies in efficiency values between calculations based on ISO standards and Schänzle and Pelz conclusions are influenced by the pump's dead volume, isentropic compressibility, and pressure.

The efficiencies provided in this article were computed using the procedures outlined in the ISO standards. Schänzle and Pelz's efficiency determinations were not included in the research findings because the measurements were conducted prior to the publication of their report. The pumps and fluids that were measured are no longer accessible, which means it is impossible to determine the bulk moduli of the fluids.

3. EFFICIENCY CHARACTERISTICS OF A SWASH-PLATE TYPE VARIABLE DISPLACEMENT AXIAL PISTON PUMP

The study's measurements aimed to investigate how different operating parameters affect the performance of the axial piston pump. Figure 4 displays the main image of the measurement system, showing two proportional directional control valves of different sizes connected in parallel. These valves were employed as loading devices to precisely control the outlet pressure at both low and high flow rates. The specific parameter values and their corresponding ranges utilized in the measurements are detailed in Table 2. Measurements were conducted using a mineral oil-based fluid with a viscosity classification of 46.

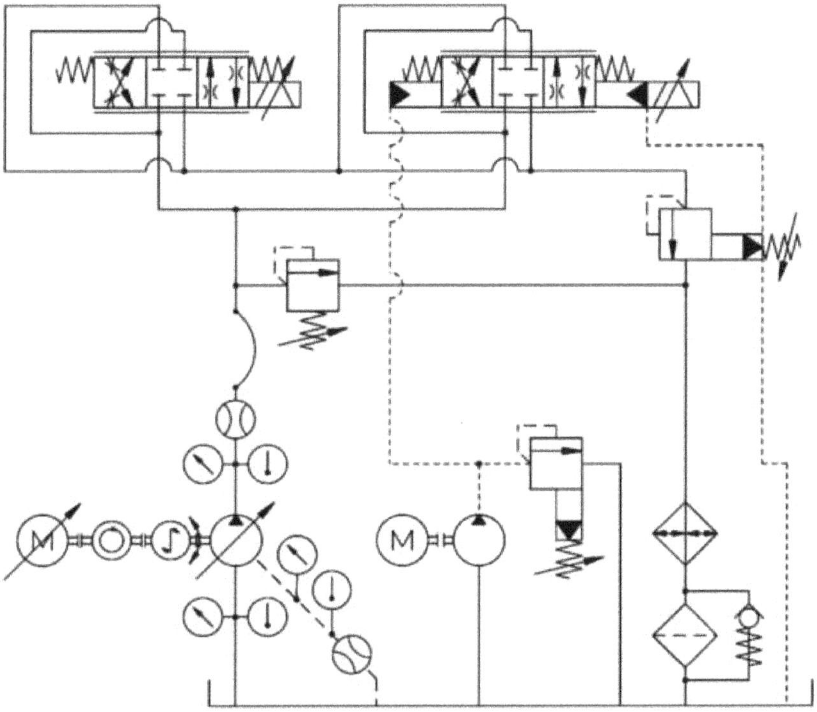

Figure 4. Measurement system.

Table 2. Values of the measurement parameters. Pressure values indicated in relation to atmospheric pressure. Maximum displacement of the measured nine piston pump is 45 cm³/r, nominal pressure 280 bar, and maximum rotational speed 2600 r/min.

Parameter	Symbol	Value	Step	Unit
Rotational speed	n	500 – 2000	250	r/min
Inlet pressure	p_1	0	-	bar
Outlet pressure	p_2	25–250	25	bar
Derived capacity	V_i	45, 33, 23, 12	-	cm^3/r
Inlet temperature	θ_i	25, 60	-	°C

The inlet pressure was maintained stable and at very close to 0 bar (overpressure) by placing the pump below the tank that had an oil column height of one meter, and by using a large diameter inlet tube to achieve very low flow velocity, thus achieving negligible flow friction. The resultant variation in inlet pressure in measurements was of class 0.04 bar. Measurement uncertainties for each measured parameter presented in **Table 3** were determined from the data of transducers, amplifiers, and data acquisition units.

Table 3. Uncertainty values of measurement parameters. Coverage factor used in calculating expanded uncertainty k = 2.

Parameter	Combined Uncertainty	Expanded Uncertainty	Unit
Rotational speed	±1.70	±3.50	r/min
Torque	±0.41	±0.82	Nm
Flow rate	±0.33	±0.66	L/min
Inlet pressure	$\pm 7.20 \times 10^{-3}$	$\pm 1.40 \times 10^{-2}$	bar
Outlet pressure	±0.60	±1.20	bar
Inlet and outlet temperature	±0.61	±1.22	°C

The main frame of the measurements was to start from the largest value of derived capacity and lowest inlet temperature, followed by unchanged capacity and highest inlet temperature. Then, adjustment of the derived capacity to the next smaller value and repeating the inlet temperature sequence until all the selected derived capacities had been passed. The inner measurement loop at each step of capacity and inlet temperature was at each selected value of rotational speed. The outlet pressure was varied through the selected range at selected steps.

The values of the derived capacity listed in **Table 2** are nominal values; the actual values used in calculating pump efficiencies were determined using measurement data and the Wilson-II method. In measurements, the pump was stripped of controllers and the pressure control channel of the pump's displacement control piston was disconnected from the pump's outlet and connected to a zero-pressure tank instead. This, as well as using the two swash-plate angle limiting screws for maximum and minimum displacement to lock the swash-plate in placc, inhibited the angle from altering during measurements, thus keeping the value of derived capacity constant at the set value.

In the following, the measurement results are presented from several points of view and commented on briefly. A wider discussion of the results is presented in **Section 4**.

Figure 5 gives an overall picture of pump efficiencies' behavior and dependence on pressure difference between outlet and inlet ports and rotational speed at two values of derived capacity and inlet fluid temperature. In volumetric efficiency, the high fluid viscosity produces high efficiency values, but in hydromechanical efficiency this phenomenon is achieved only at combinations of high pressures, low rotational speeds, and high values of derived capacity.

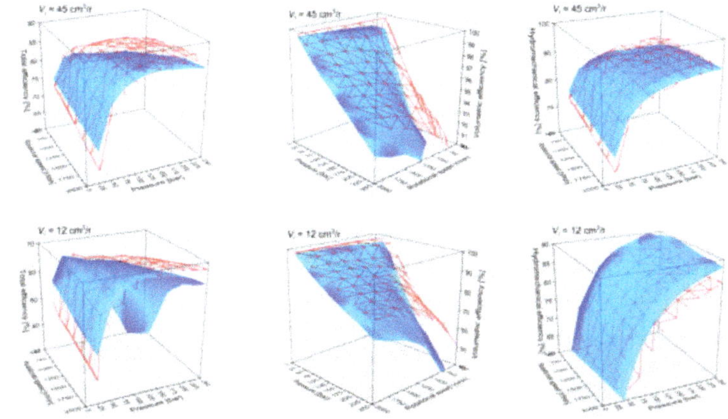

Figure 5. Efficiencies of a variable displacement axial piston pump as a function of pressure and rotational speed with derived capacities of 45 cm³/r and 12 cm³/r, and with inlet temperatures of 25 °C [ν = 84 cSt] (red wireframe) and 60 °C [ν = 21 cSt] (blue surface). Note the varying efficiency scales.

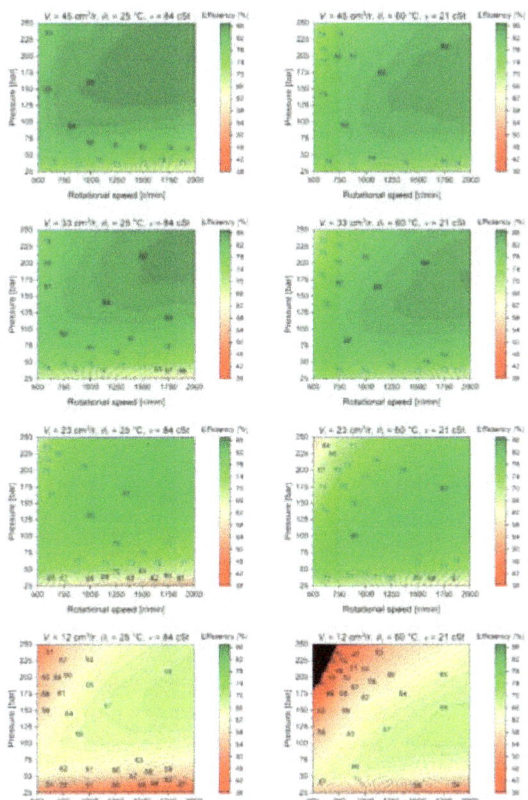

Figure 6. Total efficiency of an axial piston pump as a function of pressure and rotational speed with four values of derived capacities and two inlet temperatures. Contour lines were created from surface plots using the mesh determined by the step values presented in Table 2.

Figure 6 presents the dependence of the pump's total efficiency on pressure and rotational speed at four values of derived capacity and two values of fluid temperature at the pump's inlet in the form of contour plots. Reducing the value of derived capacity decreases the efficiency throughout the operational ranges, and rising temperature, i.e., decreasing fluid viscosity, moves the topography of high efficiencies to the direction of lower pressures.

In the following, the general results of **Figure 5** and **Figure 6**, above, are presented in more detail with 2D diagrams. When interpreting the effect of fluid inlet temperature on the pump efficiencies, it must be noted that the fluid viscosities used in the measurements were well within the range recommended by the pump manufacturer. Using very high or very low viscosities will change the behavior of the efficiencies drastically and may lead to damage to the pump. It also must be noted that the presented results are related to one certain variable displacement axial piston pump construction and cannot be generalized to apply to any kind of axial piston pump. The purpose of this presentation is to illustrate the effects of various operational parameters on the performance of a hydraulic pump.

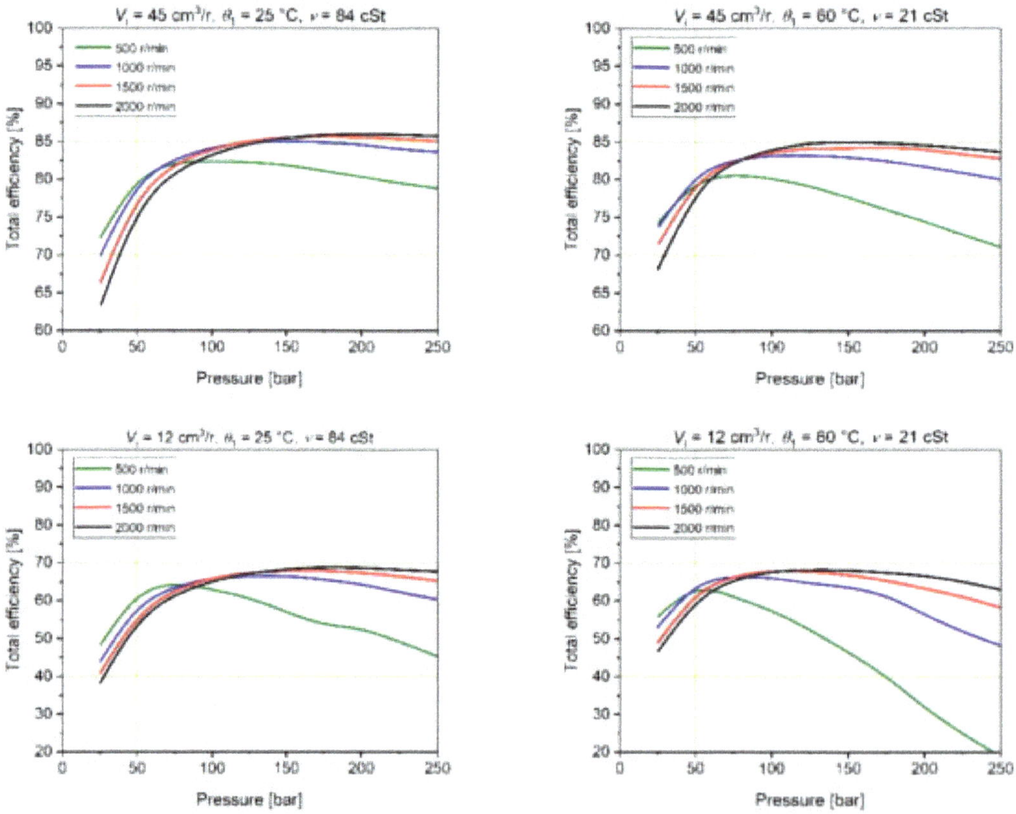

Figure 7. Total efficiency of an axial piston pump as a function of pressure with four constant rotational speeds. The ten-calculation-point efficiency curves have been smoothed with B-splines. Note the varying efficiency scales.

Figure 7, **Figure 8** and **Figure 9** present the dependence of the pump's total, volumetric, and hydromechanical efficiency on the pressure difference between outlet and inlet ports. The results are shown for four constant rotational speeds, two derived capacities, and two inlet temperatures (i.e., inlet fluid viscosities).

The effect of pressure on total efficiency is twofold. When moving from low pressures to high pressures, the efficiency first increases strongly and then decreases slowly. At low pressures, low rotational speeds result in higher efficiencies than high rotational speeds, but the situation turns the other way at high pressures. Decreasing derived capacity diminishes total efficiency all over, while increasing temperature (i.e., decreasing viscosity) has a milder effect on the total efficiency, increasing the efficiency at lower pressures and decreasing it at higher pressures, the effect being dependent on the rotational speed.

The effect of pressure on volumetric efficiency is highly linear, the value of the efficiency decreases with rising pressure. The effect of the other affecting parameters is conveyed merely in the slope of the efficiency alteration with pressure. Lowering the derived capacity and increasing the temperature (i.e., lowering the viscosity) result in decreased volumetric efficiency.

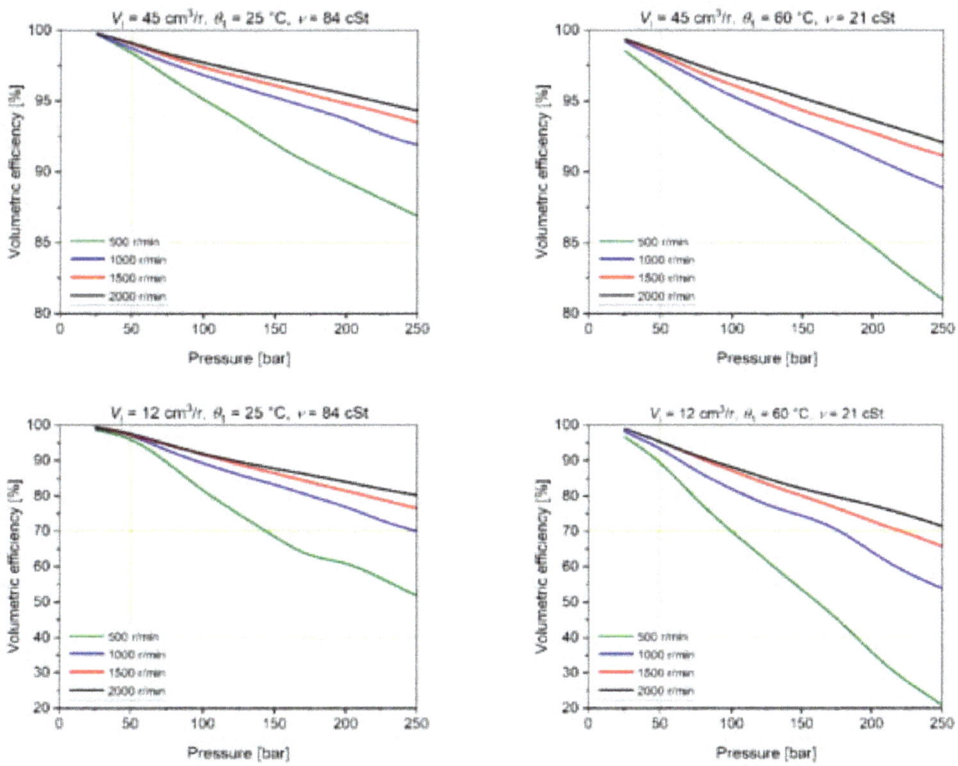

Figure 8. Volumetric efficiency of an axial piston pump as a function of pressure with four constant rotational speeds. The ten-calculation-point efficiency curves have been smoothed with B-splines. Note the varying efficiency scales.

The effect of pressure on hydromechanical efficiency resembles the behavior of total efficiency, although the decrease in hydromechanical efficiency does not take place at high pressures. Likewise, with volumetric efficiency, the pressure dictates the general behavior of hydromechanical efficiency, while the effects of the other affecting parameters manifest themselves in the vertical positioning of the efficiency curves. Reducing the derived capacity decreases the efficiency, while raising the temperature (i.e., decreasing inlet fluid viscosity) increases the efficiency. Low rotational speeds tend to increase the efficiency, although this effect is lesser in larger derived capacities.

Figure 10, Figure 11 and **Figure 12** present the above shown results from an inverse point of view, showing the dependence of the pump's total, volumetric, and hydromechanical efficiency on rotational speed with selected constant pressures. Results are shown for five pressures, two derived capacities and two inlet temperatures (i.e., inlet fluid viscosities).

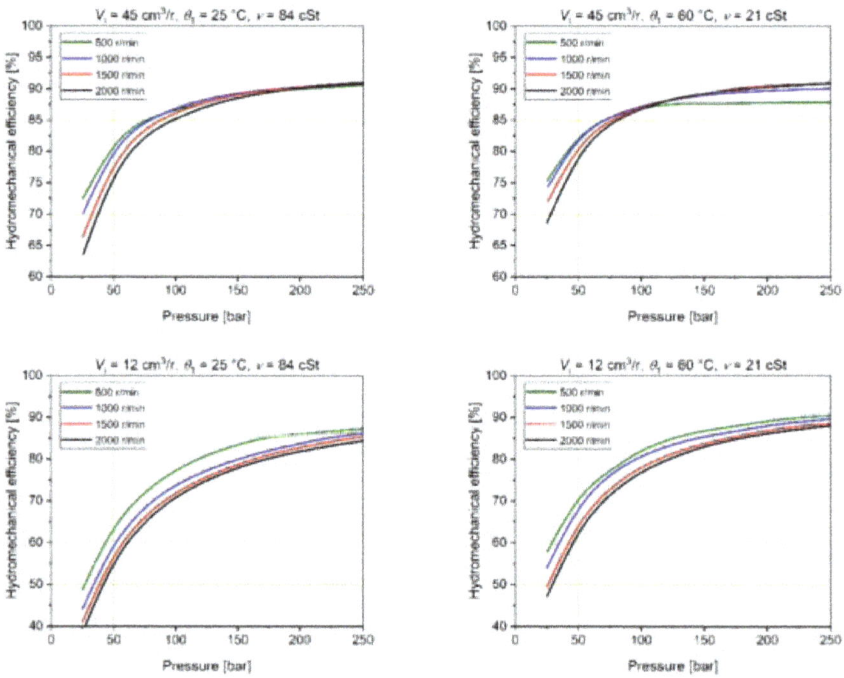

Figure 9. Hydromechanical efficiency of an axial piston pump as a function of pressure with four constant rotational speeds. The ten-calculation-point efficiency curves have been smoothed with B-splines. Note the varying efficiency scales.

The effect of rotational speed on total efficiency depends on the prevailing pressure; at low pressures the efficiency decreases with increasing rotational speed, whilst at high pressures the efficiency rises. As noted above, the decreasing derived capacity results in diminishing efficiency, as does the rise of inlet temperature (i.e., inlet fluid temperature), the effect of the latter being strongest at combinations of high pressures and low rotational speeds.

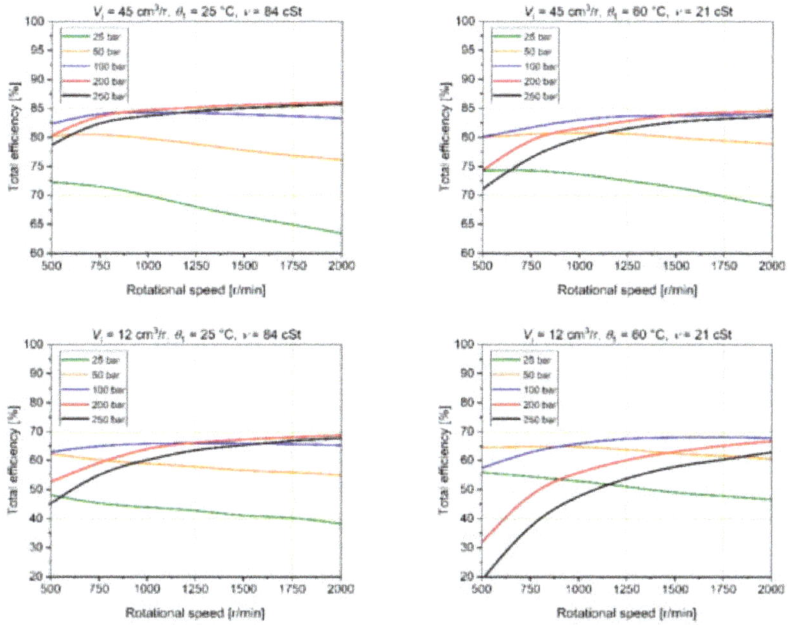

Figure 10. Total efficiency of an axial piston pump as a function of rotational speed with five constant pressures. The seven-calculation-point efficiency curves have been smoothed with B-splines. Note the varying efficiency scales.

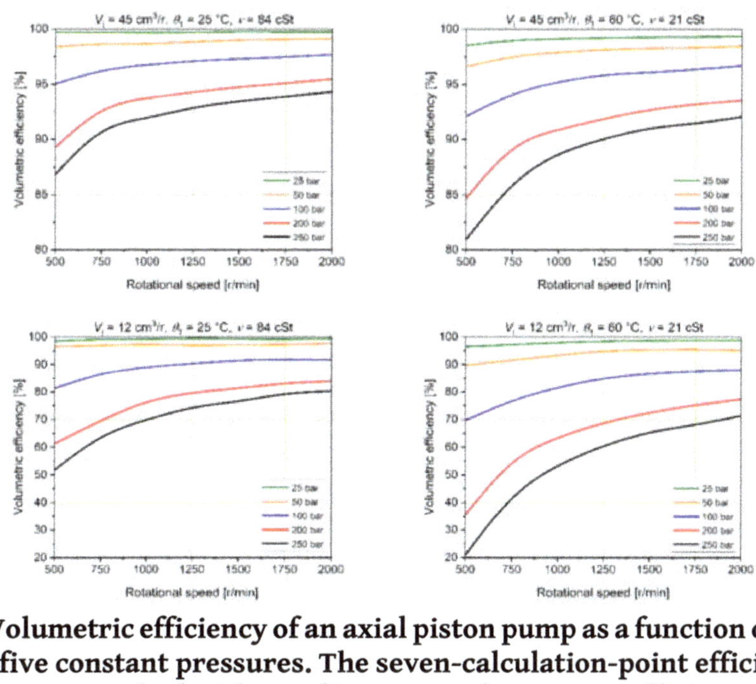

Figure 11. Volumetric efficiency of an axial piston pump as a function of rotational speed with five constant pressures. The seven-calculation-point efficiency curves have been smoothed with B-splines. Note the varying efficiency scales.

The effect of rotational speed on volumetric efficiency is generally elevating. The rate of change in efficiency is the strongest the higher the pressure, although at low pressures the efficiency is already at a high level starting from low rotational speeds. In addition, here, a decrease of the derived capacity leads to diminishing efficiency, and the same effect has also the increasing temperature (i.e., inlet fluid temperature).

The effect of rotational speed on hydromechanical efficiency is generally linear, rising speed results in decreased efficiency, although the effect is strongest at low pressures, whilst at high pressures the effect is minimal or even non-existing. As with the other efficiencies presented above, decreasing the derived capacity leads to diminishing efficiency, and the rate of change is largest at low pressures. The temperature rise (i.e., decreasing inlet fluid viscosity) increases the efficiency in general.

Figure 13, Figure 14 and **Figure 15** present the above shown results from a third angle, showing the dependence of the pump's total, volumetric, and hydromechanical efficiency on derived capacity with selected constant pressures. Results are shown for six pressures, one rotational speed, and two inlet temperatures (i.e., inlet fluid viscosities).

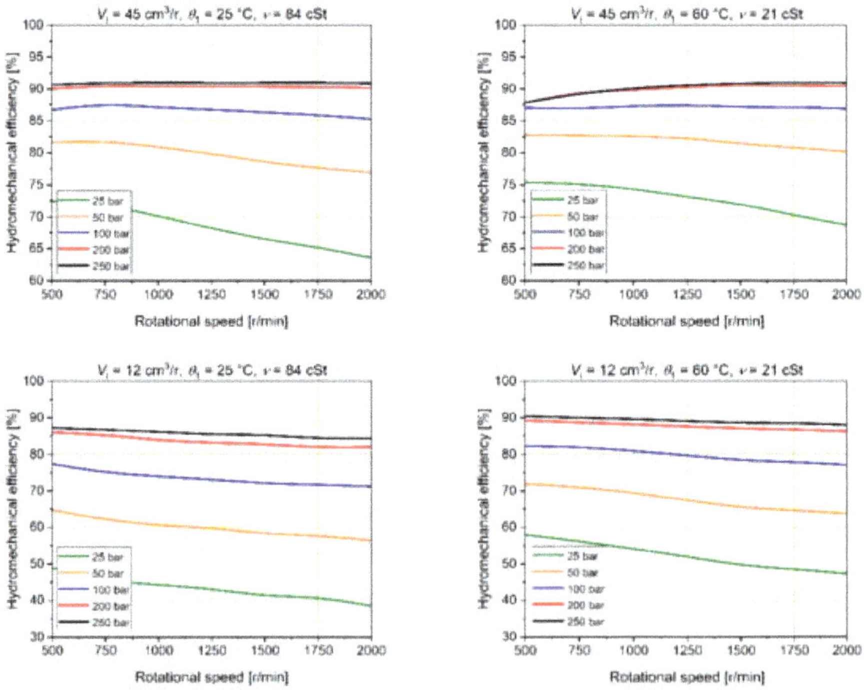

Figure 12. Hydromechanical efficiency of an axial piston pump as a function of rotational speed with five constant pressures. The seven-calculation-point efficiency curves have been smoothed with B-splines. Note the varying efficiency scales.

As already demonstrated above, the total efficiency is higher the closer the derived capacity of a variable displacement pump is to its maximum value. Rising pressure basically increases the efficiency, but from the middle pressure region up, the effect is non-existent, and with high temperatures (i.e., inlet fluid temperatures) even the opposite.

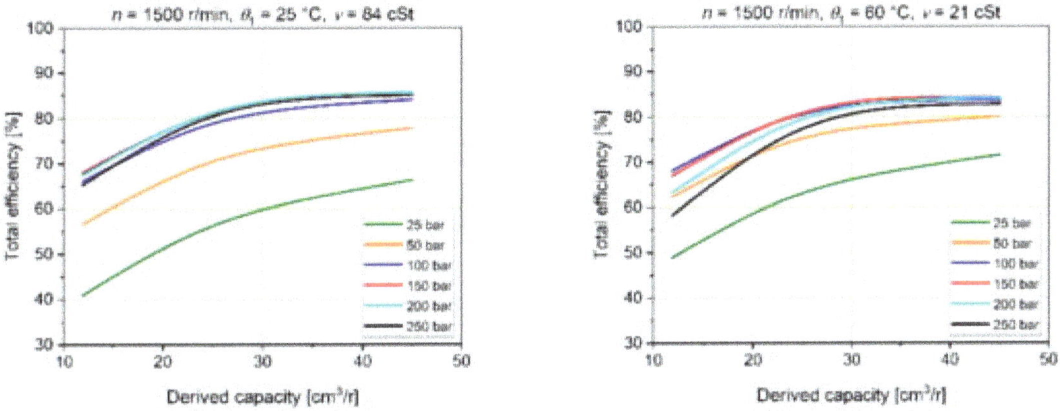

Figure 13. Total efficiency of an axial piston pump as a function of derived capacity with six constant pressures. The four-calculation-point efficiency curves have been smoothed with B-splines.

Increasing derived capacity has a similar rising effect on the volumetric efficiency, but here the efficiency is coherently: the lower the pressure, the higher the efficiency. At low pressures, however, the effect of derived capacity on efficiency is minimal. Rise of temperature (i.e., inlet fluid temperature) decreases the efficiency throughout, but here the effect diminishes with decreasing pressure.

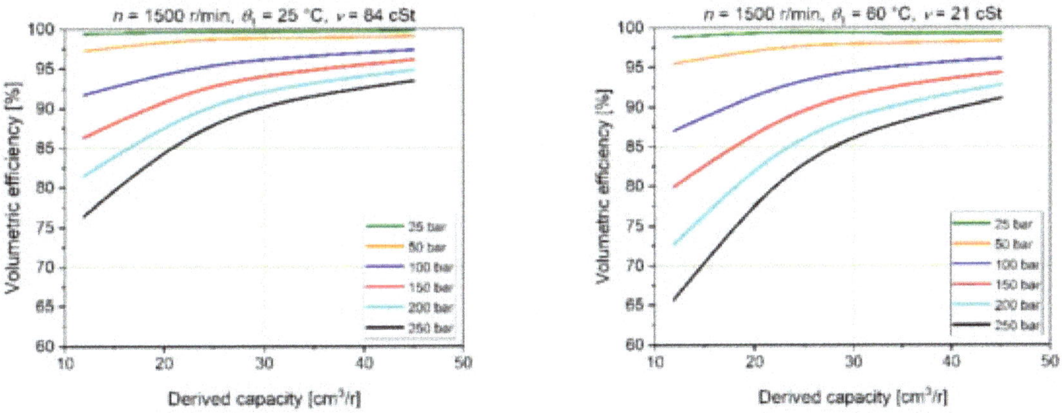

Figure 14. Volumetric efficiency of an axial piston pump as a function of derived capacity with six constant pressures. The four-calculation-point efficiency curves have been smoothed with B-splines.

When it comes to hydromechanical efficiency, the impact of increasing derived capacity on efficiency is not as consistent as with the efficiencies mentioned earlier. Efficiency increases with derived capacity at low pressures, but decreases with capacity at high pressures. An increase in temperature at the inlet enhances efficiency, but this impact lessens as pressure and derived capacity rise.

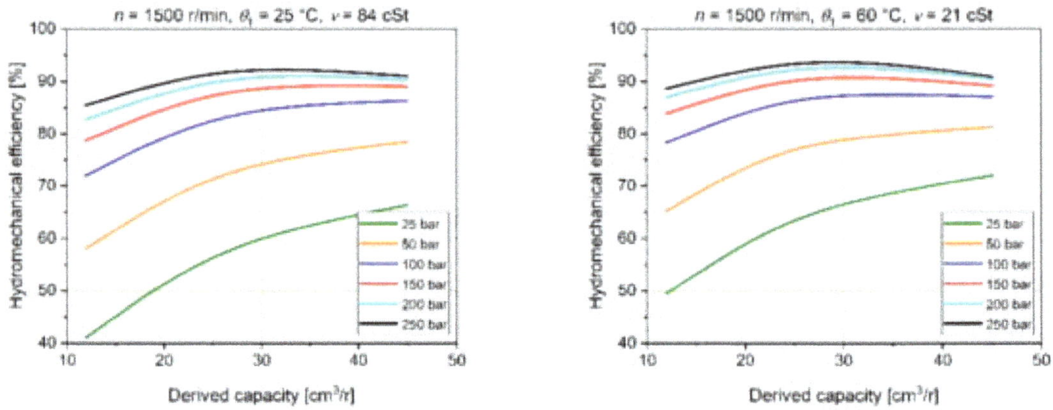

Figure 15. Hydromechanical efficiency of an axial piston pump as a function of derived capacity with six constant pressures. The four-calculation-point efficiency curves have been smoothed with B-splines.

4. DISCUSSION

The pump's performance relies on various phenomena, which collectively impact the pump and are usually quantified in relation to the pump's overall efficiency. This can be categorized into two subefficiencies: volumetric efficiency and hydromechanical efficiency. The relevant variables are determined by the parameter values at the pump's operating point. The point is often determined by the pressure variance between the pump's intake and exit ports, the pump's rotating speed, and the temperature of the fluid entering the pump, which impacts the fluid's viscosity and density. The derived capacity of variable displacement pumps is a quantity that defines the pump's operating point.

Volumetric efficiency is influenced by various flow losses occurring within the pump, as shown in Figure 3, whereas hydromechanical efficiency is affected by small losses and friction losses within the pump. Minor losses in the pump are caused by flow resistance in its intricate internal flow channels, and they become more significant as the volume of flow passing through the pump grows. Friction losses occur when parts move in relation to each other and are separated by a lubricating layer, known as viscous friction. Both losses are influenced by fluid viscosity, density, and temperature. Minor losses depend on flow rate, while friction losses are affected by pressure, which impacts the lubrication layer's thickness between moving parts. As pressure increases, the lubricating layer becomes thicker, but it is constrained by the physical restrictions of the pump mechanism and cannot expand indefinitely.

The temperature that impacts the operating point of a pump is the fluid inlet temperature. However, this does not refer to the temperature within the pump, where small and friction losses convert into power losses, increasing the temperature of both the fluid and pump structure. As a result, changes in fluid characteristics and pump part dimensions impact pump performance in a way that is difficult to distinguish from the impacts of pressure, rotational speed, and capacity. Thus, the consequences caused by the increase in temperature in the pump are not considered in this presentation.

Factors that are often overlooked when considering pump performance at an operating point include fluid parameters beyond viscosity and density. The lubricating qualities and pressure durability of the fluid can greatly affect the pump's performance, either within its working range or once it reaches a limitation point related to pressure or temperature. It is important to follow the pump manufacturer's recommendations based on the fluid's qualities. The text discusses the impact of key parameters on the performance of a variable displacement axial piston pump in terms of volumetric, hydromechanical, and total efficiency.

The findings drawn are only applicable to the measured value ranges of the measurement parameters. The study could not measure the whole pressure and rotational speed ranges of the pump due to the underpowered prime mover. The examined parameters have a consistent effect on volumetric efficiency. A change in a parameter value in one direction would either decrease or increase the efficiency value. An increase in pressure difference between the inlet and outlet ports of the pump leads to higher flow through leakage paths and greater fluid compressibility, ultimately reducing the pump's effective outlet flow rate compared to its theoretical flow delivery and causing a decrease in volumetric efficiency.

The relationship between efficiency and pressure change is predominantly linear, suggesting that the leakage is mainly laminar in nature. The impact of pressure on efficiency is influenced by rotational speed, with higher rotational speeds resulting in a smaller relative drop in efficiency as pressure increases. The leakage amount is mainly influenced by pressure and only little impacted by rotational speed, as evidenced by the efficiency's behavior in relation to rotational speed. Higher rotational speed enhances the pump's theoretical flow. When the rotational speed is relatively constant and has no impact on leakage, increasing the rotational speed leads to improved volumetric efficiency. At high pressures, leakages increase compared to low pressures. Additionally, the variations in leakage flow rate relative to theoretical flow rate are more pronounced with an increase in rotating speed. Increasing the derived capacity has a similar effect on efficiency as increasing rotational speed. The theoretical flow increases with an increase in derived capacity, while pressure-dependent leakage remains relatively constant, leading to an improvement in volumetric efficiency. Efficiency increases as pressure rises due to a more significant reduction in pressure-dependent leakage compared to low-pressure conditions. Temperature rise decreases fluid viscosity and density, which in turn reduces volumetric efficiency by affecting pressure-dependent laminar and turbulent leakage. The impact of the analyzed parameters on hydromechanical efficiency is intricate, as it is influenced by two main factors: minor losses and viscous friction losses. These factors are dependent on fluid characteristics, flow rate, and pressure. Increasing pressure enhances hydromechanical efficiency by decreasing viscous friction losses between the pump's moving elements. The impact of pressure on small losses is considered insignificant due to the minimal changes in the diameter of the pump's flow channels. Increasing pressure in lubrication sites raises the lubrication gap height and the lubrication layer thickness, reducing viscous friction between moving parts and improving hydromechanical efficiency.

The relationship between efficiency and pressure increase is not linear. This means that lubrication gap heights cannot increase infinitely with higher pressure due to limitations imposed by the pump's design and materials. Higher rotational speed often decreases hydromechanical efficiency by increasing flow through the pump, which in turn raises the pump's small losses. The pump's efficiency decreases as the rotational speed increases, with the extent of reduction being influenced by the pressure level. The drop is most significant at low pressures and becomes less noticeable at high pressures, and may even be absent. In this pump, the rotational speed does not significantly affect the hydromechanical efficiency. The decrease in efficiency at higher speeds is mainly caused by viscous friction losses. The faster the rotational speed, the thinner the lubricating layer due to decreased pressure.

Enhancing the derived capacity boosts hydromechanical efficiency, with the effect being highly dependent on pressure. Efficiency changes more significantly with capacity at low pressures compared to high pressures. The efficiency's dependency on derived capacity is mostly due to viscous friction losses rather than small losses, which only become noticeable at high flow rates. An example of this phenomenon is the reduction in efficiency as the derived capacity increases at high pressures. An increase in temperature leads to lower viscosity and density values, which in turn enhances hydromechanical efficiency by lowering minor and viscous friction losses. This phenomenon has a threshold, particularly at certain temperatures, when the lubricating ability of the fluid diminishes, leading to increased friction losses. Total efficiency is the result of combining volumetric and hydromechanics efficiency, taking into account the parameter interdependence of both. The major effect on total efficiency varies on the operating point. Domination in this context denotes a severe decline in overall efficiency. At low pressures, hydromechanical efficiency is the main factor affecting total efficiency, while volumetric efficiency becomes more significant as pressure increases. Rotational speed, derived capacity, or fluid temperature do not significantly affect the overall relationship between total efficiency and pressure. The peak total efficiency occurs within the pressure range advised by the pump manufacturer, between the lowest and highest values. The impact of rotational speed on overall efficiency is ambiguous. None of the subefficiencies definitively outperforms the overall efficiency when the pump is used at low or medium pressures and high derived capacity. Significant domination is observed when the pump operates at low rotational speeds, with the volumetric efficiency having the primary impact. This is particularly highlighted when the derived capacity is low. An increase in temperature leads to a constant drop in overall efficiency across all rotational speeds. The peak overall efficiency occurs at the upper limit of the pump manufacturer's recommended rotating speed range.

The impact of derived capacity on overall efficiency varies based on the values of other influencing elements. At low pressures, hydromechanical efficiency is the primary factor affecting total efficiency, but at higher pressures, volumetric efficiency becomes the dominant factor. At low temperatures, hydromechanical efficiency is more significant, but at higher temperatures, the difference in impact between the two subefficiencies decreases. Changes in rotational speed do not have a substantial impact on the relationship between subefficiencies. The peak total efficiency is found at the upper limit of the calculated capacity range. The efficiency dependencies on several operating point characteristics identified in the measurements are summarized in Table 4 as a crude generalization.

Table 4. Rough generalization of empirical changes in the efficiencies of the studied swash-plate type axial piston pump in relation to the increase in the operating point parameter values.

Rise of	Volumetric Efficiency	Hydromechanical Efficiency	Total Efficiency
Pressure	Decreases	Increases	Not unambiguous
Rotational speed	Increases	Decreases slightly	Not unambiguous
Derived capacity	Increases	Increases slightly	Increases
Temperature	Decreases	Increases slightly	Decreases

Although this study did not include examination of pump models, the generalized observations presented in **Table 4** are quite well aligned with the structures of common and quite simple pump models like the Wilson and Schlösser models and the affecting parameters included in those. Review of these models will be done in following studies.

5. CONCISE SUMMARY

The data above clearly demonstrate the significant impact of the operating point on the performance of a swash-plate type axial piston pump. Pressure, rotating speed, derived capacity, and temperature individually impact the volumetric and hydromechanical efficiency, thus affecting the overall efficiency. Furthermore, these effects are interconnected, with the impact of one parameter on efficiencies being reliant on at least one other parameter. Understanding the pump's efficiency characteristics is crucial for achieving greater system energy efficiencies, despite the complex relationship between efficiency and operating point factors. The pump, functioning as an energy exchanger, significantly influences energy consumption in hydraulic systems. Operating the pump below its peak efficiency range can reduce system efficiency, even if the system is otherwise tuned for high efficiency.

The study's results indicate that the pump's maximum total efficiencies are observed at the upper limit of the rotating speed range, with their placement in the pressure range being influenced by the determined capacity and viscosity values. Decreasing each of these factors pushes the regions of maximum efficiencies towards lower pressure levels. This behavior is specific to certain pump types and is influenced by the pump's construction type, as demonstrated in the measurement data released by Innas. This study's results and findings are used to develop methods for determining the optimal combination of a pump's pressure, rotational speed, and derived capacity to achieve maximum total efficiency, while still meeting the pump's power and flow rate performance criteria.

6. AT A GLANCE

To enhance the energy efficiency of hydraulic systems, focus on minimizing power losses in energy conversion components and energy controlling and conveying components. To do this, you need to use the components in their most energy-efficient operating range. Several operational parameters impact the energy conversion efficiency of a pump, the fundamental energy converter in a hydraulic system. Typically, pressure and rotational speed are the main factors examined, although fluid temperature and capacity derived from variable displacement pumps also significantly impact efficiency. Omitting these factors may lead to running the pump outside its most efficient operation range and cause high energy losses. Operating the pump efficiently requires a thorough understanding of its performance parameters, which are typically not disclosed by pump makers. This study displays the efficiency findings of a variable displacement axial piston pump based on pressure, rotating speed, derived capacity, and inlet fluid temperature. All these parameters significantly affect the energy conversion efficiency of the pump and should be considered when using a hydraulic pump.

REFERENCES

1. Performance of Hydrostatic Machines. Extensive Measurement Report. Innas BV. 2020. Available online: **https://www.google.com/search?client=firefox-b-d&q=1.+Performance+of+hydrostatic+machines.+Extensive+measurement+report** (accessed on 31 January 2022).

2. ISO 8426:2008; Hydraulic Fluid Power—Positive-Displacement Pumps and Motors—Determination of Derived Capacity. International Organization for Standardization: Geneva, Switzerland, 2008.

3. ISO 4409:2019; Hydraulic Fluid Power—Positive-Displacement Pumps, Motors and Integral Transmissions—Methods of Testing and Presenting Basic Steady-State Performance. International Organization for Standardization: Geneva, Switzerland, 2019.

4. ISO 9110-1:2020; Hydraulic Fluid Power—Measurement Techniques—Part 1: General Measurement Principles. International Organization for Standardization: Geneva, Switzerland, 2020.

5. ISO 9110-2:2020; Hydraulic Fluid Power—Measurement Techniques—Part 2: Measurement of Average Steady-State Pressure in a Closed Conduit. International Organization for Standardization: Geneva, Switzerland, 2020.

6. Post, W.J.A.E.M. Models for steady-state performance of hydraulic pumps: Determination of displacement. In Proceedings of the Ninth Bath International Fluid Power Workshop, Bath, UK, 9–11 September 1996; pp. 339–352. **[Google Scholar]**

7. Toet, G.; Johnson, J.; Montague, J.; Torres, K.; Garcia-Bravo, J. The determination of the theoretical stroke volume of hydrostatic positive displacement pumps and motors from volumetric measurements. Energies **2019**, 12, 415. **[Google Scholar] [CrossRef] [Green Version]**

8. ISO 4391:1983; Hydraulic Fluid Power—Pumps, Motors and Integral Transmissions—Parameter Definitions and Letter Symbols. International Organization for Standardization: Geneva, Switzerland, 1983.

9. Achten, P.; Mommers, R.; Nishiumi, T.; Murrenhoff, H.; Sepehri, N.; Stelson, K.; Palmberg, J.-O.; Schmitz, K. Measuring the losses of hydrostatic pumps and motors—A critical review of ISO4409:2007. In Proceedings of the ASME/BATH 2019 Symposium on Fluid Power and Motion Control, FPMC2019, Sarasota, FL, USA, 7–9 October 2019. **[Google Scholar]**

10. Li, P.Y.; Barkei, J.H. Hydraulic effort and the efficiencies of pump and motors with compressible fluid. In Proceedings of the ASME/BATH 2020 Symposium on Fluid Power and Motion Control, FPMC2020, Virtual, Online, 9–11 September 2020. **[Google Scholar]**

11. Schänzle, C.; Pelz, P.F. Meaningful and physically consistent efficiency definition for positive displacement pumps—Continuation of the critical review of ISO 4391 and ISO 4409. In Proceedings of the ASME/BATH 2021 Symposium on Fluid Power and Motion Control, FPMC2021, Virtual, Online, 19–21 October 2021. [**Google Scholar**]

04 USING LOAD-HOLDING VALVES TO EVALUATE POWER REQUIREMENTS OF A HYDRAULIC SYSTEM THROUGH A GRAPHIC METHOD

1. INTRODUCTION: AN OVERVIEW

Load Holding Valves (LHV) are essential components of mobile machinery. They are utilized in hydraulic systems to prevent uncontrolled movements of cylinders and motors caused by overrunning or gravity-assisted loads. They are also commonly used in positioning circuits and regeneration circuits. LHVs have had a substantial technical-commercial influence and are legally mandated due to their enhanced safety features as outlined in directives and regulations. They offer secure load holding, leak-proof operation, shock absorption, cavitation prevention during load decreasing, overload protection, and function as line rupture safety valves. They need to demonstrate specific functions to ensure optimal dynamic performance of the hydraulic system, including preventing overflow under stepped pilot pressure, rapid closure when pilot pressure is lost, independent opening action regardless of load pressure, and effective flow control capabilities. Various types of machinery utilize LHV valves, including excavators, cranes, aerial platforms, inspection vehicles, concrete pumping equipment, telescopic handlers, forklifts, fruit pickers, recreational facilities, industrial presses, drills, and industrial winches, among others.

LHVs valves are modulating devices that allow free flow from the valve to the cylinder and block reverse flow from the cylinder to the valve until load pressure, pilot pressure, or a combination of both opens them (traditional pressure–spring balance principle, see **Figure 1**). Though they are simple in design, their application can often frustrate technicians and engineers. The known problems of LHVs are the excessive power consumption associated with over-pressurization of the flow supply and their tendency to introduce instability. The solutions to both issues are opposed to each other since what works well for stability is harmful for energy savings.

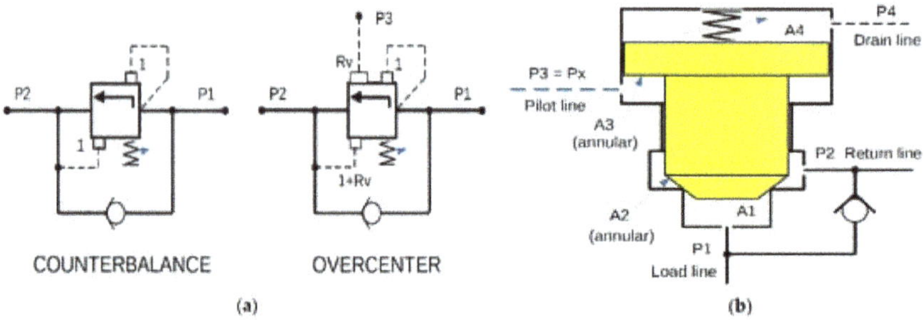

Figure 1. The traditional principle of pressure–spring load-holding valve model. (a) ISO 1219 symbols; (b) Functional scheme of the load holding valve, LHV.

Application engineers interested in energy consumption must determine how to compute the power requirements of an LHV under certain operating conditions. The research aimed to introduce a simplified graphical way for analyzing the power needs of hydraulic systems using Load Holding Valves (LHVs) for controlling overrunning loads. The paper is structured as follows: Section 1 introduces the work discussed in this paper. Section 2 provides a concise diagram of load-holding valves and demonstrates a common use for regulating the motion of a linear actuator. Special attention is also focused on the graphical depiction of the unique curves of both components. Section 3 provides a brief overview of the current state of the art. Section 4 introduces the graphical way for doing a power balancing. Section 5 describes the experimental tests and provides examples of using this analysis methodology. Section 6 discusses conclusions on the proposed technique.

2. REVIEW OF LOAD-HOLDING VALVES

The initial designs were created by the Vickers engineers in the 1930s, and the first valves introduced to the market were designed by Racine. Subsequently, these valves have been categorized according to their design and operation, including counterbalance valve, overcenter valve, holding valve, load-control valve, pilot-assisted load-control valve, load-holding valve, and motion-control valves, among others.

The main debate revolves around the terminology of counterbalance and overcenter valve names. Counterbalance valves are pressure relief valves combined with a check valve to establish a one-way back pressure in a hydraulic system, preventing the actuator from moving uncontrollably under increasing loads. Two pressures and a spring balance the valve poppet. The load pressure (P1) force opens the valve orifice, while the spring force and back pressure force (P2) close it. The overcenter valve can be viewed as a counterbalance valve with an external pilot line (P3) and is more energy-efficient in systems with fluctuating loads. The authors are inspired by recent advancements in this valve type to seek a more universal nomenclature, like load-holding valve (LHV). Figure 1b illustrates the diagram of a load-holding valve. The P4 typically connects to the P2, however it can also be directly connected to an atmospheric reference.

According to **Figure** 1b, where $A4=A1+A2+A3$, the force balance on the LHV poppet (Equation (1)) allows for calculating the pressure required to open the valve and starts to move the load. In order to hold the load, the valve cracking pressure (or setting pressure) must be set higher than the load pressure. Because of the hysteresis of the moving parts of the valve (mainly due to friction), there is a difference between cracking and reseat pressure (the pressure at which the valve closes) with the reseat pressure being lower than the cracking pressure. Hysteresis is one of the reasons why the typical LHV valve setting is approximately 30% higher than the load pressure to ensure that the reseating pressure will be high enough to maintain the load.

$$P_1 + P_3R_v = P_M + P_2(R_v + 1) \quad (1)$$

where, P_1 is the load pressure, P_3 is the pilot pressure, $PM=Fspring/A1$ is the spring pressure, P_2 is the return pressure, and Rv is the pilot ratio. The pilot ratio is given by the ratio between the pilot pressure area and the relief area.

$$R_v = \frac{A_3}{A_1} \quad (2)$$

In **Figure** 2a, a typical application is shown. A directional valve is used in conjunction with an LHV to control the movement of an actuator. The pilot port of the LHV is connected to the opposite side of the actuator.

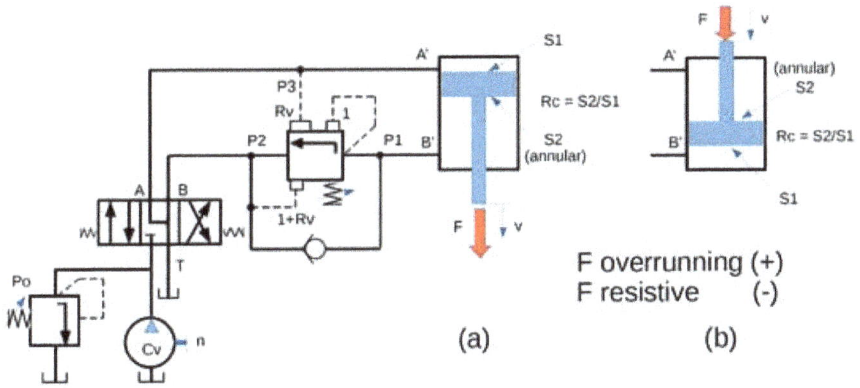

Figure 2. Typical application hydraulic circuit. (a) LHV valve is connected to the chamber on the rod side; (b) LHV valve is connected to the chamber on the piston side.

The force balance also makes it possible to determine the pressures in the hydraulic cylinder chambers (piston and rod side) as a function of a given load. This balance is established for two situations of the hydraulic cylinder, that is, when the LHV valve is connected to the chamber on the rod side (see **Figure 2**a) and when it is connected to the chamber on the piston side (see **Figure 2**b).

For case 2a

$$P_1 R_c - P_3 = P_{Load} \tag{3}$$

$$R_c = \frac{S_2}{S_1} \tag{4}$$

$$P_{Load} = \frac{F_{Load}}{S_1} \tag{5}$$

where, P_1 is the rod side chamber pressure, P_3 is the pilot pressure (equal to piston side chamber pressure), P_2 is the return pressure, and Rc is the hydraulic cylinder section ratio. The sectional area ratio of the cylinder is given by the ratio between the effective sections of rod and piston side.

The load, F, can be considered resistive load (opposite direction for the actuator velocity "sign -"), or overruning load (same direction concerning the actuator speed "sign +"), as depicted in **Figure 2**.

Assuming that P_2 is very small and combining Equations (1) and (3), the pressures of the hydraulic cylinder chambers can arise from the pressures of the hydraulic cylinder chambers as a function of the spring pressure, P_M, and the equivalent load pressure, PLoad.

$$P_1 = \frac{1}{1 + R_c R_v} P_M + \frac{R_v}{1 + R_c R_v} P_{Load} \tag{6}$$

$$P_3 = \frac{R_c}{1 + R_c R_v} P_M - \frac{1}{1 + R_c R_v} P_{Load} \tag{7}$$

Analogously, for case 2b

$$P_1 - P_3 R_c = P_{Load} \tag{8}$$

where, P_1 is the piston side chamber pressure, P_3 is the pilot pressure (equal to the rod side chamber pressure), and P_2 is the return pressure.

In the same way, assuming that P_2 is very small and combining Equations (1) and (8), the pressures of the hydraulic cylinder chambers can be expressed as a function of the spring pressure, PM, and the equivalent load pressure, PLoad.

$$P_1 = \frac{R_c}{R_c + R_v} P_M + \frac{R_v}{R_c + R_v} P_{Load} \tag{9}$$

$$P_3 = \frac{1}{R_c + R_v} P_M - \frac{1}{R_c + R_v} P_{Load} \tag{10}$$

Two types of operating curves can be found in the technical documents related to LHVs. On the one hand, the so-called "steady-state operating curve" expresses the load pressure as a function of the pilot pressure (P_3 vs. P_1). **Figure 3** shows, for cases (2a) and (2b), the plots of the hydraulic actuator/LHV system based on Equations (1), (2) and (8), corresponding to the steady-state operating curves. The red lines are the characteristic curves of the LHV with different pilot ratios, while the solid black line corresponds to the unloaded hydraulic cylinder, and the dashed gray line corresponds to the hydraulic cylinder with an overrunnig load equivalent to 100 bar. Consequently, intersection point A corresponds to the LHV opening point and intersection point B to the LHV set pressure. This shows the effect of the setting pressure PM and the pilot ratio Rv on the operation of the LHV. It is significant to note that this representation is used by some manufacturers to highlight differential aspects in the performance of the latest LHV versions, such as the two-stage valve, the multiple pilot ratio (adaptive) valve, and the setting self-adjusting (load match) valve.

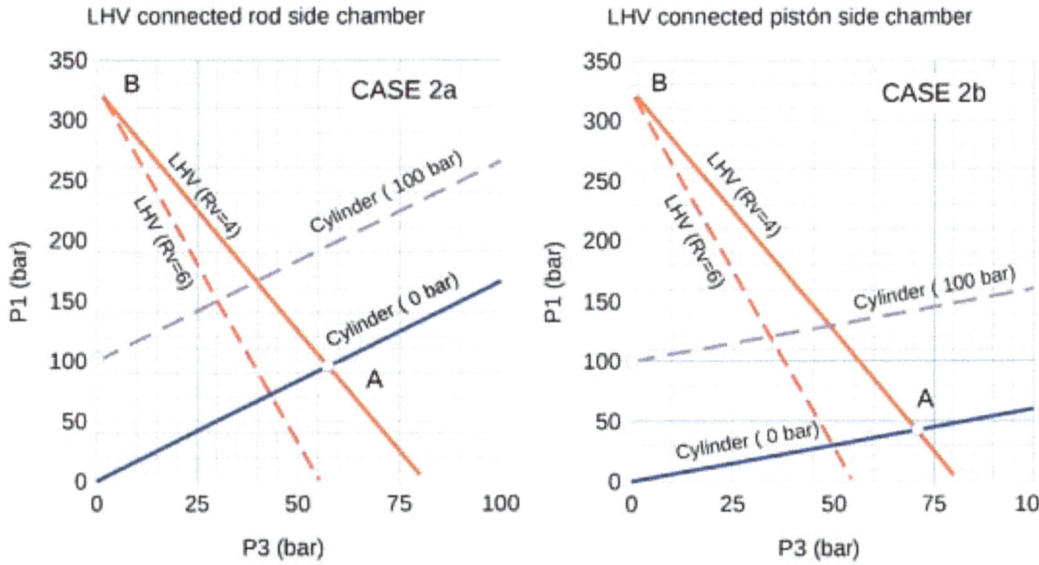

Figure 3. Steady-state operating curves $P_1 = f(P_3)$ for the two cases (2a) and (2b). Point A corresponds to the LHV opening point and point B to the LHV set pressure.

Another variant of the operating curve for cases (2a) and (2b) is shown in **Figure 4**. These plots express the pressures P_1 and P_3 of the system hydraulic actuator/ LHV in the cylinder chambers as a function of the equivalent load pressure, PLoad, which is the graphical representation of the two systems of linear Equations (6), (7), (9) and (10). The positive axis direction refers to overrunning loads, while the negative direction corresponds to resistive loads. The red line corresponds to the P_1 pressure and the black line to the pilot pressure, P_3. It can be seen that when the equivalent load pressure decreases, the pilot pressure increases.

The plots in **Figure 4** were first introduced by Professor Nicola Nervegna [**1**] and Professor Luca Zarotti [**2**]. Ritelli and Vacca used these graphs to evaluate energy aspects in [**3**], considering the hydraulic power consumption equal to the product of the pressure and the flow rate through the valve. If the flow rate is constant, the hydraulic power is proportional to the pressure in the actuator chamber P_1. Although these graphs have a high conceptual and qualitative value from the energetic point of view, the steady-state operating curves $P1=f(P3)$ shown in **Figure 3** can be used to evaluate the power requirements of the hydraulic system with LHV valves, as will be discussed in the following sections.

Table 1 shows the specifications of the essential hydraulic components that make up the hydraulic circuit used. These specifications were to draw **Figure 3** and **Figure 4**, while **Table 2** shows the coordinates of the most significant points of the curves shown in **Figure 4.**

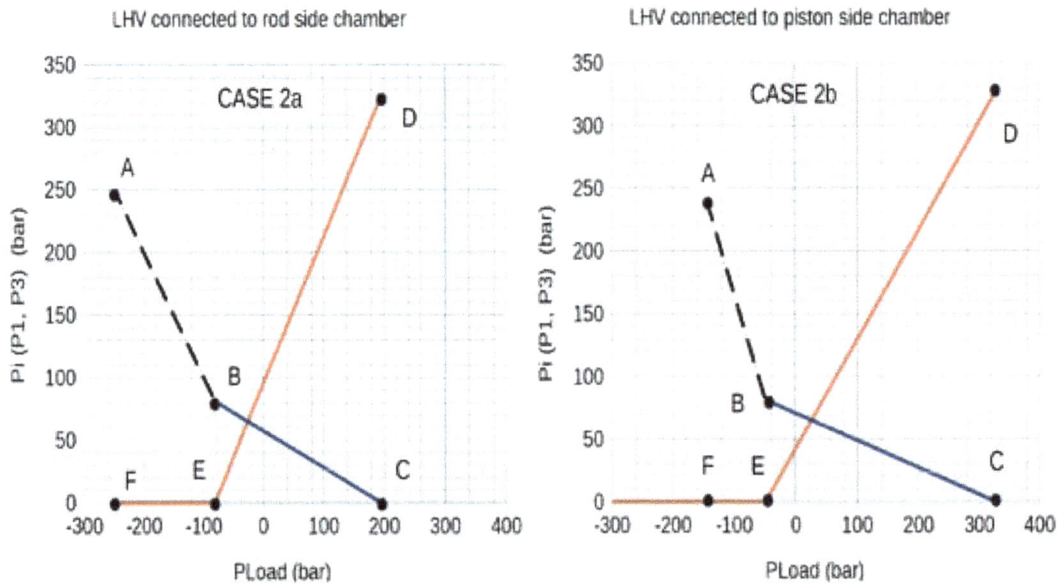

Figure 4. Steady-state operating curves $P(P1, P3)=f(Pload)$ for the two cases (2a) and (2b). The meanings of points A–F are defined in Table 2.

Table 1. The specifications of the primary hydraulic components of the hydraulic circuit imperative to depict graphs included in Figure 3 and Figure 4.

Cylinder		LHV		System/Load	
Piston diameter	100 mm	Pilot ratio	4	Relief valve	250 bar
Rod diameter	63 mm	Setting pressure	325 bar	P_{Load} (min)	0 bar
Section ratio	0.603	Load max	270 bar	P_{Load} (example)	100 bar
		Nominal flow	90 L/min		

Table 2. List of critical points in the diagrams in Figure 4 in bar.

Case (2a)	*x*	*y*	*x*	*y*
A	P_{vlp}	P_{vlp}	250	250
B	$(1/R_v)\cdot P_M$	$1/R_v\ P_M$	81	81
C	$R_c\cdot P_M$	0	196	0
D	$R_c\ P_M$	P_M	196	325
E	$(1/R_v)\cdot P_M$	0	81	0
F	P_{vlp}	0	250	0
Case (2b)	***x***	***y***	***x***	***y***
A	$R_c\ P_{vlp}$	P_{vlp}	151	250
B	R_c/R_v	$1/R_v\ P_M$	49	81
C	P_M	0	325	0
D	P_M	P_M	325	325
E	$(R_c/R_v)\cdot P_M$	0	49	0
F	$R_c\ P_{vlp}$	0	151	0

3. THE STATE OF THE ART

Figure 5 displays a historical collection of papers categorized as overcenter, counterbalance, and load-holding valves. This list is not comprehensive, but it adequately demonstrates the progression of scientific and technological elements.

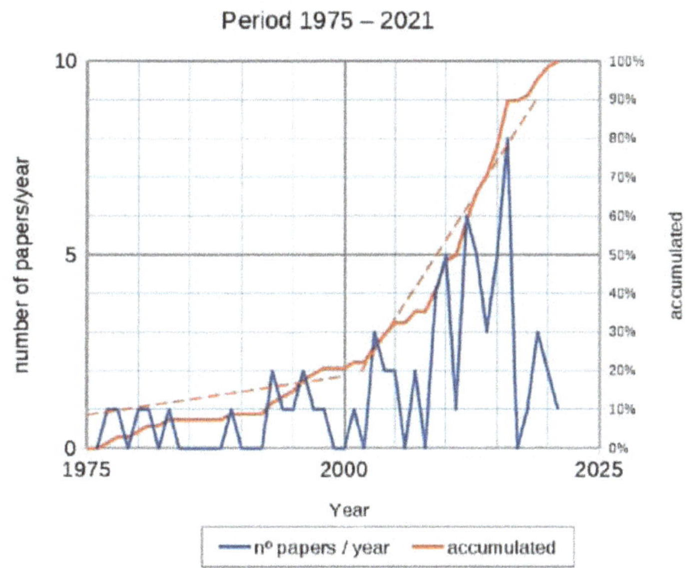

Figure 5. The number of papers/year (in blue) published related to load-holding valve issues shows an accumulated behavior (in red) with a tendency of 80% in the last 20 years.

The figure shows two distinct periods: the first period, from the early 2000s to the mid-2010s, focused on investigating the causes of instabilities, while the second period concentrated on developing designs and circuits to reduce or eliminate these effects. In a subsequent phase, which spans about the last ten years, new valve designs have emerged. Modeling is conducted using concentrated parameter methods and/or dispersed parameter methods, with a focus on research concerning energy usage.

3.1. About Instability

The severity of the oscillations is influenced by numerous parameters, including the external load on the actuator, the mechanical structure's properties, the damping and hysteresis of the LHV, the operator input, and the volumes and restrictions in the hydraulic lines. All of these have undergone thorough investigations by several referees [**4,5,6,7,8,9,10,11,12,13,14**].

The stability issue of LHV is thought to be caused by a phase lag between the pilot pressure and the actuator's outlet pressure. Separating the pilot pressure and the actuator intake pressure can help prevent oscillations. Ref. [16] proposed using a pressure feedback system to enhance the dynamic response of LHV, which can help prevent oscillations and enhance the stability of the hydraulic system. Sorensen et al. [17] proposed an innovative solution using a low-pass filtered value of the secondary circuit load pressure for the pilot connection of the LHV. Groof [18] presented a straightforward criterion for predicting instability from a conceptual perspective. The criterion is derived from a third-order equation and determines the stability of a system. It just concentrates on the hydraulic system, disregarding the mechanical stiffness of the entire system. Equation (11) is used to check the stability of the system when the LHV is installed on the rod side of the cylinder (Figure 2a), whereas Equation (12) is utilized when the LHV is mounted on the piston side of the cylinder (Figure 2b).

$$\frac{G_{relief}C_{hB}}{R_c} > \frac{G_{pilot}}{C_{hA}} \quad (11)$$

$$\frac{G_{relief}C_{hB}}{C_{hA}} > \frac{G_{pilot}}{R_c} \quad (12)$$

where, Gpilot, gradient of pilot function, describes the opening characteristic as a function of the pilot pressure

$$G_{pilot} = \frac{dQ_{throughLHVvalve}}{dP_{pilot}} \quad (13)$$

Grelief, gradient of relief function, describes the opening behavior of the valve depended on the load pressure

$$G_{relief} = \frac{dQ_{throughLHVvalve}}{dP_{relief}} \quad (14)$$

Ch, hydraulic capacitance is given by the increase in effective stored fluid volume dV per change in pressure dP. The subscripts A and B indicate the hydraulic conduction and the volume of the cylinder hydraulic chamber on the corresponding side.

$$C_h = \frac{dV}{dP} \quad (15)$$

By way of illustration and for scenario (2b), the above criterion (Equation (12)) can be expressed by Equation (16).

$$\frac{V_2}{V_1} > R_v R_c \left(\frac{2P_1 A_1}{1 + 2P_1 + k \cdot x_{LHV} - k_f \cdot x_{LHV} \cdot P_1} \right) \quad (16)$$

where,

- k, elastic constant of the spring of LHV
- kf, flow force constant
- V_1, hydraulic cylinder piston side chamber volume
- V_2, hydraulic cylinder rod side chamber volume
- xLHV, lifting height of the spool/poppet with respect to the seat surface of the LHV

The key factors affecting stability are the setting pressure and the pilot ratio, as mentioned in the preceding paragraphs. Decreasing the pilot ratio leads to increased operational stability, but also results in higher pressure levels and thus more energy consumption, as referenced in [3]. This is particularly important for minor external loads. Incorporating dampening into the hydraulic circuit design is another effective method. By raising volumes, adding orifices, and using logic valves, more damping is incorporated into the system. The pilot line has received special attention. Positive results have been produced by altering the pressure in various ways on the LHV pilot port line, such as introducing delays and generating a variation in the journey back and forth. Sciancalepore and Vacca (2019) suggested a method involving LHVs to regulate actuator speed and decrease energy usage. The system involves regulating the pilot port of the LHV using an external pressure source with an adjustable pilot. Two control strategies are outlined: the "Smart LHV" focuses on minimizing system pressure without controlling actuator velocity, while the "Smart System" efficiently manages actuator velocity using the LHV in overrunning load situations.

3.2. About Modeling

The majority of these studies focus on modeling the static and changing features of load-holding valves in both the time and frequency realms through a lumped parameter method. Yet, efforts to create physical or semi-physical models of these valves have faced numerous obstacles, such as friction, hysteresis, non-linear discharge area traits, fluctuating discharge coefficients, and variable flow forces.

Currently, there are various commercially accessible software programs like MATLAB/Simulink™, AMESim™, Dymola™, Maple/MapleSim™, and 20-Sim™ that greatly facilitate tasks. These packages can be categorized into two modeling approaches: a basic semi-physical model and a non-physical model (black box). The initial method utilizes static modeling based on force balance and the Bernoulli orifice pressure-flow equation to enable the valve to manage load velocity. It does not consider phenomena like as friction and flow forces. Alternatively, the dynamic modeling method relies on Newton's second law and the fluid continuity equation in some instances. Experimental verification was necessary in both instances. The second approach of the model utilizes two distinct pressure ratios to calculate the flow through the valve along with several factors that need to be determined through experimentation. However, LHVs are seldom correctly represented because of the challenge in acquiring fundamental model parameters and the intricacy of determining equations for flow forces and friction.

Computational fluid dynamics (CFD) can help understand these phenomena, but sometimes experimental work and modeling approaches are needed for time-domain simulation. The works of the references are particularly intriguing and noteworthy [**14,20**].

3.3. About Energetic Aspects

Examining the energy efficiency of fluid power systems is now essential for machinery makers due to the increasing electrification of machines and the constraints of battery capacity. Several researchers are engaged [21,22,23,24,25], and some are specific to studying the different hydraulic load retention systems in mobile and industrial applications [**15,26,27**].

Ritelli and Vacca [**3**] focused on the analysis of LHV energy savings. They quantified the energy consumption of LHVs with different pile ratios, showing that in a crane application, it is possible to achieve 59% energy savings with high pile ratios.

Regarding the energy efficiency of LHVs, there are some relevant previous efforts. LHVs with external and mixed control types were considered by ref. [**13**]. Significant examples are also the LoadMatch™ model, recently commercialized by Sun Hydraulics, which dynamically changes the valve setting as a function of the load pressure. In contrast, other works focus on introducing an external control to the LHV to influence its opening based on the instantaneous loading conditions. The most important contribution in this direction is the work in ref. [**17**], which proposed to use an auxiliary hydraulic circuit to control the pilot of the LHV, and ref. [**28**], which presented a method to control the pilot port of the CBV through an external pressure source (adjustable pilot). For both cases of "Smart CBV" and "Smart System", remarkable energy savings of 75% and up to 90%, respectively, were observed.

Despite scientific and technological advances reported and the wide variety of models and algorithms, it can be said that the use of this type of valve is far from easy. This wide range of tools is often not available to application engineers. In addition, the use of these tools still require a high level of application-specific knowledge. The impact outside academia remains limited. One reason may be that the design of hydraulic systems depends to some extent on the application.

In this paper, the authors highlighted the importance of understanding how LHVs work, not only in the critical opening condition but also outside of this condition, with the aim of helping to select the most suitable valve for the application according to its characteristics (extracted from the corresponding technical catalogs). Of course, experimentation is the best way to judge the system's power consumption but, the availability of a simple graphical method in the course to estimate the impact of the LHV on the overall power consumption may be very useful. Of course, experimentation is the best way to judge the system's power consumption. However, the availability of a simple graphical method in the course to estimate the impact of the LHV on the overall power consumption may be very useful. In the next section, a brief background overview of the different types of performance curves is presented.

3.4. Performance Curves

Selecting the appropriate LHV valve for a particular application can be a difficult endeavor. To apply an LHV accurately, one must comprehend its functionality and possess adequate technical data regarding its performance. After thoroughly examining the catalogs and various technical publications, such as release and technical papers, handbooks, user manuals, spreadsheets, prospectuses, and brochures, the following comments can be summarized:

- They all provide a simple and generic description of how it works, symbols, valve cavity, specifications or technical data, and some performance graphs or characteristic curves. Typical technical data include maximum operating pressure, setting pressure (cracking pressure) interval, nominal flow, pilot ratio, internal leakage, and hysteresis, among others.

- Although graphical presentations of performance should help viewers quickly and easily understand the critical information, there is no unanimous agreement on which are the most appropriate. Of the different ways of expressing its performance, the next conditions may be pointed out:
- Pressure drop in pilot operation condition (valve fully opened by the pilot pressure), (from port 1 to port 2, see **Figure 1**).
- Pressure drop in check valve operating condition (free flow) (from port 2 to port 1) and
- Pressure drop in pressure relief working condition (valve opened by the load). Unfortunately, some technical documents or specification sheets do not provide this minimum information, which is necessary for applying LHVs in a system.
- Generally, the curves shown at the end of **Section 2** of this document and in **Figure 3** and **Figure 4** are not usually included in the catalogs, but only in technical papers and to highlight the significant improvements in the performance of different valves.

4. GRAPHIC METHOD TO ESTIMATE THE POWER BALANCE

The method is attractive because only the pressures in the three ports and the relief function curve of the valve are sufficient to evaluate the power consumption. The method has been inspired by the work done at the Agder University, Norway, led by Prof. M. Hansen [**11**]. It consists of a simple semi-physical model.

4.1. Description of the Four-Quadrant Diagram

For the description of the graphical method, the four-quadrant diagram of the Cartesian plane is a good description of the proposed method (**Figure 6**). The following variables are represented on the axes of this diagram:

- x-axis (+): pilot pressure (P_3) in bar which corresponds to the pressure that prevails in the actuator rod chamber
- x-axis (-): flow through the load holding valve (Q_1) in L/min, equal to the flow leaving the piston side actuator chamber
- y-axis (+): load pressure (P_1) in bar, corresponding to the pressure in the actuator piston chamber
- y-axis (-): flow rate entering the piston side actuator chamber (QS) in L/min according to the configuration of the hydraulic circuit shown in scenario (2b), **Figure 2**.

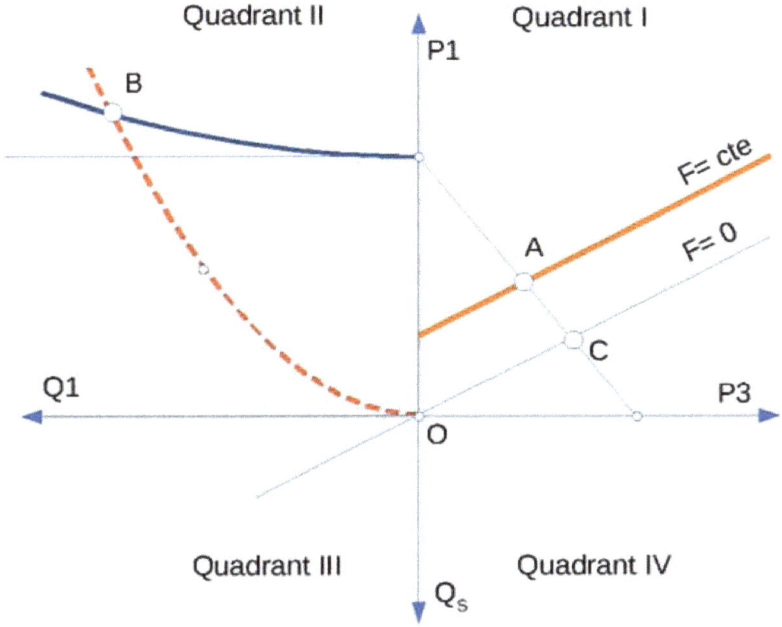

Figure 6. The cartesian plane is divided into four quadrants and characteristic curves of the hydraulic actuator and the load-holding valve. A represents the intersection between the curve of the cylinder function and that of the LHV when the force is constant; B is the intersection between the curve of the operating point of the LHV acting as a relief valve and the flow curve; C is the same as A but zero for the case of zero force.

The four-quadrant diagram allows us to plot the characteristic curves of an actuator and holding valve as follows:

- First quadrant "I": Upper right region. It is used to represent the characteristic curve of the hydraulic actuator and the characteristic curve $P1=(P3)$ of the load-holding valve.
- Second quadrant "II": Upper left region. It is used to show the characteristic curve of the load holding valve acting as a pressure-limiting valve (relief function).
- Third quadrant "III": Lower left region. It is used as an auxiliary two-dimensional space.
- Fourth quadrant "IV": Lower right region. It is used as an auxiliary two-dimensional space.

Alternatively, it can be used to depict the characteristic curve of the load-holding valve that expresses the pilot pressure (P_3) as a function of the flow rate (Q_1) through the load-holding valve for a predetermined load pressure value (P_1)

4.1.1. Steady-State Operating Curve and Hydraulic Cylinder Characteristic Curve

Equations (1) and (3) have been plotted in quadrant I. The intersection between the two curves gives the operating point of the hydraulic system (actuator/LHV valve) when the cylinder is subjected to a load, overrunning or resistive, as in **Figure 3.**

4.1.2. Characteristic Curve P1=(Q) of an LHV Valve Acting as a Pressure Limiting Valve (Relief Function)

Considering the simplified diagram of the holding valve poppet shown in **Figure 7** and assuming that P_3 is not working and the flow is in a steady-state and incompressible fluid (ρ is constant), the application of the principle of conservation of momentum to the control volume allows us to write:

$$I_2 - I_1 = P_1 A_1 - P_2 A_2 - F_{solid-fluid} \qquad (17)$$

$$F_{fluid-solid} = -F_{solid-fluid} \qquad (18)$$

where: P1 A1 and P2 A2 denote the net force due to the pressure distribution in the inlet and outlet sections, respectively, I_1 and I_2 are the momentum fluxes of the flow entering and leaving the control surface,

$$I_1 = \frac{\rho Q^2}{A_1} \qquad (19)$$

$$I_2 = \frac{\rho Q^2}{A_2(x)} \cos(\alpha) \qquad (20)$$

where

$$A_2(x) = \pi \cdot d \cdot sen(\alpha) \cdot x = K_2 \cdot x \qquad (21)$$

$$K_2 = \pi \cdot d \cdot sen(\alpha) \qquad (22)$$

and Ffluid-solid is the force on the poppet exerted by the fluid.

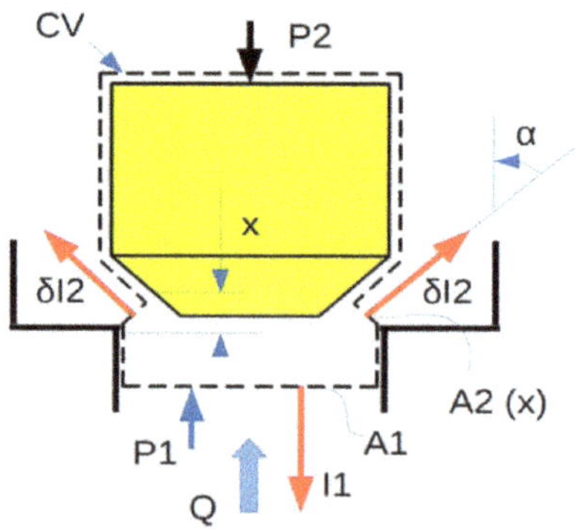

Figure 7. Control volume of the poppet/seat surface system of an LHV valve.

Applying Newton's second law of motion to the valve poppet (considered a free solid)

$$F_{fluid-solid} - (F_{spring} + M_{solid}\, g) = M_{solid}\dot{v} \tag{23}$$

Accepting an equilibrium position defined by the distance, x, considering that Msolid is negligible then:

$$F_{fluid-solid} = F_{spring} \tag{24}$$

$$F_{spring} = K_1(x + x_0) \tag{25}$$

where, K_1 is the constant of the spring

$$F_{precompresion} = K_1 x_0 \tag{26}$$

$$P_M = \frac{F_{precompresion}}{A_1} \tag{27}$$

Combining the previous Equations (17)–(27), it follows that the relief function curve

$$P_1 = P_M + \frac{K_{spring}}{A_1}x + K_2 Q^2 \tag{28}$$

with

$$K_2 = \frac{\rho}{A_1^2} - \frac{\rho}{A_1 \cdot (\pi d) \cdot x \cdot \tan\alpha} \tag{29}$$

On the other hand, for a predefined distance between poppet and seat surface, the flow rate can be calculated by the equation

$$Q_{out}(y, P) = K_2 \cdot x \cdot C_d \cdot \sqrt{2\frac{P_1}{\rho}} \tag{30}$$

or alternatively

$$P_1 = K_3 Q^2 \tag{31}$$

with

$$K_3 = \left(\frac{1}{K_2 \cdot x \cdot C_d \cdot \sqrt{\frac{2}{\rho}}}\right)^2 \tag{32}$$

In quadrant II, see **Figure 8**, the relief function curve (Equation (28)) and the flow rate curve (Equation (31)) are represented (not scaled). The point of intersection between the two curves (point B) indicates the operating point of the LHV acting as a relief valve; that is, the valve opens at pressure P and allows fluid flow Q.

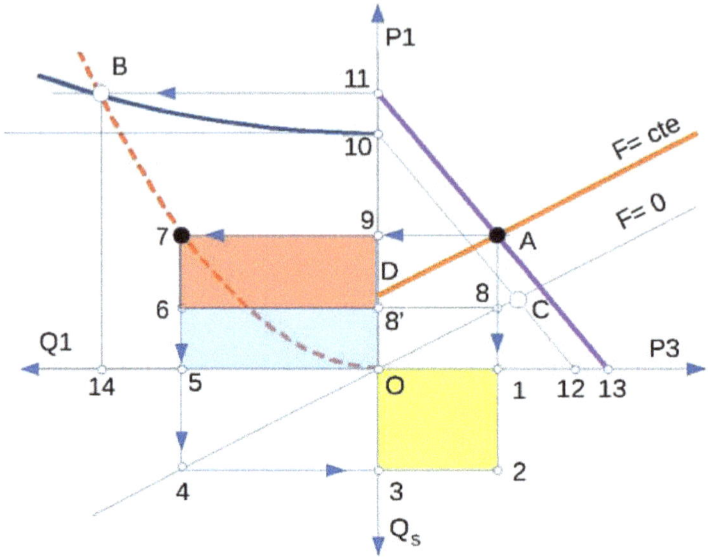

Figure 8. Scheme to display the described graphics approach. NLHV = ABlue + ARed = AO_{579}, NS = AYellow = ABlue = AO_{123} = $A_{0568'}$ and NM = ARed = $A_{6798'}$.

4.1.3. Power Balance Applied to the Actuator/LHV Hydraulic System

Before explaining how the power balance is proposed, it is helpful to explain the different curves shown in the four-quadrant diagram (**Figure 8**), as well as the points that are considered most representative:

- Curve 10–12: characteristic curve $P_1 = f(P_3)$ of the LHV valve in the static balance position (closing condition). It corresponds to the representation of Equation (1). Load force *fload*
- Curve 11–13: characteristic curve $P_1 = f(P_3)$ of the LHV valve in the permanent regime when a flow rate Q_1 flows through the valve, as a consequence of the opening of the obturator by the action of pressure P_1 and pilot pressure P_3. Load force constant $F1>0$
- Curve O–C: characteristic curve $P_1 = f(P_3)$ of the hydraulic actuator when it is subjected to zero force. It corresponds to the representation of Equation (3).
- Curve D–A: characteristic curve $P_1 = f(P_3)$ of the hydraulic actuator when it is subjected to an overrunning force, F = constant.
- Curve 10–B: characteristic curve P = f(Q) of the LHV valve acting as a pressure limiting valve (relief function). See Equation (28).
- Curve O–7–B: LHV poppet pressure drop in the open position as a result of the shutter opening due to the action of the pressure P_1 and the pilot pressure P_3. See Equation (31).

Assuming that the operating point of the hydraulic system (hydraulic actuator/ LHV valve) is defined by point A, that is, the intersection of curves (11–13) and (D–A), this point corresponds to the operating point of the LHV when a flow rate Q_1 flows as a result of the opening of the obturator under the action of pressure P_1 and pilot pressure P_3, thus defined by the coordinates (P_3, P_1).

Other points of interest are:

- Point 7: This point is defined by the coordinates (Q_1, P_1), intersection of the curve (0–B), with the line of constant pressure equal to P_1.
- Point 5: defined by the coordinates (Q_1, 0)
- Point 1: defined by the coordinates (P_3, 0)
- Point 3: defined by the coordinates (0, Q_S)

Based on the hydraulic circuit (e.g., scenario 2b) outlined in **Figure 2**, the following powers involved in movements are defined:

NM power due to the movement of the actuator as a consequence of being subjected to the overruning force F.

$$N_M = F_{overrunning} \cdot v_{actuator} \tag{33}$$

NS, power supplied by the pump feeding the hydraulic actuator chamber on the rod side

$$N_S = P_S Q_S = P_{S(:point\ 1:)} Q_{S(:point\ 3:)} \tag{34}$$

NLHV, power related to the fluid flow through the LHV valve as a result of the combined action of pressure P_1 and pilot pressure P_3

$$N_{LHV} = P_1 Q_1 = P_{1(:point\ 9:)} Q_{1(:point\ 5:)} \tag{35}$$

A simple power balance allows us to establish the following equation:

$$N_{LHV} = N_S + N_M \tag{36}$$

N_S power is defined by the rectangle (0123) (yellow area), where side 03 has magnitude QS and side 01 has magnitude equal to P_3. Note that the area (0123) is equal to the area (0568'). Using the hydraulic actuator curve (for condition F=0=0) extended from Quadrant I to Quadrant III, represented by the "40C" line, the following relationships are established:

$$P_{1(:point\ 8':)} = R_c P_{3(:point\ 1:)} \tag{37}$$

$$Q_{1(:point\ 5:)} = \frac{Q_{S(:point\ 3:)}}{R_c} \tag{38}$$

multiplying Equations (37) and (38), it turns out

$$P_{1(:point\ 8':)} Q_{1(:point\ 5:)} = P_{3(:point\ 1:)} Q_{S(:point\ 3:)} \tag{39}$$

Equation (39) highlights the aforementioned equality, that is the area (0568') = area (0123). On the other hand, note that the area of the rectangle (0579) (ABlue + ARed), where side 05 has magnitude Q_1 and side 09 is equal to P_1, represents the power NLHV, i.e., the power related to the fluid flow through the LHV valve. Then, by virtue of Equation (36), it can be deduced that the area of the rectangle (6798') (ARed), represents the power due to the movement of the actuator as a consequence of being subjected to the overrunning force F.

Based on these equations, the ratio ϕ is defined between the power provided by the pump and the power related to the flow through the LHV valve, see Equation (40). This power ratio is an index that indicates the percentage of the pump power needed to open the LHV valve and, therefore, an index of the energetic goodness of the LHV valve.

$$\phi = \frac{N_S}{N_{LHV}} = \frac{area(O568')}{area(O579)} = \frac{Area_{Blue}}{(Area_{Blue} + Area_{Red})} \tag{40}$$

5. EXPERIMENTAL VALIDATION

5.1. Lab Testing

To validate the proposed methodology, first of all, the performance of an LHV valve was tested. Specifically, the reference valve tested at the LABSON Laboratory (Terrassa-Barcelona) was the LHV-4 valve (note: there was no particular reason for the choice, only its availability in the laboratory). **Figure 9** shows the diagram of the hydraulic circuit used.

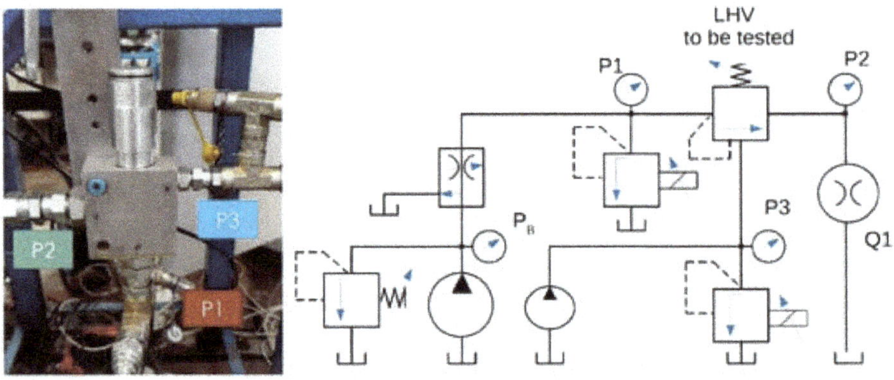

Figure 9. LHV-4 picture. Diagram of the hydraulic circuit used where: P_1: upstream water pressure; P_2: downstream pressure; P_3: pilot pressure.

Figure 10 presents the results obtained. **Figure 10**a shows the steady-state operating curve of LHV valve, while **Figure 10**b shows of relief function curve. On the one hand, the relief functions with setting pressure of 150 and 250 bar are shown. These curves were evaluated based on the adjustment curves shown in **Figure 10**a. On the other hand, the relief curve with setting pressure 100 bar, obtained by varying the pressure P_1 with the help of a proportional pressure relief valve, showed its hysteresis phenomenon.

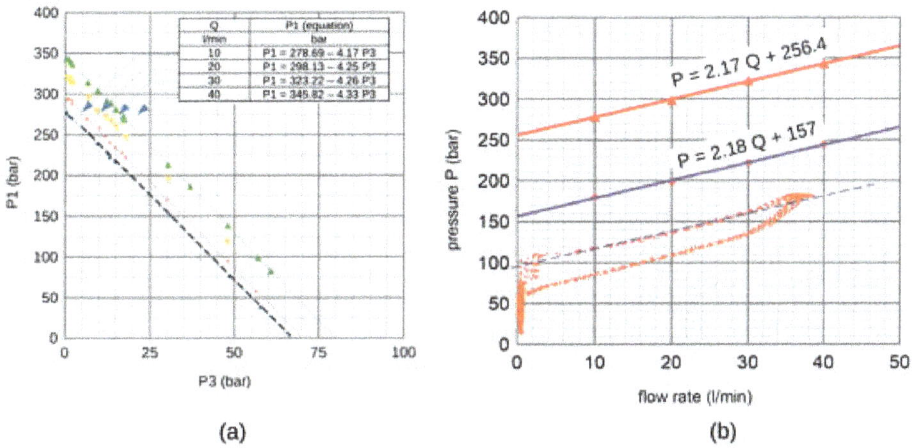

Figure 10. LHV experimental curves: (a) steady-state operating curves and (b) relief function curve.

5.2. Field Testing

This part outlines the experimental tests conducted to confirm the technique described in the preceding section. Multiple tests were conducted on the hydraulic lifting cylinder of a telescopic handler known as M1. The initial testing involved equipping the M1 telehandler with the LHV4-M1 valve, a well-known component, to replace the original valves in the commercial version.

Figure 11 shows panoramic images of the experimental tests carried out. The tests were carried out with different load conditions of the telehander.

Figure 11. The telehandler machine conveniently instrumented.

The hydraulic cylinder was actuated by means of a manual directional control valve. The field test was equipped with multiple pressure sensors, a flow sensor, and a cylinder position sensor. Displacement and angle sensors were also fitted to the boom.

Sensor data acquisition and electronic actuation were carried out with the help of a multi-axis motion controller RMC200, equipped with input and output modules (Delta Computer Systems). It is easy to use, and the software provided with it has powerful plotting capabilities.

Sensors and their main characteristics are listed in **Table 3**. Calibration values were obtained from the manufacturer's information and from comparison with a standard when the manufacturer's information was not available. All values were checked in preliminary tests. Sensor values were measured and recorded at a frequency of 1 kHz. Higher rates were not considered necessary for the energy analysis. Digital filtering at 100 Hz was required for the calculation of the speeds, especially with the angle sensor. The data were processed with commercial software (DIAdem, Excel) to calculate the total energy consumption.

Table 3. List of main characteristics of used sensors.

Magnitude	Manufacturer	Model	Range	Accuracy
Pressure	WIKA	MH2	0–250/0–400 bar	>0.5% span
Flow	HYDAC	EVS3199TF	6–60 L/min	>2% act. val.
Position	Micro Epsilon	WDS-1500_PS60-SR-U	0–1500 mm	>+/-1.5 mm
Angle (tilt)	SICK	TMM 56E-PMH045	+/-45°	+/-0.3°
Temperature	Omega	PT100	10–100°	>0.36 °C

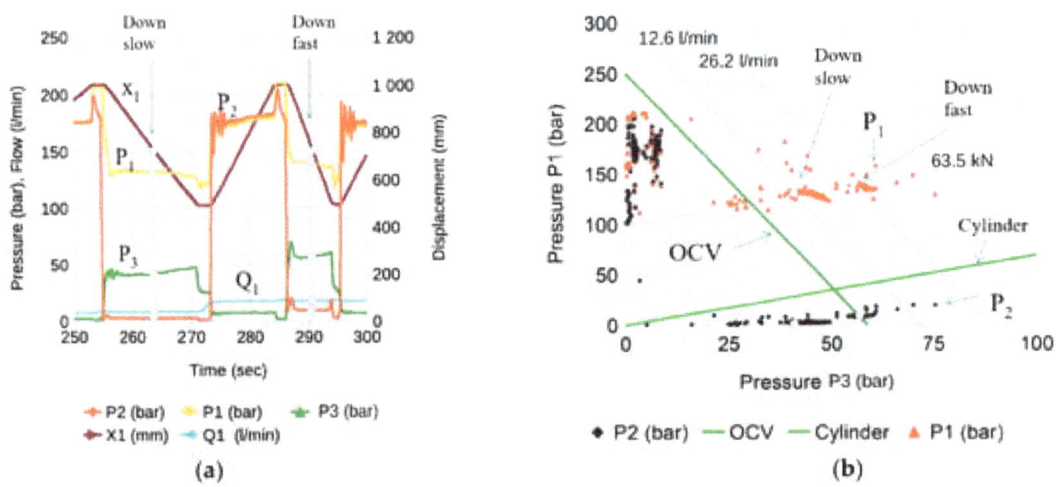

Figure 12. Example of experimental results obtained on M_1 machine using LHV_4-M_1 valve where: (a) shows the temporal evolution of the experimentally measured variables. (b) experimental results plotted on steady-state operating curves for $P_1 = f(P_3)$.

Figure 12 shows the results obtained with machine M_1 in which a 90 × 50 hydraulic cylinder and the LHV4 valve were installed to move the telescopic arm.

Figure 12a shows the temporal history of the load pressure P_1, the return pressure P_2, and pilot pressure P_3, together with the hydraulic cylinder position x_1 and pump flow QS companion. It is observed that there is a time interval in which the values P_1, P_2, P_3, and Q_1 remained essentially constant, which corresponds to the downward cylinder movement, x_1, and, consequently, to the time interval in which the valve is working. The companion **Figure 12**b shows the working pressure space of rod side P_1 versus the pilot P_3 pressures, respectively. The average values are represented by a small circle in **Figure 12**a,b.

Procedure

The method is presented over the description set out in **Figure 13** with real values (and **Figure 8** as reference).

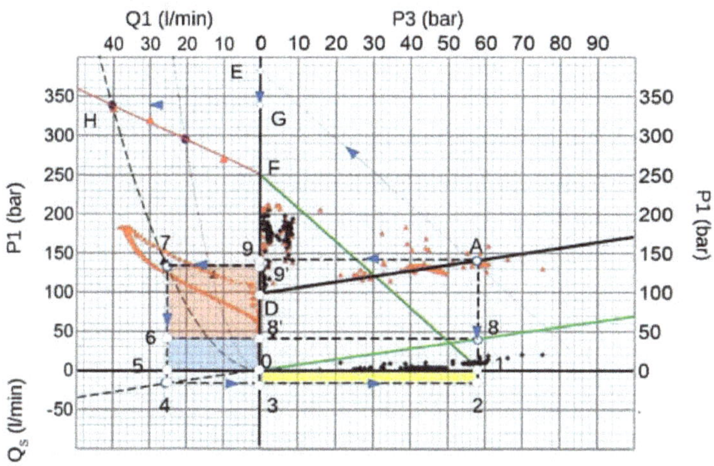

Figure 13. Graphical representation of the power balance for a real case.

Actions to be carried out over the first quadrant:

- Plot the working point A (e.g., P_1 = 142 bar and P_3 = 58 bar);
- Plot LHV curve based on its setting point Pm = 250 bar and the pilot ratio 4.25 as it is shown in **Table 1** (e.g., straight-line 1F);
- Plot an auxiliary line parallel to 1F at the A point projecting it to the pressure axis (e.g., E point);
- Calculate the effect of the return pressure using the equation P2·(Rv + 1), which is represented by the segment EG. G point represents P_M pressure shown in Equation (1);
- Trace a horizontal line crossing at A point to obtain the working pressure (e.g., point 9);
- Obtain the differential pressure P_1-P_2 by plotting 9′ position (e.g., segment length between 0–9′).

Actions in the second quadrant:

- Plot the relief LHV function curve taking the setting pressure as 250 bar (e.g., the experimental curve shown in **Figure 10**b);
- Draw a horizontal line starting at G and crossing the LHV function curve (e.g., H point is obtained). Therefore, the K_3 constant can be calculated from Equation (3) obtained using H coordinated, $K3=PHQ2H$;
- The parabolic performance following Equation (31);
- Trace a horizontal line across 9` to obtain the coordinates of intersection with the parabolic curve OH to bring point 7, which corresponds to the orifice working conditions;
- Draw a vertical segment from point 7 to the flow axis to obtain the flow Q_1 through the valve (point 5);
- Extend the cylinder actuator characteristic curve from the first quadrant to the third quadrant (segment "04", as an extension of segment "08");
- The pump flow rate QS is represented by point 3, which is calculated from point 4 by crossing the vertical axis;
- Extend line "4–3" to the intersection with the vertical line through point 1, obtaining point 2;
- The graphics method provides flow rate (represented by points 3 and 5 compared with experimental ones, as shown in **Table 4**).

Table 4. Working values.

			Go down Slow	Go down Fast
load pressure	P_1	bar	132	142
return pressure	P_2	bar	3	9.21
pilot pressure	P_3	bar	42	58
supply flow	Q_S	l/min	9	18
valve flow	Q_1	l/min	13	26
setting pressure	P_m	bar	250	250

Power ratios ϕ can be obtained applying Equation (40).

- "(0123)" area represents the power, NS (yellow area)
- "(0568`)" area has an identical area of "(0123)" (blue area)
- "(679`8`)" area is equivalent to the power load, F·v (red area)
- "(0579`)" area is the power dissipated in the valve, $NLHV=P1Q1$, which is equal to the sum of "(0568`)" and "(679`8`)"

The yellow area, identical to the blue area, shows that using LHV valves in a hydraulic system introduces additional energy consumption.

To conclude this research work, it is essential to apply and validate the proposed methodology in more cases. For this purpose, two models of telehandlers (M_1 and M_2) were equipped with seven different LHV valves. The mixed sets of valves and machines were distinguished with LHV_i-M_j where i = {1 ... 7} and j = {1,2}. The relevant data, setting pressure, and relief characteristic curve were obtained from the commercial catalogs of the valves.

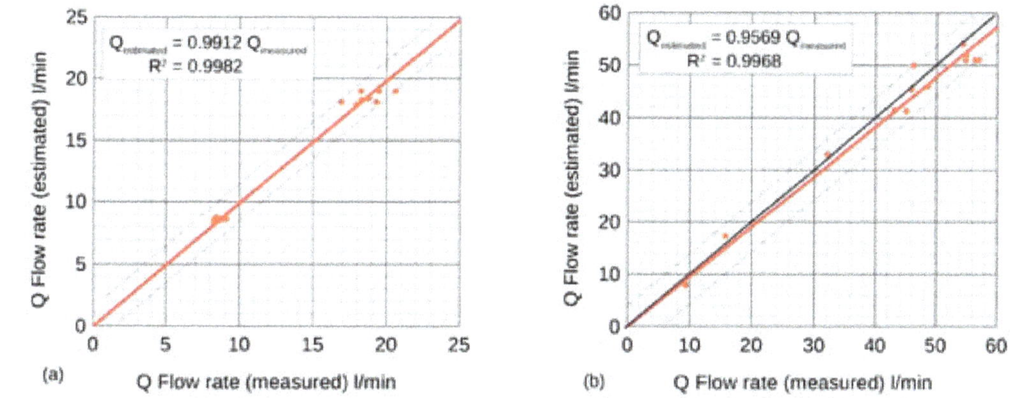

Figure 14. Comparative graphs between measured and estimated flows (Q_S) for the two machines: 14 (a) machine M_1 and 14 (b) machine M_2.

Figure 15 shows the power ratio, ϕ, as a function of the net force to which the lifting cylinder is subjected for machines M_1, **Figure 14**a, and M_2, **Figure 14**b.

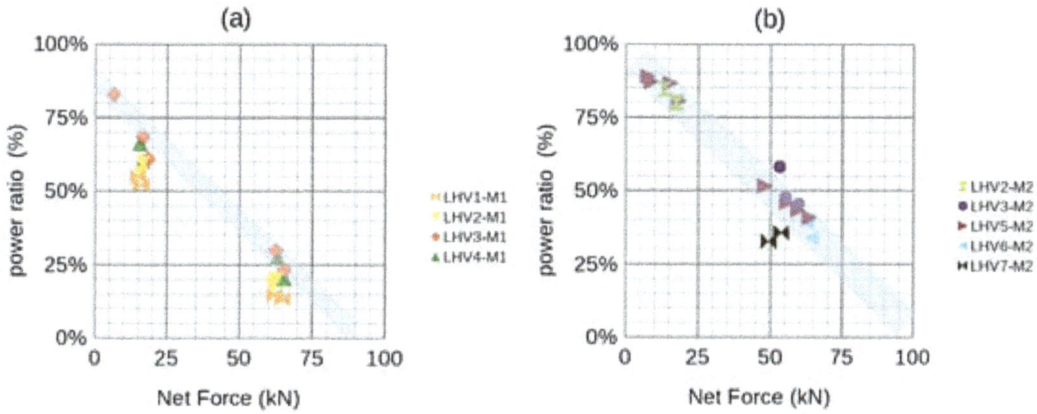

Figure 15. Graph power ratio as a function of the net force (for all cases). (a) M_1 machine; (b) M_2 machine.

Figure 14 shows the high correlation between the flows estimated by applying the proposed method and the experimental tests carried out in the field with the two machines presented and the set of seven valves used. The high correlation measured through the coefficient $R_2 \approx 0.99$ is notable. Taking into account that the proposed method allows estimating the flows and powers involved in the operation of the hydraulic cylinder/holding valve system, it seems reasonable to compare the calculated values with the actual values measured in the experimental tests in order to obtain the accuracy of the method. The estimated and measured flow rates for machine M_1 are shown in **Figure 14**a, where the dashed lines define an error range of ±7.5%. On the other hand, the values for the machine M_2 are shown in **Figure 14**b, where the dashed lines define an error range of +8%/-12%.

Figure 16 shows the order of magnitude of the errors made in the estimation of power according to the following definition:

$$error(\%) = \frac{N_{supply}(estimated) - N_{supply}(mesured)}{N_{supply}(mesured)} \tag{41}$$

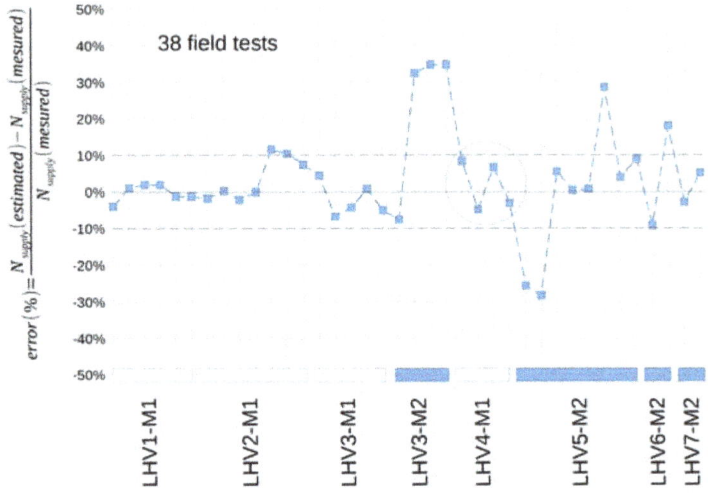

Figure 16. Bounds of the power estimation errors of the applied method.

6. CONCISE SUMMARY

The authors emphasized the significance of comprehending the operation of LHVs in the critical opening situation in this chapter. Outside of this region, it assists in choosing the best appropriate valve for the application based on its properties. To further understand these valves, instruments such as software and models are necessary. However, it requires a significant amount of specialized knowledge related to a particular application, typically developed in academic settings. The methodology in this paper is ad hoc and designed for immediate usage by end-users. The main benefit of this method is that the data from a catalog and the pressure measurements from the three valve ports are enough to calculate the power ratio. Although DAQ and high-level equipment were utilized to demonstrate the effectiveness of this methodology, they are not necessary for end-users to assess power. End-users can utilize basic gauge-pressure sensors to implement the suggested technique and achieve favorable outcomes.

The visual technique accurately estimates the flow rate passing through the valve and the pump's flow rate. The powers involved are therefore computed. The functional link between the power ratio and the load's net force is determined secondly. Less load results in more substantial power output. Figure 15 outlines the performance variations seen among the various valves that were evaluated. Recent design changes have been implemented to enhance performance and generate greater energy savings, resulting in contrasting outcomes. Figure 16 demonstrates the validity of the strategy in relation to the suggested simplifications and hypotheses. The approach often calculates power ratios within an error range of ±10%, which is quite acceptable.

7. AT A GLANCE

The utilization of a load-holding valve (LHV) in a hydraulic system is widely recognized to result in increased energy consumption. The study introduced a simpler graphical technique for evaluating the power needs of hydraulic systems that have load-holding valves for controlling overrunning loads. This method aids in comprehending the load-holding valves' function when the actuator is in motion. It also enables viewing of how the key parameters (pilot ratio, set pressure) and additional factors like flow rate, back pressure, and load force impact the overall system consumption. This method is appealing because it allows for the evaluation of energy consumption and the definition of the power ratio using only the pressures at the three ports and the valve relief function curve. The power ratio serves as an index reflecting the percentage of energy required to open the LHV valve. The approach was utilized in practical scenarios, specifically on two categories of elevating mobile machinery. It was confirmed through multiple outside experiments on two mobile machines, resulting in experimental data. Both machines were fitted with a set of seven distinct performance LHV valves for the tests. This strategy could help hydraulic machine manufacturers choose an appropriate valve for energy-efficient applications, particularly in the context of the growing trend towards electrification.

REFERENCES

1. Nervegna, N. Oleodinamica e Pneumatica; Politeko, T., Ed.; Politeko: Arlington, VA, USA, 2003. **[Google Scholar]**

2. Zarotti, G.L.; Nervegna, N. Rational Design of Mobile Hydraulics by Digital Computer; SAE Ed.: Warrendale, PA, USA, 1979. **[Google Scholar]**

3. Ritelli, G.F.; Vacca, A. Energetic and Dynamic Impact of Counterbalance Valves in Fluid Power Machines. Energy Convers. Manag. **2013**, 76, 701–711. **[Google Scholar]** **[CrossRef]**

4. Overdiek, G. Design and Characteristics of Hydraulic Winch Controls by Counterbalance Valves. In European Conference on Hydrostatic Transmissions for Vehicle Applications; Mechanical Engineering Publications Limited for the Institution of Mechanical Engineers: Aachen, Germany, 1981. [**Google Scholar**]

5. Persson, T.; Krus, P.; Palmberg, J.-O. The Dynamic Propertles of Over-Center Valves in Mobile Systems. In Proceedings of the 2nd International Conference on Fluid Power Transmission and Control, Hangzhou, China, 20–22 March 1989; Zhejiang University: Hangzhou, China, 1989. [**Google Scholar**]

6. Miyakawa, S. Stability of a Hydraulic Circuit with a Counter-Balance Valve. Bull. JSME **1978**, 21, 1750–1756. [**Google Scholar**] [**CrossRef**]

7. Handroos, H.; Halme, J.; Vilenius, M. Steady State and Dynamic Properties of Counterbalance Valves. In Proceedings of the 3rd Scandinavian International Conference on Fluid Power (SICFP'93), Linköping, Sweden, 25–26 May 1993. [**Google Scholar**]

8. Chapple, P.J.; Tilley, D.G. Evaluation Techniques for the Selection of Counterbalance Valves. In Proceedings of the Expo and Technical Conference for Electrohydraulic and Electropneumatic Motion Control Technology, Anaheim, CA, USA, 23–24 March 1994. [**Google Scholar**]

9. Rahman, M.M.; Porteiro, J.L.F.; Weber, S.T. Numerical Simulation and Animation of Oscillating Turbulent Flow in a Counterbalance Valve. In Proceedings of the IECEC-97 Thirty-Second Intersociety Energy Conversion Engineering Conference (Cat. No.97CH6203), Honolulu, HI, USA, 27 July–1 August 1997; Volume 2, pp. 1525–1530. [**Google Scholar**]

10. Lisowski, E.; Stecki, J.S. Stability of a Hydraulic Counterbalancing System of a Hydraulic Winch. In Proceedings of the Sixth Scandinavian Intl. Conf. on Fluid Power, Tampere, Finland, 26–28 May 1999; pp. 921–933. [**Google Scholar**]

11. Andersen, T.O.; Hansen, M.R.; Pedersen, P.; Conrad, F. The Influence of Flow Force Characteristics on the Performance of over Centre Valves. In Proceedings of the 9th Scandinavian International Conference on Fluid Power, Linköping, Sweden, 1–3 June 2005. [**Google Scholar**]

12. Liu, X.; Liu, X.; Wang, L.; Chen, J. The Dynamic Analysis and Experimental Research of Counter Balance Valve Used in Truck Crane. In Proceedings of the 2010 International Conference on Electrical and Control Engineering, Wuhan, China, 25–27 June 2010; College of Mechanical Science and Engineering: Tampere, Finland, 2010. [**Google Scholar**]

13. Zähe, B.; Anders, P.; Ströbel, S. A New Energy Saving Load Adaptive Counterbalance Valve. In Proceedings of the 10th International Conference on Fluid Power, Dresden, Germany, 8–10 March 2016; pp. 425–436. **[Google Scholar]**
14. Jakobsen, J.H.; Hansen, M.R. CFD Assisted Steady-State Modelling of Restrictive Counterbalance Valves. Int. J. Fluid Power **2020**, 21, 119–146. **[Google Scholar] [CrossRef]**
15. Andersson, B.R. Energy Efficient Load Holding Valve. In Proceedings of the 11th Scandinavian International Conference on Fluid Power SICFP, Linköping, Sweden, 2–4 June 2009; Volume 9. **[Google Scholar]**
16. Kjelland, M.B.; Hansen, M.R. Numerical and Experiential Study of Motion Control Using Pressure Feedback. In Proceedings of the 13th Scandinavian International Conference on Fluid Power, SICFP2013, Linköping, Sweden, 3–5 June 2003. **[Google Scholar]**
17. Sørensen, J.K.; Hansen, M.R.; Ebbesen, M.K. Novel Concept for Stabilising a Hydraulic Circuit Containing Counterbalance Valve and Pressure Compensated Flow Supply. Int. J. Fluid Power **2016**, 17, 153–162. **[Google Scholar] [CrossRef]**
18. de Groof, A.G. Instability of Counterbalance Valves; Hogeschool: Rotterdam, The Netherlands, 2014. **[Google Scholar]**
19. Sciancalepore, A.; Vacca, A.; Weber, S. An Energy-Efficient Method for Controlling Hydraulic Actuators Using Counterbalance Valves with Adjustable Pilot. J. Dyn. Syst. Meas. Control **2021**, 143, 111007. **[Google Scholar] [CrossRef]**
20. Christensen, M. Design Af Hydraulisk Lastholdeventil. Design Af Mekaniske Systemer. Master's Thesis, Aalborg Universitet, Aalborg, Denmark, 2007. **[Google Scholar]**
21. Vukovic, M.; Leifeld, R.; Murrenhoff, H. Reducing Fuel Consumption in Hydraulic Excavators—A Comprehensive Analysis. Energies **2017**, 10, 687. **[Google Scholar] [CrossRef] [Green Version]**
22. Mahato, A.C.; Ghoshal, S.K. Energy-Saving Strategies on Power Hydraulic System: An Overview. Proc. Inst. Mech. Eng. Part J. Syst. Control Eng. **2021**, 235, 147–169. **[Google Scholar] [CrossRef]**
23. Roquet, P.; Raush, G.; Berne, L.J.; Gamez-Montero, P.-J.; Codina, E. Energy Key Performance Indicators for Mobile Machinery. Energies **2022**, 15, 1364. **[Google Scholar] [CrossRef]**

24. Berne, L.J.; Raush, G.; Gamez-Montero, P.J.; Roquet, P.; Codina, E. Multi-Point-of-View Energy Loss Analysis in a Refuse Truck Hydraulic System. Energies **2021**, 14, 2707. **[Google Scholar]** **[CrossRef]**

25. Rydberg, K.-E. Energy Efficient Hydraulics—System Solutions for Minimizing Losses; National Conference on Fluid Power, "Hydraulikdagar'15"; Linköping University: Linköping, Sweden, 2015. **[Google Scholar]**

26. Lin, T.; Chen, Q.; Ren, H.; Huang, W.; Chen, Q.; Fu, S. Review of Boom Potential Energy Regeneration Technology for Hydraulic Construction Machinery. Renew. Sustain. Energy Rev. **2017**, 79, 358–371. **[Google Scholar]** **[CrossRef]**

27. Axin, M. Fluid Power Systems for Mobile Applications with a Focus on Energy Efficiency and Dynamic Characteristics. Ph.D. Thesis, Fluid and Mechatronic Systems Department of Management and Engineering, Linköping University, Linköping, Sweden, 2013. **[Google Scholar]**

28. Sciancalepore, A.; Vacca, A.; Pena, O.; Weber, S.T. Lumped Parameter Modeling of Counterbalance Valves Considering the Effect of Flow Forces. In Proceedings of the ASME/BATH 2019 Symposium on Fluid Power and Motion Control, FPMC 2019, Longboat Key, FL, USA, 7–9 October 2019. **[Google Scholar]**

05 THEORETICAL ANALYSIS OF ACTIVE FLOW RIPPLE CONTROL IN POSITIVE DISPLACEMENT PUMPS

1. INTRODUCTION: AN INTRODUCTION

Positive displacement machines are frequently used in hydraulic systems due to their capacity to deliver high power density, as well as their strong and dependable nature. One significant disadvantage is the fluctuation of the flow rate, leading to pressure ripple, noise, and vibrations that might damage the components in the hydraulic system. This phenomena can have adverse effects on the durability and performance of hydraulic components [1];. Additionally, vibrations can impact both humans and the environment [2].

Researchers studying the decrease in flow and pressure fluctuations caused by positive displacement pumps have concentrated on various solutions, with many including the geometric optimization of components. A meticulous design of the port plate in axial piston pumps facilitates a seamless shift of fluid pressure between suction and delivery phases and vice versa[3,4,5]. Side bushings in gear pumps are constructed with specific grooves to enhance machine performance in this aspect [6,7,8,9,10,11]. Other methods involve devices external to the pump and could be classified as passive or active methods.

Passive approaches involve installing external components, such resonators, outside the pump. Passive devices attached to a pump supply line have a significant limitation in that they are typically constructed and calibrated to function under specific conditions, making them useful only within limited pressure ranges or against specific frequencies [12,13,14]. For instance, if the pump shaft speed varies, these systems become less effective, as the pressure ripple frequency is directly influenced by the pump's rotational speed. A passive solution is suggested in reference [13] to decrease pressure fluctuations on both the suction and delivery sides. A specific flow ripple attenuator is introduced and linked to the suction and delivery lines. Solutions utilizing active control can adjust to fluid pressure and machine shaft speed variations, surpassing the constraints of passive methods. Researchers have explored combining passive and active approaches, as discussed in references [12,14]. Passive pipes are used to reduce high frequencies, while an active actuator targets lower frequencies.

The technique described in [12,14] involves using a variable volume actuator on the pump's delivery side. This actuator is controlled by a piezoelectric actuator through an adaptive filter with a feedforward algorithm.

Active approaches involve creating a flow signal that is added to the pump delivery flow rate to reduce its oscillation. These solutions must produce a signal with the appropriate amplitude and phase to counteract the original signal. Active systems have a significant disadvantage in that they are complex devices that need to be attached outside to the pump, resulting in a somewhat intricate and large unit compared to the pump by itself. An active fluid-borne noise reduction system has been implemented on an aviation hydraulic pump in reference [15]. The system consists of an in-house rotary valve design that is attached to the pump shaft, taking into account the impact of speed.

The literature discusses studies on reducing structure-borne noise affecting swash plate vibrations in axial piston pumps controlled by a servo valve. Other studies focus on compensating for the uneven flowrate typically generated. In reference [19], a comparable approach is suggested where the swash plate is manipulated to modify the instantaneous displacement. A second method described in reference [19] involves utilizing a piezo stack to activate a cylinder that is linked to the pump outlet.

This research introduces an active approach to reduce flow ripple in a positive displacement pump. This approach utilizes an actuated piston positioned on the pump's delivery side to counteract the main flow ripple, effectively reducing the primary harmonic of the flow oscillation. This study aims to initiate the research process by presenting a theoretical analysis. The two solutions' models were created using the Simcenter AmesimTM software. Two strategies are suggested for minimizing flow ripple by utilizing a piston that, through oscillation, produces a consistent flow. The distinction between the two solutions is in the method by which the piston is activated. This study provides an analytical comparison of two options, highlighting their pros and cons. An adaptive control logic based on the LMS (Least Mean Square) algorithm was built to drive both active systems, allowing them to adapt to changing pump working conditions. The control algorithm was created using SimulinkTM and then cosimulated with the Amesim physical model.

2. DESCRIPTION OF THE ACTIVE METHODS

2.1. Piston Actuated by Piezo-Stack Actuator (1st Solution)

This method involves using a cylindrical actuator attached to the pump's delivery line to reduce flow ripple. The schematic may be found in Figure 1.

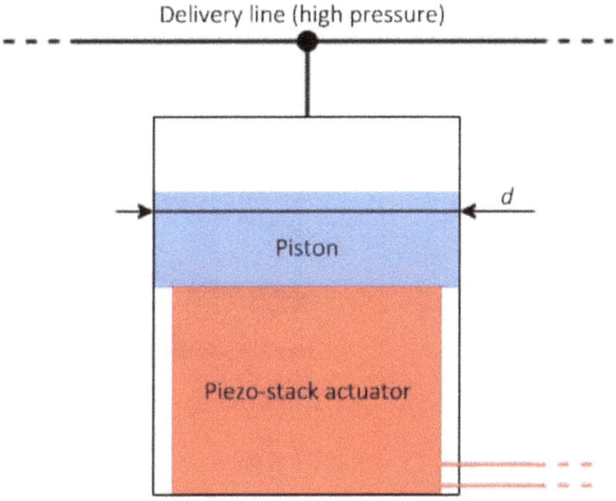

Figure 1. Schematic of piezo-stack actuated piston active method.

The piston is moved by a piezo-stack actuator controlled by a voltage amplifier that uses a feedforward logic signal determined by the LMS algorithm. The system functions as an adaptive filter that adjusts its response based on the varying operating conditions of the volumetric machine, such as changes in delivery pressure, flow rate, and pump rotation speed, due to this configuration.

2.2. Hydraulically Actuated Piston (2nd Solution)

The second proposed active approach is similar to the previous one, but it utilizes a different solution to activate the piston, as illustrated in Figure 2. The piston is moved by fluid pressure controlled by a three-way proportional valve, which is operated by a piezo-stack actuator. The piston features two cylindrical portions with varying sizes.

The piston is positioned to come into contact with a fluid at varying pressures. The larger diameter face of the piston, denoted as D, is in direct contact with volume A, which is connected to the delivery line of the hydraulic circuit. A spring is placed to counteract the force exerted by the fluid in A at delivery pressure on the piston. The chamber containing the spring is linked to the tank, resulting in the absence of hydrostatic stresses on surfaces touching this chamber. The face with the lower diameter d is touching volume B. The volume can also be linked to the delivery or tank line via a three-way proportional valve. The valve is controlled by a piezo-stack actuator powered by a voltage amplifier. The first system utilizes an LMS algorithm to provide a control signal for the piezo actuator, implementing a feedforward logic. The system functions actively in the following manner. When the piezo-stack actuator maintains the valve in position 1, volume B is linked to the delivery line, allowing high-pressure fluid to flow into it and drive the piston upwards.

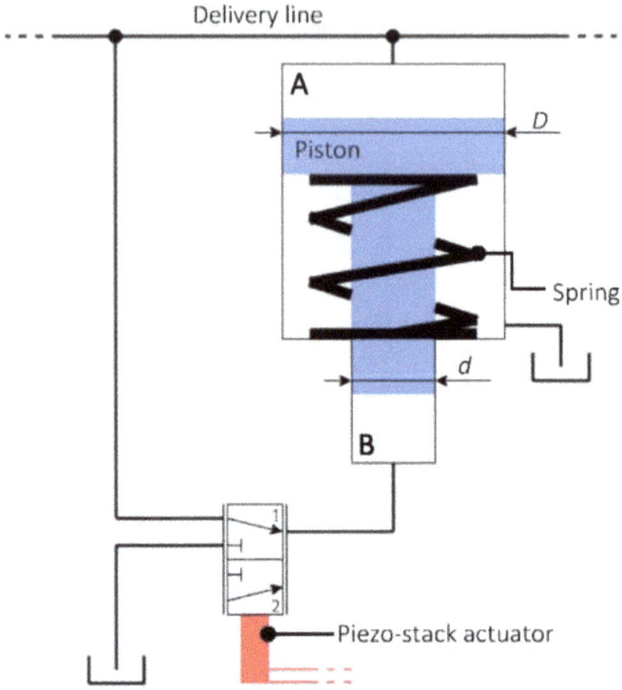

Figure 2. Schematic of hydraulically actuated piston active method.

The piston has a larger diameter on the upper side than on the lower side, causing fluid to be released towards the delivery line during this phase. When the piezo stack actuator moves the valve to position 2, chamber B is linked to a low-pressure tank line, allowing the fluid to flow as the piston descends due to the delivery pressure acting on the larger side of the piston. When this active system is functioning, it is anticipated that a specific quantity of fluid will be drained. During each switching cycle of the piezo-actuated valve, a little amount of pressurized fluid is redirected from the pump delivery to volume B. This action allows the piston to be activated and then released towards the tank to reduce the pressure in the volume. The drainage acquired is limited, as revealed by the advanced model.

3. DETERMINING AND DEFINING ACTIVE SYSTEM OPERATING PARAMETERS

3.1. Piston Actuated by Piezo-Actuator (1st Solution)

The piston and piezo-actuator were dimensioned based on the analysis of the flow ripple produced by the pump. Integrating the oscillating flow rate component in the time domain allows for determining the fluid volume required for the piston to move during each pulse of the piezo-actuator to eliminate the flow ripple entirely. The volume V is calculated by determining the peak-to-peak value of the integral of flow ripple, as seen in Figure 3.

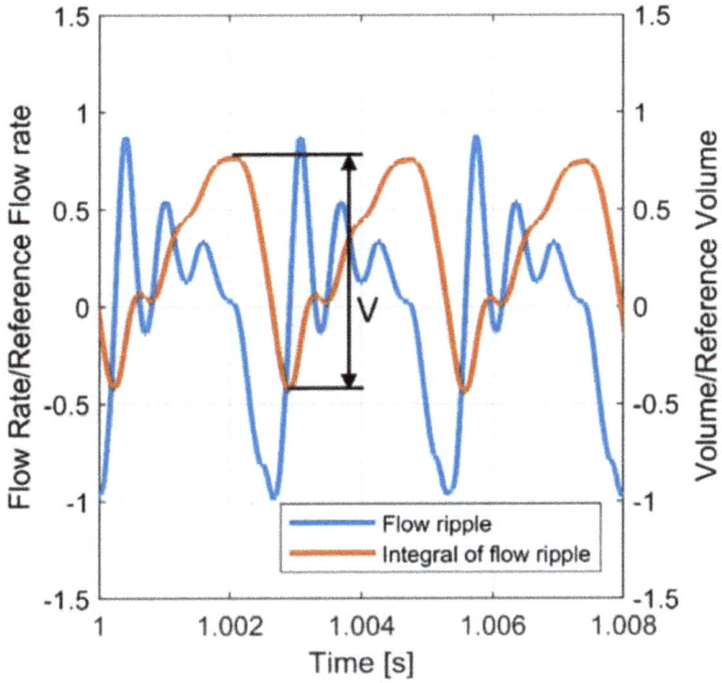

Figure 3. Simulated pump delivery flow ripple and its integral, 250 bar, 2500 rpm.

The volume V was determined based on the flow ripple produced by the pump at its most challenging operating conditions. To guarantee the active system's effectiveness across the whole operating range of the pump, even under extreme conditions. Naming d the piston diameter and s the maximum displacement of the piezo-stack actuator, which corresponds to the stroke of the piston, it is possible to relate these two parameters thanks to the fluid volume V that must be handled by the piston:

$$V = s \cdot \pi \frac{d^2}{4} \tag{1}$$

from which d can be written as a function of s:

$$d = \sqrt{\frac{4V}{\pi s}} \tag{2}$$

The diameter d depends the load F on the piston and the actuator at a fixed pressure p. When the pressure is at its maximum, pmax, the load is also at its maximum Fmax, and writing d as a function of s as in the (2) it is possible to obtain

$$F_{max} = p_{max} \cdot \frac{V}{s} \tag{3}$$

The piezo-stack actuator used to actuate the piston should then be able to withstand such a load.

3.2. Hydraulically Actuated Piston

Sizing and defining operational parameters for the active technique with a hydraulically operated piston are challenging due to the intricate interplay among the system's components. The sizing was conducted through iterative optimization stages utilizing the optimization tools within the Amesim program. The parameters subject to optimization are the upper diameter of the piston (D), the lower diameter of the piston (d), and the stiffness of the spring connected to the piston. The optimization aims to maximize the reduction of the first harmonic of the flow ripple while ensuring efficient fluid drainage throughout the device's operation. The optimization process is carried out using the Amesism program, which utilizes the NLPQL (Non-Linear Programming by Quadratic Lagrangian) algorithm.

4. MATHEMATICAL MODEL

4.1. Fluid Model

A simple fluid model was used to represent the hydraulic working fluid, gas/vapor release is not taken into account since the pressure at the delivery side of the pump always stays well above the cavitation threshold. The temperature was considered constant and equal to T = 50 °C. Density was calculated as a function of pressure variation considering the liquid bulk modulus β:

$$\rho = \rho_0 \cdot [1 + 1/\beta \cdot (p - p_0)] \tag{4}$$

4.2. Pump Model

A pump model was utilized to create a flow ripple for testing two active solutions. This study utilizes a mathematical model of the pump based on the axial piston pump Amesim submodel. The authors were able to concentrate their research solely on the active solutions for reducing flow ripple due to the availability of this existing model. A concise overview of the model is provided. The pump is considered a sequence of interconnected open volumes based on the Filling and Emptying principle.

The pressure in every chamber is obtained by integrating the pressure time derivative considering the volumes as capacitive elements:

$$\frac{dp}{dt} = \frac{\beta}{\rho} \frac{1}{V(t)} \left(\sum \dot{m} - \rho \frac{dV(t)}{dt} \right) \tag{5}$$

Where ρ is the fluid density, β is the bulk modulus, V(t) the instantaneous volume of the chamber, $m^{\cdot}$ is the flow rates from the inlet port and to the outlet port, respectively. Flow rates going through inlet and outlet ports for every volume chamber are computed with the orifice law:

$$\dot{m} = c_q A(t) \rho \sqrt{\frac{2|\Delta p|}{\rho}} \quad (6)$$

Where cq is the discharge coefficient and A(t) is the flow area. The geometrical features of the simulated pump were used to define all geometrical parameters needed by the Amesim axial piston pump mathematical submodel.

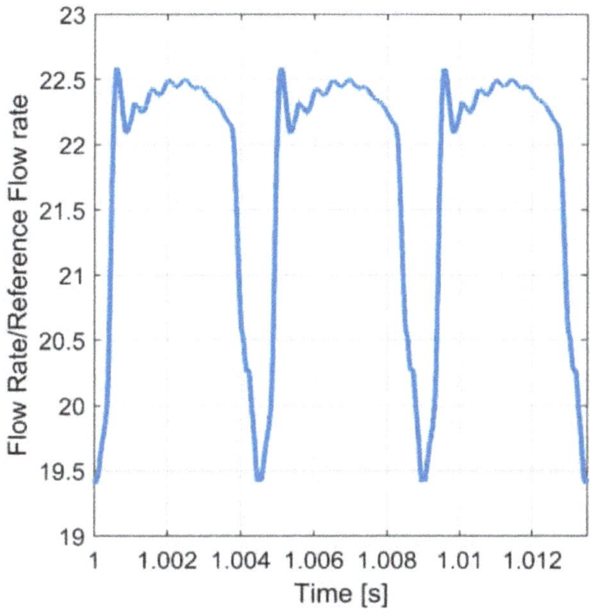

Figure 4. Flow rate at 1500 r/min, 250 bar.

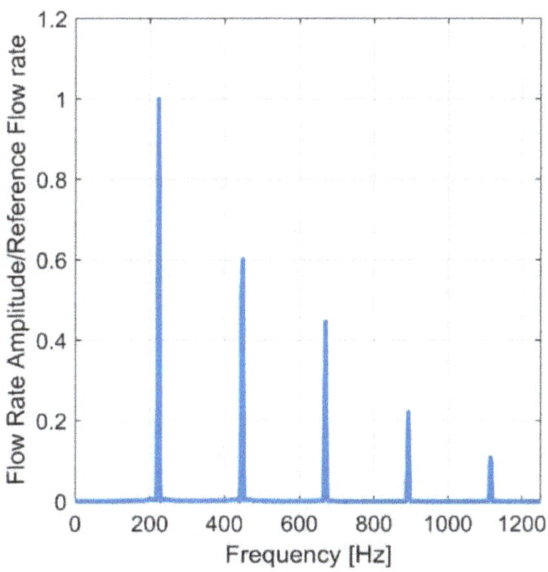

Figure 5. FFT of the flow rate at 1500 r/min, 250 bar.

The target of the pump model is only to generate a typical flow ripple of a positive displacement pump, in this particular case axial piston pumps were considered. **Figure 4** and **Figure 5** report the output flow ripple of the model and its Fast Fourier Transform (FFT), respectively.

4.3. Piezo-Stack Actuator Model

The piezoelectric actuator model was simulated using the Amesim linear piezoelectric stack actuator submodel.

In this model, the force F and the voltage U applied to the piezo-stack actuator are related to the displacement dZ and the electric charge dQ by the following relation:

$$\begin{pmatrix} F \\ U \end{pmatrix} = \begin{pmatrix} K_{11} & K_{12} \\ K_{21} & K_{22} \end{pmatrix} \cdot \begin{pmatrix} dZ \\ dQ \end{pmatrix} \tag{7}$$

Thanks to typical the piezo-stack actuator specifics (electrical capacitance, closed circuit stiffness, blocking load), it is possible to obtain the parameters K11, K12, K21, and K22 that determine the functioning.

The model described is less sophisticated than a real piezo-stack actuator. The model does not consider hysteresis effects or the influence of temperature and operation frequency on the piezo-stack electrical capacitance. If hysteresis effects are insignificant, the change in electrical capacitance should be taken into account for this investigation. An increase in electrical capacitance leads to a decrease in the piezo-stack actuator's responsiveness to high-frequency control signals, as shown in Section 4.4. The model utilizes an electrical capacitance that is overestimated by a safety factor for this purpose.

4.4. Piezo-Stack Actuator Power Supply

The high-performance piezo-stack actuators in this work are defined by an expected operating voltage range of 0–1000 V. A high-power amplifier is required to achieve this voltage. The power amplifier's output current must be restricted to prevent exceeding its maximum threshold, causing delays and reducing its response to the control signal. The piezo-stack actuator's high electric capacitance causes a delay in voltage buildup at its ends when there is a restricted charge current. The piezo-stack actuator is connected to an electrical circuit that functions similarly to an RC circuit, as seen in Figure 6.

An electrical resistance connected in series with the piezo-stack actuator is necessary to keep the current below the maximum threshold set by the power amplifier. The resistance should be minimized to reduce the power amplifier output current, which might lead to increased response time and loss of high frequency performance. Figures 7 and 8 demonstrate how electrical resistance affects the current of the power amplifier and the voltage applied to the piezo-stack actuator when a 0–1000 V step is utilized as the control signal in the system.

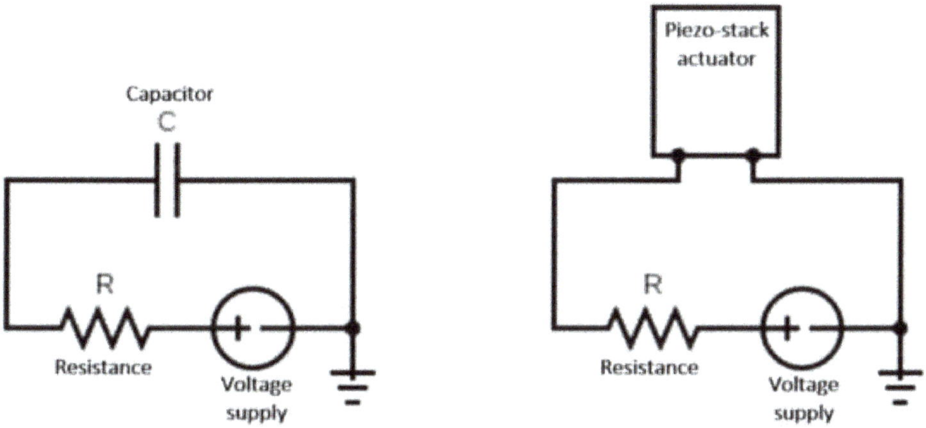

Figure 6. Analogy between RC circuit and circuit with piezo-stack actuator.

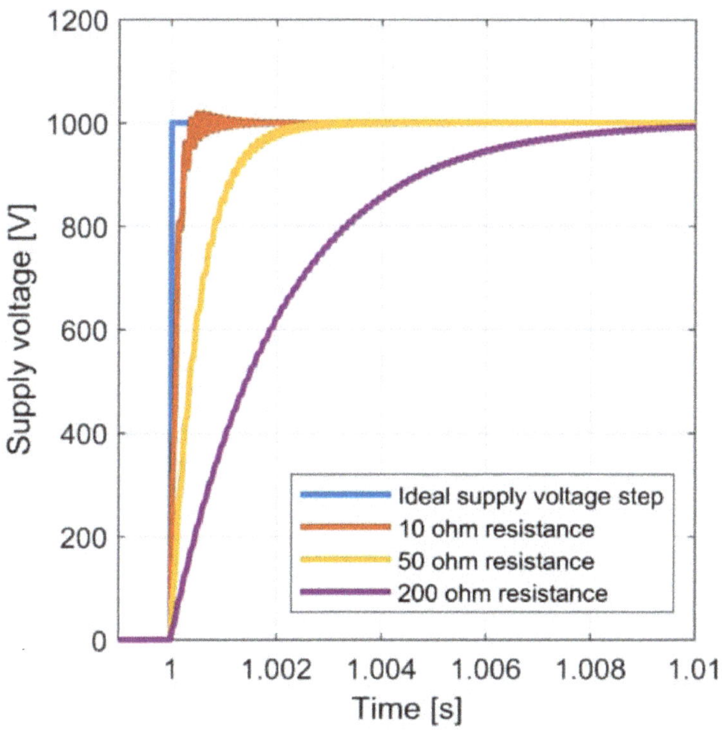

Figure 7. The 0–1000 V step response of the circuit with a piezo-stack actuator and different resistance values.

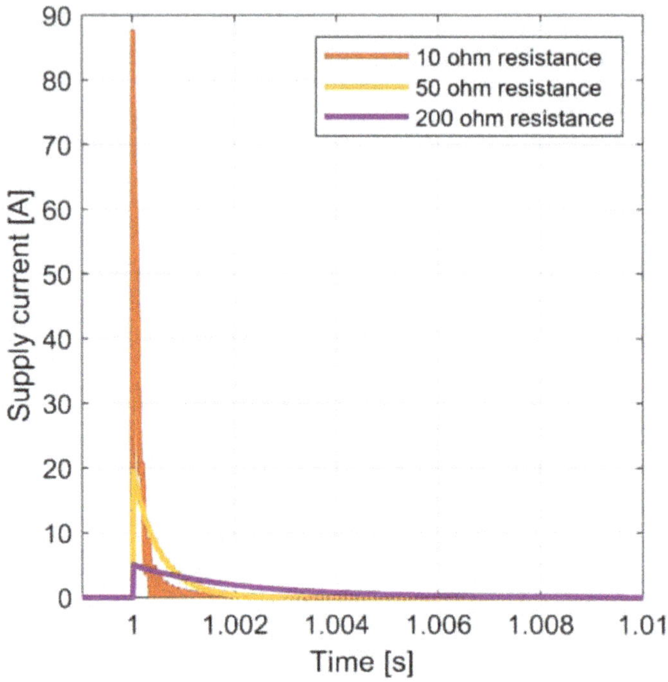

Figure 8. Electrical current in the piezo-stack actuator circuit, 0–1000 V step input for different resistance values.

5. CONTROL ALGORITHM MODEL

The active systems are programmed to function as adaptive filters, namely FIR-type filters, with parameters that are continuously adjusted using a specific algorithm. The algorithm utilized in this scenario is the LMS (Least Mean Square) algorithm. Figure 9 displays the layout of a typical adaptive filter.

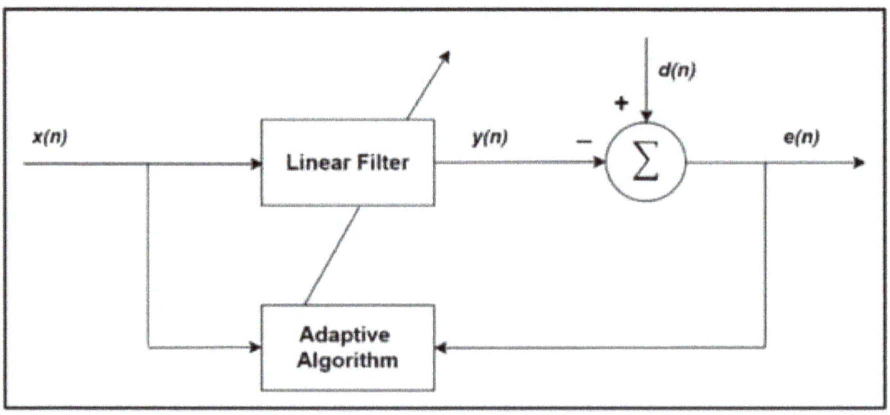

Figure 9. Scheme of a generic adaptive filter.

The adaptive method receives an input signal x(n) that is a sinusoid with the same frequency as the error signal e(n) that needs to be minimized. The adaptive method calculates the parameters for the linear filter in order to reduce the error signal. The linear filter described receives an input signal x(n) and produces an output cancellation signal y(n). The actual pump flow rate, d(n), is adjusted by subtracting y(n), which contains both the desirable average value and the unwanted oscillating component. The error signal e(n) for the adaptive filter is derived from the pressure ripple at pump delivery, which is then filtered to remove the average value using a high-pass filter. The LMS method was used in the Simulink environment through two blocks: the LMS update block, which provides the weights for the Linear Filter block illustrated in Figure 10. The FIR linear filter has a length of 32, consisting of 32 filter weights w(n) that are continuously updated by the LMS block.

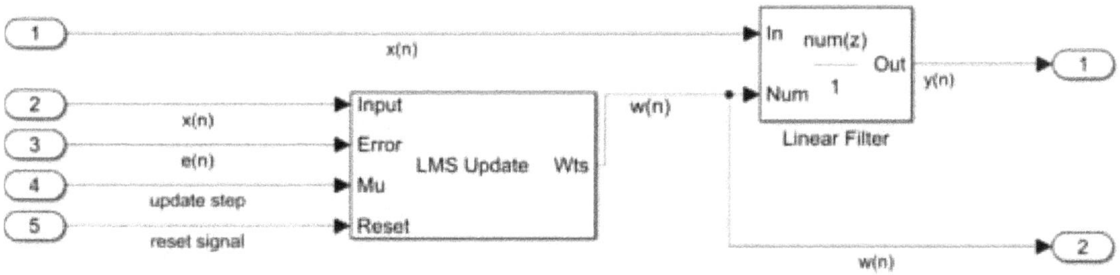

Figure 10. Scheme of the Simulink control algorithm model.

In LMS algorithm, the updating of the N filter weights w(n) happens as follows:

$$w(n+1)_k = w(n)_K + \mu e(n)x(n-K), \ \ K = 0,\ 1,\ \ldots,\ N-1 \quad (8)$$

where K is the index of the filter weight and μ is the update step size that must be large enough to provide fast convergence of the filter Simulations were conducted to test the effectiveness of two active flow ripple cancellation devices under various operating circumstances of the positive displacement pump. The goal was to assess the pump's capability to minimize flow fluctuations and adjust to various conditions such as delivery pressure, flow rate, and rotation speed.

Simulations involve applying several working cycles to the pump, causing changes in operational circumstances. Two distinct cycles were examined, each consisting of three stable parts distinguished by varying pressure, pump speed, and swash plate tilt.weights and at the same time small enough to ensure algorithm stability.

6. SIMULATIONS RESULTS

Simulations were conducted to test the effectiveness of two active flow ripple cancellation devices under various operating circumstances of the positive displacement pump. The goal was to assess the pump's capability to minimize flow fluctuations and adjust to various conditions such as delivery pressure, flow rate, and rotation speed.

Simulations involve applying several working cycles to the pump, causing changes in operational circumstances. Two distinct cycles were examined, each consisting of three stable parts distinguished by varying pressure, pump speed, and swash plate tilt.

6.1. Piston Actuated by Piezo-Actuator (1st Solution)

In the first run, the pump working cycle was set as shown in **Table 1**.

Table 1. Pump working conditions in first simulation cycle.

Cycle Section	1	2	3
Average delivery pressure [bar]	250	135	30
Pump rotation speed [rpm]	2500	2500	2500
Swash plate angle/max. angle [%]	100	100	50
Average flow rate [L/min]	200	200	100

Between every steady-state section, a short transition section was also inserted to allow the working conditions to change smoothly.

In **Figure 11**, **Figure 12** and **Figure 13**, the FFT of pump delivery flow rate in the three steady-state conditions is reported. It is possible to see the effect of flow ripple reduction produced by the active system comparing the main harmonic amplitude with and without active system intervention.

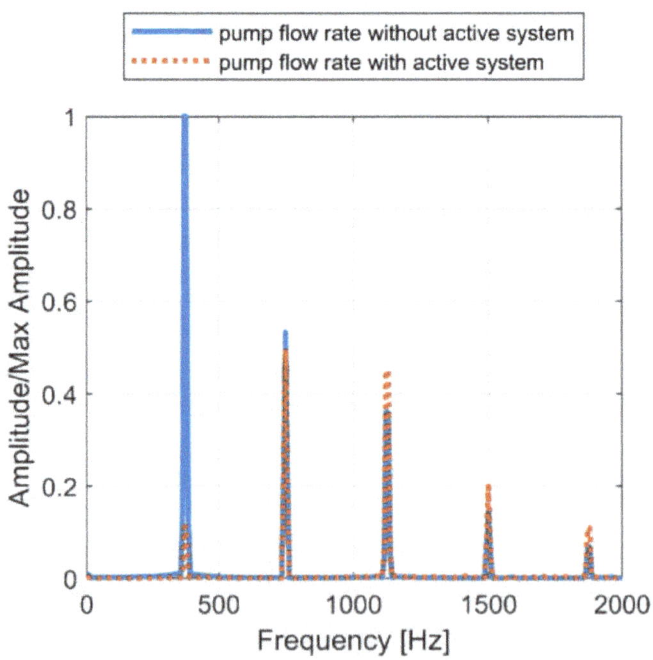

Figure 11. Simulation cycle 1, section 1: flow ripple FFT at 250 bar, 2500 rpm, 100% swash plate inclination.

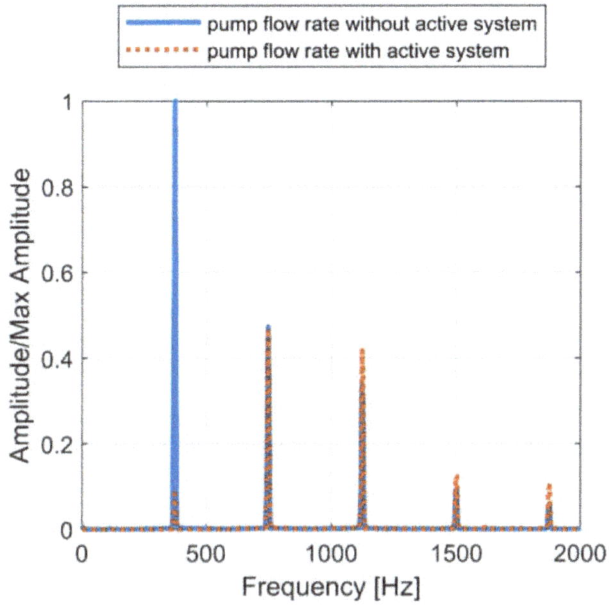

Figure 12. Simulation cycle 1, section 2: flow ripple FFT at 135 bar, 2500 rpm, 100% swash plate inclination.

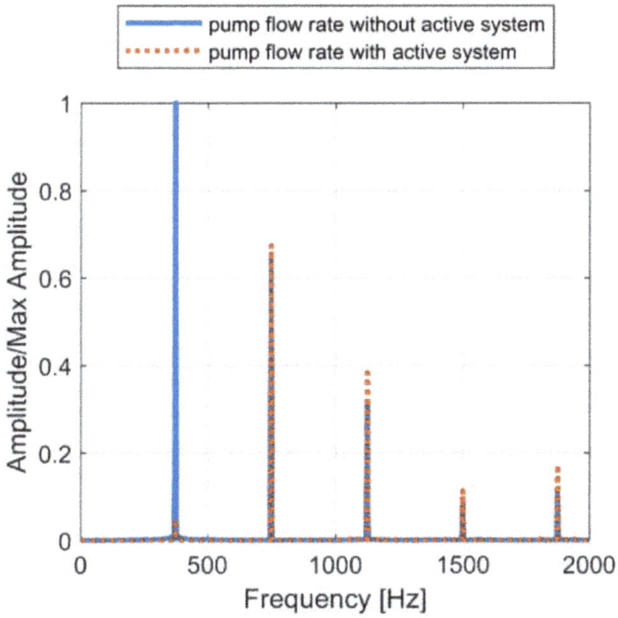

Figure 13. Simulation cycle 1, section 3: flow ripple FFT at 30 bar, 2500 rpm, 50% swash plate inclination.

In each one of the three cycle sections, the first order harmonic of the flow ripple is clearly reduced. The reduction amounts to 89% for cycle section 1, to 92% for section 2, and to 95% for section 3.

It is possible to note that higher order harmonics may show a small increase or decrease even if the active system is meant to act only on the first main one. This is attributable to the introduction of the active system that brings with it small changes in the delivery line characteristics, like fluid volume for example, modifying the frequency response of the hydraulic system. This phenomenon can be seen in all the other simulated conditions, for which the aforementioned explanation remains valid.

The active method also shows good performances in autonomously updating the control system parameters to adapt to pump working condition changes. In **Figure 14**, 4 of the 32 weights of the FIR filter updated with the LMS algorithm during simulation are shown. Only 4 weights are reported for simplicity. As it can be seen from the graph, each weight converges rapidly to a new value after the working condition changes. Convergence time is in the order of tenths of a second.

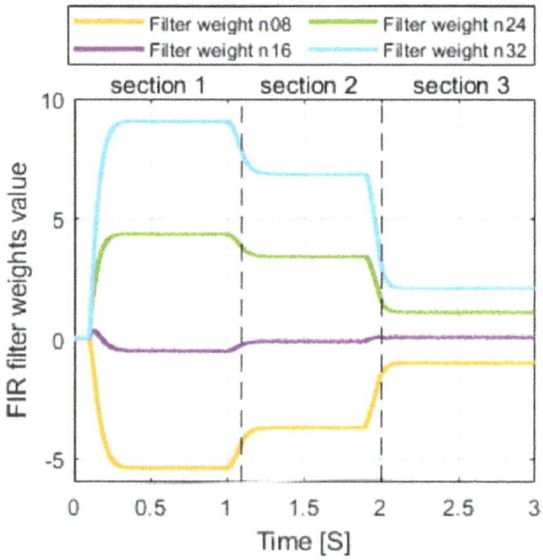

Figure 14. Simulation cycle 1: weight trend in FIR discrete filter.

Since in the first run cycle only constant speed and decreasing pressure situations were tested, the second run cycle contemplates a decrease in pump speed between sections 1 and 2 and an increase in pressure between sections 2 and 3, as it can be seen in **Table 2.**

Table 2. Pump working conditions in second simulation cycle.

Cycle Section	1	2	3
Average delivery pressure [bar]	250	90	220
Pump rotation speed [rpm]	2500	1500	1500
Swash plate angle/max. angle [%]	100	100	100
Average flow rate [L/min]	200	120	120

In **Figure 15, Figure 16** and **Figure 17**, results referring to the simulated cycle indicated in **Table 2** are reported. Results regarding section 1 are not reported because the section 1 of the second cycle has the same parameters of the section 1 of the first cycle. From the results, it is possible to state that also in conditions of variable pump speed and increasing delivery pressure the control algorithm can allow the active system to adapt to varying pump working conditions. Flow ripple first harmonic reduction for sections 1–3 amounts, respectively, to 89%, 82%, and 74%.

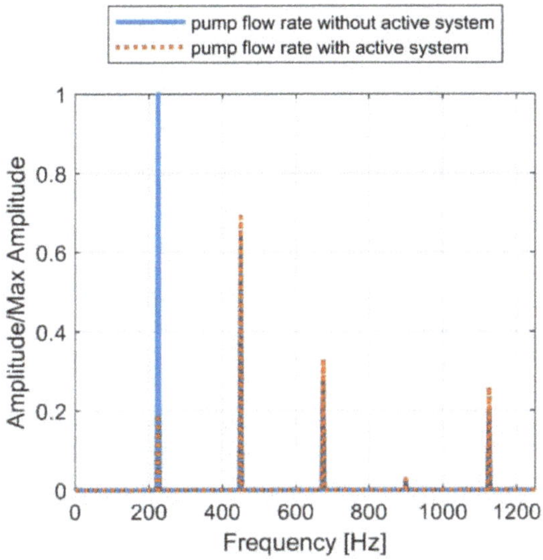

Figure 15. Simulation cycle 2, section 2: flow ripple FFT at 90 bar, 1500 rpm, 100% swash plate inclination.

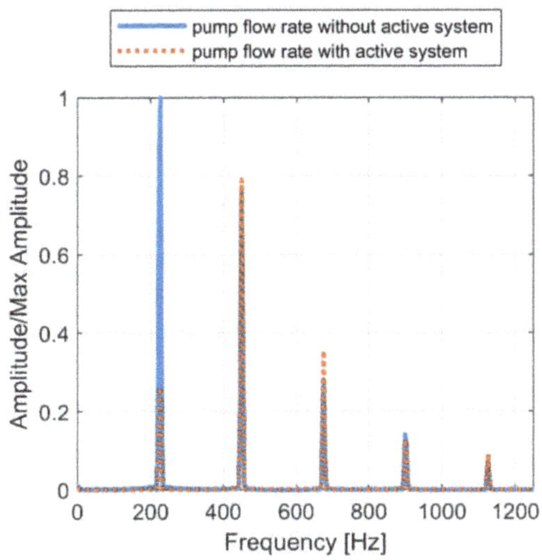

Figure 16. Simulation cycle 2, section 3: flow ripple FFT at 220 bar, 1500 rpm, 50% swash plate inclination.

During functioning of the active system, the supply current of the piezo stack actuator reaches very high values of over 8 A. The characteristic power supply maximum current of a piezo-actuator is usually below this value; for this reason, it was decided to limit the supply current by appropriately increasing the electrical resistance placed in series with the piezo-actuator, in order to maintain peak current below the 6 A threshold.

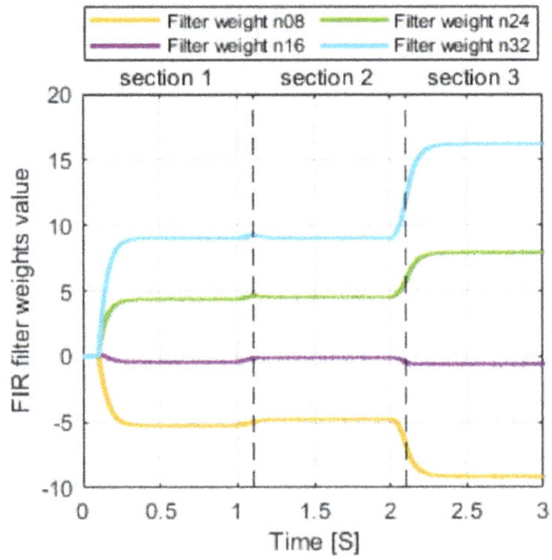

Figure 17. Simulation cycle 2: weight trend in FIR discrete filter.

A simulation was run with pump working cycle identical to the first run (**Table 1**) to compare effectiveness with and without electric current limitation. Results are reported in **Figure 18, Figure 19, Figure 20** and **Figure 21.**

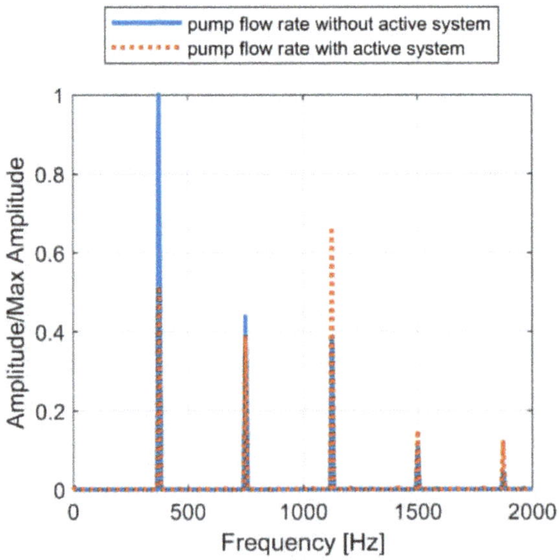

Figure 18. Limited current simulation, section 1: flow ripple FFT at 250 bar, 2500 rpm, 100% swash plate inclination.

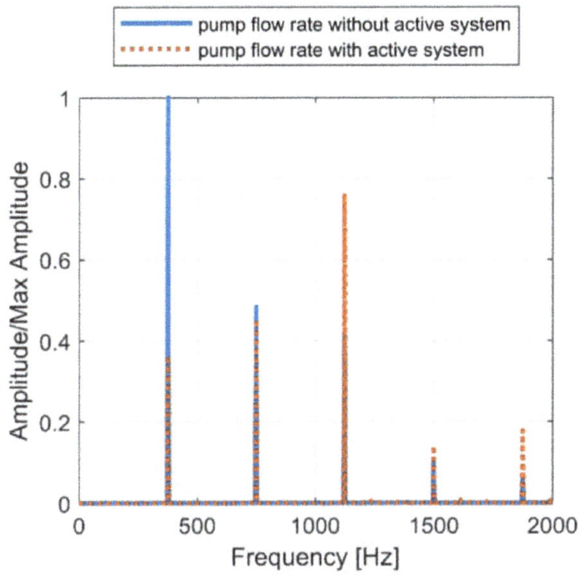

Figure 19. Limited current simulation, section 2: flow ripple FFT at 135 bar, 2500 rpm, 100% swash plate inclination.

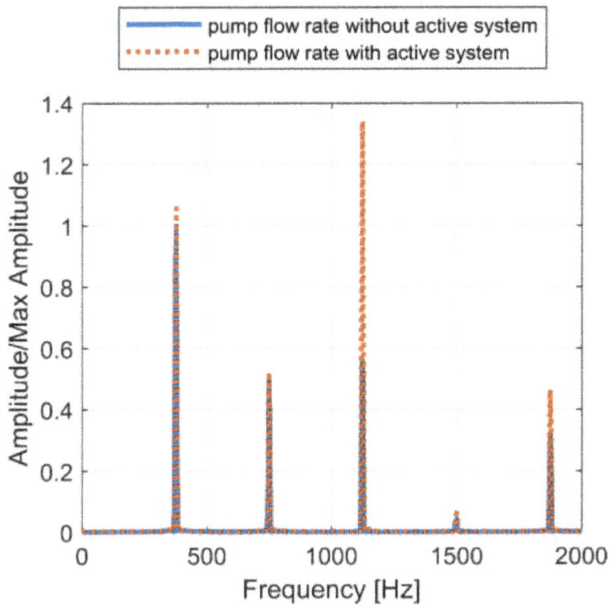

Figure 20. Limited current simulation, section 3: flow ripple FFT at 30 bar, 2500 rpm, 50% swash plate inclination.

The effectiveness of the active system is greatly diminished when the supply current is limited to 6 A compared to when a higher current is used. Furthermore, the adaptive method is unable to effectively update the FIR filter weights, leading to divergence and hindering the reduction of flow ripple.

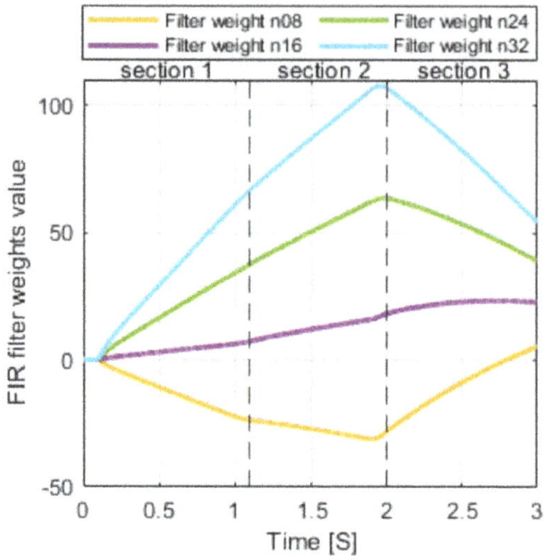

Figure 21. Limited current simulation: weight trend in FIR discrete filter.

One solution to this issue involves increasing the number of piston-actuator pairs and utilizing smaller piezo-stack actuators with a higher bandwidth that require less supply current. An alternative method is to create a specialized power supply that can deliver high peak currents to optimize the application. Both methods would be costly, leading to increased system complexity and expenses.

6.2. Piston Actuated Hydraulically via Piezo-Valve (2nd Solution)

The same working cycles used for the 1st solution were applied to the 2nd solution which relies on a piezo-valve to hydraulically actuate the piston.

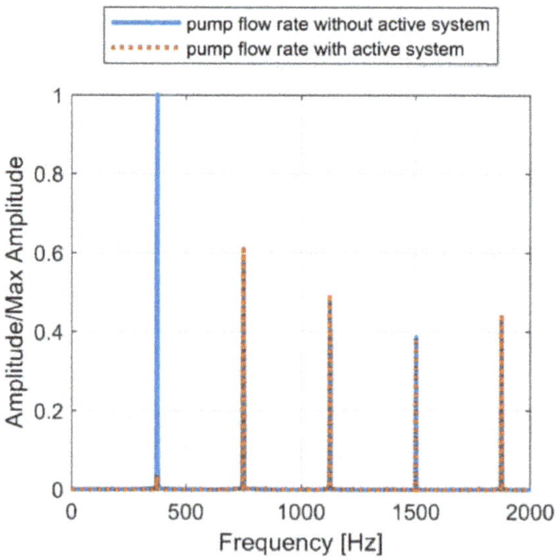

Figure 22. Simulation cycle 1, section 1: flow ripple FFT at 250 bar, 2500 rpm, 100% swash plate inclination.

In **Figure 22, Figure 23, Figure 24** and **Figure 25**, results referring to the first cycle stage (**Table 1**) are reported. The reduction in the first harmonic of flow ripple is clearly effective, reaching values of 97% in section 1 of the cycle, 96% in section 2, and 93% in section 3. However, convergence in FIR filter weights appears to be slower if compared to the 1st piezo-actuated active system case. In particular, in section 3 of the cycle, filter weight stable values were not reached in 1 s time, and therefore it is possible to see that they are still updating when the simulation stops, with their trend still sloping.

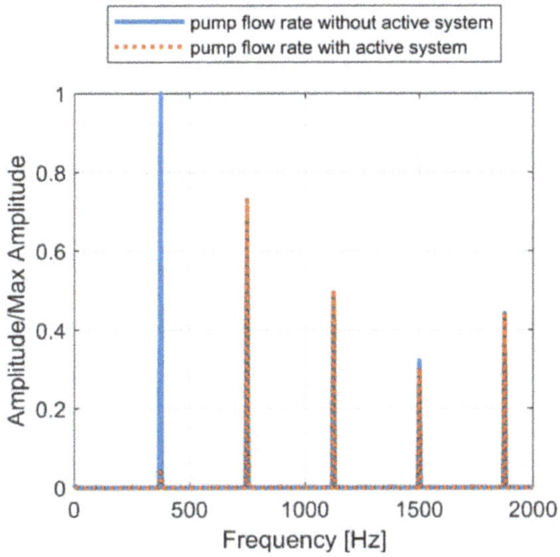

Figure 23. Simulation cycle 1, section 2: flow ripple FFT at 135 bar, 2500 rpm, 100% swash plate inclination.

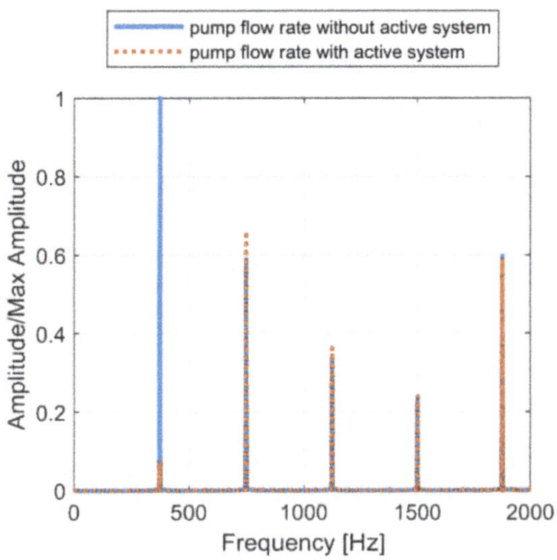

Figure 24. Simulation cycle 1, section 3: flow ripple FFT at 30 bar, 2500 rpm, 50% swash plate inclination.

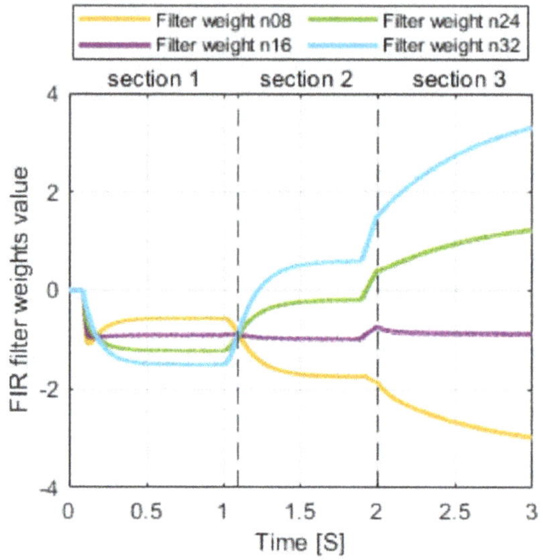

Figure 25. Simulation cycle 1: weight trend in FIR discrete filter.

In **Figure 26**, **Figure 27** and **Figure 28**, results referring to the second cycle pattern (**Table 2**) are reported. Results regarding section 1 are not reported because the section 1 of the second cycle has the same parameters of the section 1 of the first cycle. In this case, it is possible to see by the trend of the FIR filter weights that the control algorithm is not able to keep up with changing working conditions of the pump. In particular, after being able to converge in the first section of the cycle, filter weights then rapidly diverge starting from the section number 2, compromising the effectiveness of the active system.

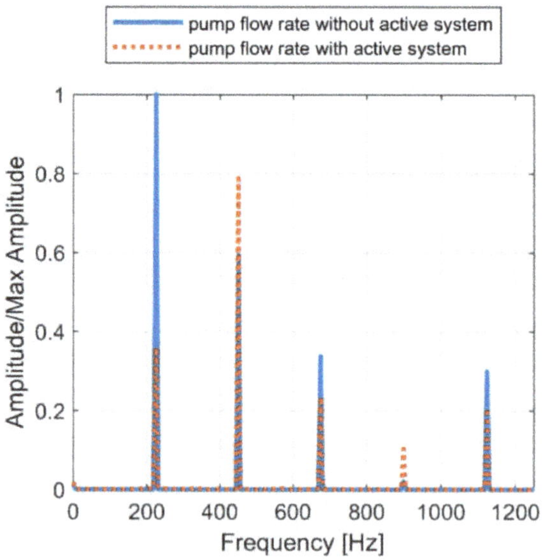

Figure 26. Simulation cycle 2, section 2: flow ripple FFT at 90 bar, 1500 rpm, 100% swash plate inclination.

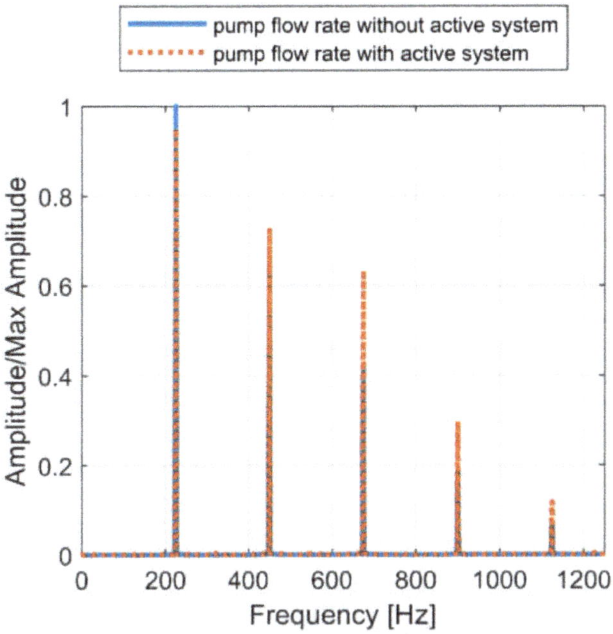

Figure 27. Simulation cycle 2, section 3: flow ripple FFT at 220 bar, 1500 rpm, 50% swash plate inclination.

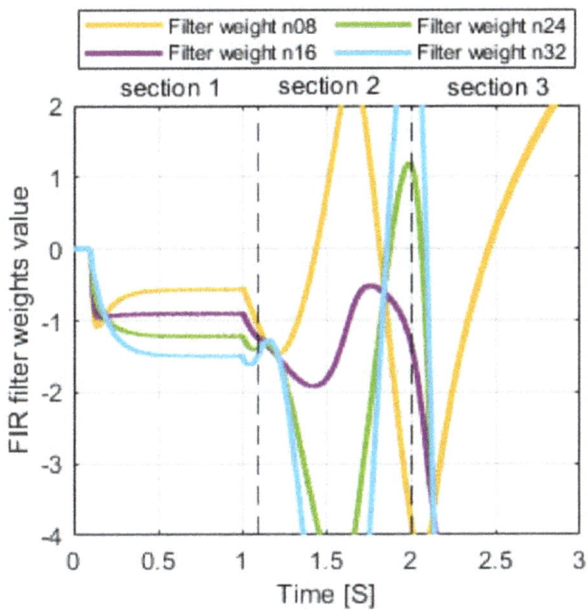

Figure 28. Simulation cycle 2: weight trend in FIR discrete filter.

The lack of strong correlation between the control signal generated by the FIR block and the actual movement of the piston may explain the undesired behavior of the 2nd active system control algorithm compared to the 1st active system. In the second system, the control signal affects a piezo-valve that impacts the piston's oscillation, however the relationship is less significant compared to a control signal that directly operates the piston using a piezo-stack actuator in the first active system. The LMS algorithm requires a high correlation between the control signal it generates and the cancelation signal in order to operate well. In this secondary active system, the piston's oscillation, responsible for generating the flow ripple cancellation signal, is greatly influenced by the average delivery pressure and the first harmonic of the flow ripple. The algorithm cannot quickly adjust to changes in the pump's functioning condition. The amount of fluid drained in the operation of this active system is consistently minimal, never surpassing 1.5% of the pump delivery flow rate under any simulated operating conditions.

7. CONCISE SUMMARY

This chapter introduced two dynamic techniques for reducing pump flow fluctuations. Both approaches use pistons to generate signals that negate flow ripple. One method uses a piston operated directly by a piezo device, while the other method involves a piston actuated hydraulically through a piezo-valve. Mathematical models of both strategies were outlined, along with a control algorithm designed to provide adaptability for both systems. Both methods are applicable to any type of positive displacement pump, however an axial piston pump was employed for simulation in this study. The numerical simulations' results were analyzed, highlighting the capabilities and challenges of the two suggested methodologies.

The first active solution can significantly reduce flow ripple and adjust to changing operating conditions if there is enough supply current. Thanks to this dynamic system, an 80-90% reduction in the primary harmonic flow ripple was obtained. The second active method helps to overcome some of the drawbacks of the first way. A smaller piezo-stack actuator is required to operate the valve, which reduces costs, dimensions, and prevents issues associated to high supply current. The control algorithm must be enhanced to enable the active system to effectively adjust to changes in pump working conditions due to the increased complexity of the approach. The active system demonstrated significant potential by reducing the primary harmonic flow ripple by over 90% when operating correctly, with a fluid drainage rate of only 1.5% of the pump delivery flow. The next phase of this research will involve designing and optimizing more robust and dependable control algorithms based on the one employed in this study. Next, prototypes of the two options will be created and an experimental campaign will be conducted to confirm the accuracy of simulation results.

8. AT A GLANCE

Positive displacement machines are commonly associated with a significant downside, which is the fluctuation in the flow rate they deliver. This work introduces two active methods to decrease the flow ripple produced by a pump using an external device controlled by a piezo-stack actuator. The study concentrates on a theoretical analysis to gather information about the performance of offered solutions and their primary advantages and disadvantages. The suggested active methods include a cylindrical actuator linked to the pump's supply line. The piston might be moved either by a piezo-stack actuator or by a proportional piezo actuated valve modulating a differential pressure. The actuators were simulated and a control algorithm utilizing the Least Mean Square method was implemented to ensure adaptability for both systems under varying operating conditions. The mathematical model that has been built allows for the identification of the significant potential of these solutions in significantly reducing flow ripple. The first architecture was deemed the most effective approach, while the second one enabled a decrease in production costs.

REFERENCES

1. Karpenko, M.; Prentkovskis, O.; Šukevičius, Š. Research on high-pressure hose with repairing fitting and influence on energy parameter of the hydraulic drive. In Eksploatacja i Niezawodnosc—Maintenance and Reliability; Polish Maintenance Society: Lublin, Poland, 2022; Volume 24, pp. 25–32. ISSN 1507-2711. **[Google Scholar]**
2. Stosiak, M. The impact of hydraulic systems on the human being and the environment. J. Theor. Appl. Mech. **2015**, 53, 409–420. **[Google Scholar]** **[CrossRef]** **[Green Version]**
3. Ye, S.; Zhang, J.; Xu, B. Noise Reduction of an Axial Piston Pump by Valve Plate Optimization. Chin. J. Mech. Eng. **2018**, 31, 57. **[Google Scholar]** **[CrossRef]** **[Green Version]**
4. Johansson, A.; Olvander, J.; Palmberg, J.-O. Experimental verification of cross-angle for noise reduction in hydraulic piston pumps. Proc. Inst. Mech. Eng. Part I J. Syst. Control Eng. **2007**, 221, 321. **[Google Scholar]** **[CrossRef]**
5. Harrison, A.M.; Edge, K. Reduction of axial piston pump pressure ripple. Proc. Inst. Mech. Eng. **2000**, 214, 53–64. **[Google Scholar]**
6. Borghi, M.; Zardin, B. Axial Balance of External Gear Pumps and Motors: Modelling and Discussing the Influence of Elastohydrodynamic Lubrication in the Axial Gap. In Proceedings of the ASME International Mechanical Engineering Congress and Exposition, Houston, TX, USA, 13–19 November 2015. **[Google Scholar]** **[CrossRef]** **[Green Version]**

7. Zhao, X.; Vacca, A. Theoretical investigation into the ripple source of external gear pumps. Energies **2019**, 12, 535. **[Google Scholar] [CrossRef] [Green Version]**

8. Zhao, X.; Vacca, A. Numerical analysis of theoretical flow in external gear machines. Mech. Mach. Theory **2017**, 108, 41–56. **[Google Scholar] [CrossRef]**

9. Zhou, J.; Vacca, A.; Casoli, P. A Novel Approach for Predicting the Operation of External Gear Pumps Under Cavitating Conditions. Simul. Model. Pract. Theory **2014**, 45, 35–49. **[Google Scholar] [CrossRef]**

10. Rundo, M.; Altare, G.; Casoli, P. Simulation of the filling capability in vane pumps. Energies **2019**, 12, 283. **[Google Scholar] [CrossRef] [Green Version]**

11. Corvaglia, A.; Ferrari, A.; Rundo, M.; Vento, O. Three-dimensional model of an external gear pump with an experimental evaluation of the flow ripple. Proc. Inst. Mech. Eng. Part C J. Mech. Eng. Sci. **2021**, 235, 1097–1105. **[Google Scholar] [CrossRef]**

12. Pan, M.; Ding, B.; Yuan, C.; Zou, J.; Yang, H. Novel Integrated Control of Fluid-Borne Noise in Hydraulic Systems. In Proceedings of the BATH/ASME 2018 Symposium on Fluid Power and Motion Control FPMC 2018, Bath, UK, 12–14 September 2018. **[Google Scholar]**

13. Shang, Y.; Tang, H.; Sun, H.; Guan, C.; Wu, S.; Xu, Y.; Jiao, Z. A novel hydraulic pulsation reduction component based on discharge and suction self-oscillation: Principle, design and experiment. Proc. Inst. Mech. Eng. Part I J. Syst. Control Eng. **2020**, 234, 433–445. **[Google Scholar] [CrossRef]**

14. Pan, M.; Johnston, N.; Plummer, A. Hybrid Fluid-borne Noise Control in Fluid-filled Pipelines. J. Phys. Conf. Ser. **2016**, 744, 012016. **[Google Scholar] [CrossRef] [Green Version]**

15. Waitschat, A.; Thielecke, F.; Behr, R.M.; Heise, U. Active Fluid Borne Noise Reduction for Aviation Hydraulic Pumps. In Proceedings of the 10th International Fluid Power Conference, Dresden, Germany, 8–10 March 2016. **[Google Scholar]**

16. Kim, T.; Ivantysynova, M. Active Vibration/Noise Control of Axial Piston Machine Using Swash Plate Control. In Proceedings of the ASME/BATH 2017 Symposium on Fluid Power and Motion Control, Sarasota, FL, USA, 16–19 October 2017; Paper No. FPMC2017-4304. p. V001T01A053. **[Google Scholar] [CrossRef]**

17. Casoli, P.; Pastori, M.; Scolari, F.; Rundo, M. Active pressure ripple control in axial piston pumps through high-frequency swash plate oscillations—A theoretical analysis. Energies **2019**, 12, 1377. **[Google Scholar] [CrossRef] [Green Version]**

18. Casoli, P.; Pastori, M.; Scolari, F. Swash plate design for pressure ripple reduction—A theoretical analysis. In Proceedings of the AIP Conference, Modena, Italy, 11 September 2019. **[Google Scholar]** **[CrossRef]**

19. Hagstrom, N.; Harens, M.; Chatterjee, A.; Creswick, M. Piezoelectric actuation to reduce pump flow ripple. In Proceedings of the ASME/BATH 2019 Symposium on Fluid Power and Motion Control 2019 FPMC, Sarasota, FL, USA, 7–9 October 2019. **[Google Scholar]**

06 SELECTION OF ACTUATOR MODE IN MULTIPLE CHAMBERS USING REINFORCEMENT LEARNING: SIMULATIONS AND EXPERIMENTS

1. INTRODUCTION: AN OVERVIEW

Each actuator in a multiple hydraulic actuator system experiences varying pressure levels due to the load it carries. The supply pressure is regulated based on the greatest needed pressure when the actuators are managed by a load-sensing system with pressure-compensation valves and a single pump.

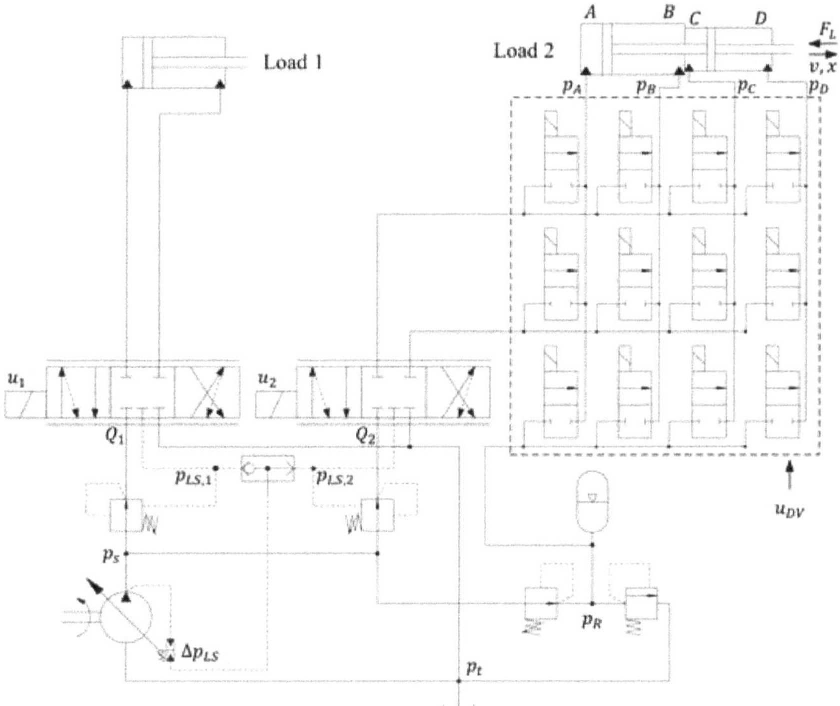

Figure 1. Hydraulic-system architecture.

There is a discrepancy in pressure between the actuators and the supply, which is rectified by adjusting the pressure-compensation valves. Resistive control losses are a significant contributor to energy losses in hydraulic systems[1,2]. To minimize these losses, consider employing a multi-chamber actuator in one of the loads. Figure 1 displays a hydraulic diagram of a pressure-compensated valve-controlled load-sensing system, with Load 2 being powered by a multi-chamber actuator.

The set of on/off valves between the proportional valve and the actuator allows for the combinations of different chambers that define the possible actuator modes to be used. The capability to select different actuator modes allows the modulation of the resultant pressure from the load to make $pLS,2$ similar to $pLS,1$. This is shown in the illustrative flow-pressure diagrams in **Figure 2.**

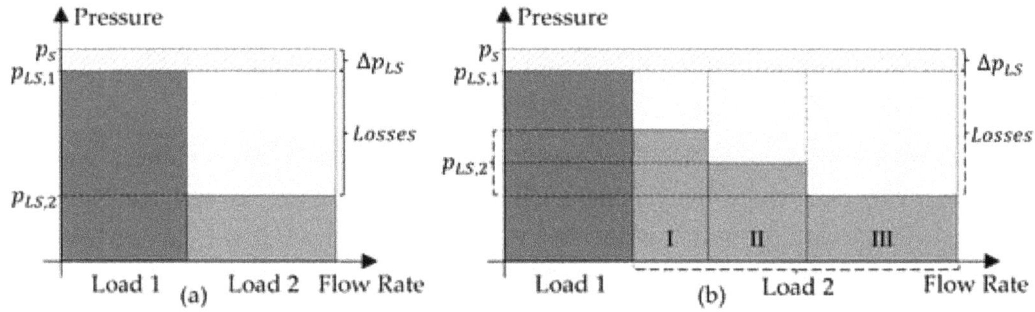

Figure 2. Flow-pressure diagram showing the effect of mode selection on resistive control losses. (a) With a conventional actuator; (b) with a multi-chamber actuator.

Modulating the pressure, as shown in Figure 2b, allows for a decrease in resistive control losses. A traditional actuator, as shown in Figure 2a, is unable to carry out this modulation, leading to increased resistive control losses. Further elaboration on this system architecture can be found in reference [3].

This architecture offers an additional degree of freedom that can be manipulated, known as mode selection. The mode that minimizes losses depends on the speed and forces of all actuators, which vary during system operation, making it a complex challenge to solve. Multi-chamber actuators are typically part of systems where the choice of modes determines the ultimate control objective, such as speed or position. The discrepancy in speed or location is converted into a target force that the multi-chamber actuator needs to apply. The force reference in [4] is compared to the available forces for each mode, and the mode with the smallest force inaccuracy is then applied to the system. The mode selection in [5] minimizes force error, similar to [4], but does not address the issue of frequent switching. Mode-selection method used in reference [6] is contrasted to a strategy that penalizes high-amplitude pressure shifts between distinct modes, similar to the one in reference [4]. The authors in reference [7] explain how a controller operates by selecting the mode that minimizes energy usage while still being able to power the current load. The mode selection in [8] is determined by a set range of force capacity for each mode. The active mode is capable of applying the desired force, with the force differential being regulated by the proportional pressure control in a specific chamber.

Reference [9] examines model predictive control (MPC) to address issues with force spikes during mode transition. The mode selection process is still carried out by minimizing the force differential between the reference and available modes, as well as minimizing the energy losses related to switching. Throttle control is used to modify the pressure in each chamber to match the computed pressure reference, thus managing the force difference. MPC aims to reduce force transients while switching. Another study utilizing MPC is discussed in reference [10], showing the superiority of MPC compared to less complex mode-selection algorithms, as demonstrated in reference [4]. The modes in [2] are selected based on their capacity to overcome the current force of the actuator while minimizing pump pressure and flow rate, even if they are not employed for a multi-chamber actuator. The authors in reference [11] explain a state machine used to manage mode selection based on factors such as load pressure and predefined pressure levels for each mode. The authors highlight potential instability and non-smooth operation of the actuator while switching modes. They propose that adjusting the proportional valve control could reduce these issues, aligning with concepts discussed in previous studies [8,9]. The mode selection of the independent metering valve system in [12] is determined by the present load force and the force capabilities of each mode.

The control issue in this case differs from the ones examined in the aforementioned studies. The mode selection enables the control end objective, which is executed by the proportional valve. The control objective for the multi-chamber actuator is to choose a mode that allows the desired motion to be achieved with minimal resistive control losses. The mode selector can be seen as facilitating energy management by aiming to minimize energy losses through the usage of the multi-chamber actuator. This work explored a controller for mode selection based on RL instead of constructing one based on analytical equations of energy losses in the system.

Reinforcement learning (RL) utilizes neural networks as actor and critic function approximators, demonstrating effectiveness in solving complicated issues like playing Atari. The primary benefits are automated learning through interaction with the environment and discovering optimized control solutions that are naturally challenging to solve analytically. Moreover, machine learning algorithms demonstrate strong performance in settings beyond their specific training, a common occurrence in the operation of mobile robots given their wide range of applications.

Ref. [14] proposes an RL-based solution using the Q-learning algorithm to control hybrid system power in mobile working machines in order to reduce fuel consumption. The authors demonstrate the potential of initially training the controller using simulation models of the system and subsequently implementing it on the actual system. Reference [15] introduces a reinforcement learning energy-management approach for hybrid construction equipment. The authors employ a blend of Dyna learning and Q-learning in their investigation, which is confined to simulation outcomes.

Additional uses of Reinforcement Learning (RL) in mobile machinery involve: Reference [16] utilizes cameras, lidar, and motion and force sensors to optimize bucket loading of fragmented rock with a multi-objective goal, focusing on maximizing bucket loading capacity. Reference [17] trains RL for motion control of a forestry crane with an emphasis on reducing energy consumption. Reference [18] applies RL for trajectory tracking control of an excavator arm, with the controller producing valve-control signals directly.

The authors note that the process of developing RL controllers typically begins with pre-training the agent in a simulation environment. The benefits of this method include avoiding negative real-world outcomes, cutting expenditures related to gaining practical experience, and running simulations faster than real-time. The examined papers demonstrate the controller's ability to identify and execute solutions for intricate problems. The paper showcases the training and use of an RL-based controller for selecting modes in a multi-chamber hydraulic actuator, using both simulation and experimental data. The actuator is a component of a multi-actuator load-sensing system powered by a single pump. Choosing alternative modes can decrease resistive losses in control valves by improving the pressure alignment between actuators. A Deep Q-Learning (DQN) agent was developed and trained to optimize mode selection in order to reduce system energy losses.

Contribution and Objectives

Our group has already published research on the topic discussed in this paper. The system design and potential efficiency enhancements are outlined in [3]. The initial use of the RL-based method for controlling this system was examined and introduced in reference [20]. This work expands on prior articles by utilizing a load-sensing pump instead of a continuous pressure source, increasing the complexity of the learning challenge and making it more closely resemble a real-world mobile system application.

This work aims to present experimental data that illustrate the utility of RL for controlling intricate hydraulic systems. This case involves choosing modes for multi-chamber actuators to minimize resistive control losses. Specifically, RL demonstrates the ability to identify an optimized control solution with less reliance on manual controller design. The study is restricted to training the agent in simulation and does not include ongoing learning once deployed to the real system.

2. AVAILABLE MODES, MODEL DESCRIPTION, AND CONTROL STRUCTURE

The multi-chamber actuator used in this study has four chambers and is connected to three supply lines (A/B/R). This gives a total of 87 possible modes. However, most modes are ruled out due to the reasons presented in [**3**]. The modes used in this study are given in **Table 1**. Mode 1 is the only mode that is not an agent's decision, because it is implemented as a rule to ensure safe operation. The steady state force displayed in **Table 1** is calculated considering a pressure of 100 bar on port A and 15 bar on ports B and R. This is done to show which modes can exert higher force.

Table 1. Available modes and connected chambers.

Mode	A	B	R	Force [kN]
1	-	-	-	-
2	AC	BD	-	77
3	ABC	D	-	70
4	A	BD	C	55
5	AB	D	C	48
6	C	BD	A	12

The model of the physical system was developed in the multi-domain system simulation tool HOPSAN [**21**] and describes the motion of the boom arm as a function of the motion of the actuators. It also models the dynamic behaviour of the pressures, flow rates, internal leakage, and closing and opening of the valves. The model was validated against experimental results, with a detailed description presented in [**20**].

The controller and training algorithm for the DQN agent were implemented in MATLAB/Simulink [**21**]. During the training in simulation, the agent learns by interacting with the system. After training, the agent controls the real system.

The actual controller of the system is not only composed of the trained agent selecting the optimal mode, it also contains other control and safety rules. An overview of the controller structure is shown in **Figure 3**. The reward branch of the agent is only used during the training phase.

In **Figure 3**, it is also seen how the interaction of the agent with the environment takes place during the learning phase and then in the controlling phase. Paraphrasing [**13**] for the present control problem: The agent interacts with the environment in a sequence of actions, observations, and rewards. At each agent time step (t), the agent observes the observations (t), selects an action $a(t$ from the set of modes (**Table 1**), and applies it to the environment. It observes the new observations (t+1) and the reward $r(t)$. It tries to maximise the reward.

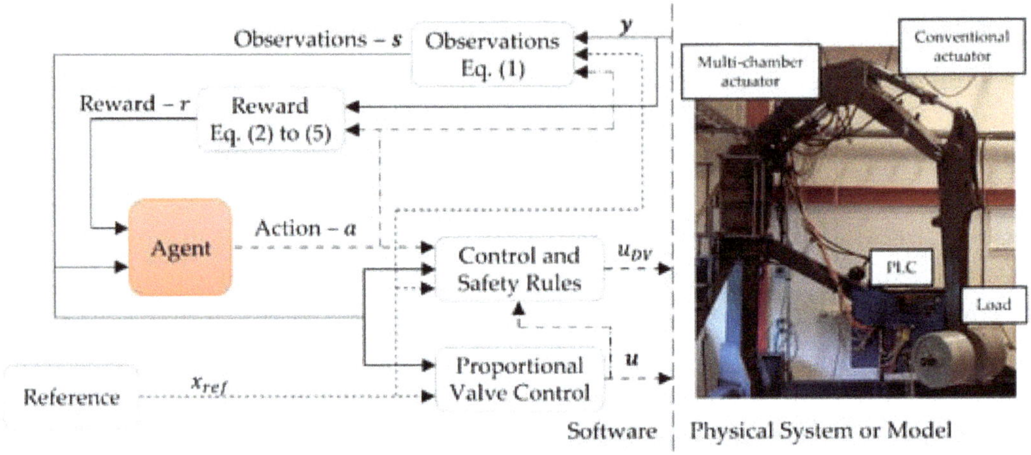

Figure 3. Structure of the controller and picture of the test bench.

From the RL perspective, everything that is not the agent is considered as the environment it is interacting with. It is important to make the environment in the simulation as close to the real environment as possible. For the control of hydraulic systems in mobile applications, along with having a sufficiently accurate model of the physical system, there is the need to include additional safety rules. Therefore, the environment is the combination of the system model and the additional rules. The added control and safety rules for this system are:

- Apply mode 1 if $|u2|$<0.002 m$|2|$<0.002 m;
- Limit the available modes according to position and external load;
- Compensate $u2$ due to the difference in areas in each mode; and
- Use mode 4 for the lowering motion ($u2$ < --0.002 m).

The first rule is used because the valve has an overlap of 2 mm, so the on/off valves can be closed. The second rule prevents the agent from choosing a mode that cannot drive the load with the available maximum pump pressure, otherwise the load could fall. For the same operator control input signal to the proportional valve, the third rule ensures approximately the same actuator speed for the different modes, see [**3,20**] for details.

The fourth rule is implemented because the return motion, due to the much larger area of the actuator connected to the return than the supply, results in excessive throttling on the meter-out edge of the valve. This causes the pump to operate at maximum pressure due to a perceived high load, which causes the pressure-compensation valves to operate fully open and, thus, the compensation losses are close to zero. Thus, the selection of modes, based on minimising the compensation losses, does not work for the returning motion. Therefore, due to this design constraint, the return motion is not an agent's decision.

While the RL agent is responsible for the mode selection, another P controller 'mimics' the machine operator controlling the proportional valves. The P controller is implemented inside the proportional valves control block in **Figure 3**.

3. LEARNING SETUP FOR THE AGENT

The application has a continuous observation space (pressures, speed, etc.) and a discrete action space (modes to be selected). A DQN agent [**13,22**] is suitable for this type of observation and action space, and it mainly consists of a neural network calculating the value for taking a certain action given the current observations. The value is an estimation of the sum of the reward that the agent can collect over a future time horizon, by following a certain sequence of control actions. In this case, the network at each agent's time step predicts the value of taking each action. A greedy function selects and implements in the system the action that has the highest estimated value, which is, in other words, the action that would lead to the best performance, according to the reward function. This agent is, thus, a non-linear map between the system states (observations of pressures, speed, position, ...) and the optimised control action (modes). The reader is referred to [**13,22**] for a description of the type of agent and training algorithm.

Each agent action corresponds to one mode (**Table 1**). Inside the 'Control and Safety Rules' block in **Figure 3**, a lookup table maps the action to the corresponding vector $uDV$of the open/closed digital valves that implements the mode in the system.

The structure and parameters of the network are presented in **Table 2**. No sensitivity analysis was performed to evaluate the size of the network on the performance of the task. However, the size of the network was chosen to be small.

Table 2. Structure of the agent network.

Layer	Size
Feature inputs (observations)	9 input features
Fully connected with Rectified Linear Unit activation function	70 neurons
Fully connected with Rectified Linear Unit activation function	35 neurons
Fully connected output (actions)	5 neurons

The observations $\boldsymbol{s}$ used as input features are

$$s(t) = [a, u_2, p_A, p_B, p_C, p_s, p_{LS,1}, v, x], \quad (1)$$

where a is the previous action, $u2$ is the proportional valve control signal, pA-B-C are the chambers' pressures, ps is the supply pressure, pLS,1 is the load-sensing pressure for the conventional actuator, v is the actuator speed, and x is the position.

The reward function r to be maximised is composed of three terms,

$$r = K_1 r_{Velocity} + K_2 r_{Power} + K_3 r_{Switch}, \tag{2}$$

$$r_{Velocity} = -1\ if\ |v| < 0.001\ \mathrm{m/s}, \tag{3}$$

$$r_{Power} = -(|Q_2(p_s - p_{LS,2})| + |\ Q_1(p_s - p_{LS,1})|)/P_{Norm}, \tag{4}$$

$$r_{Switch} = \begin{cases} r_{Switch} & if\ a_{t+1} = a_t \\ r_{Switch} - 1 & if\ a_{t+1} \neq a_t \end{cases}. \tag{5}$$

The velocity term ($rVelocity$) penalizes the agent if a motion is requested (|$u2$|>0.002|2|>0.002), and the multi-chamber actuator does not move with a minimum velocity. This encourages the agent to learn to meet a minimum control-performance requirement. The power loss term ($rPower$) is a penalty based on the hydraulic system control losses, to encourage the agent to find a mode that results in smaller energy losses due to pressure compensation. The switch term ($rSwitchh$) penalizes the agent for frequent mode switching.

Table 3 presents the parameters and load cases used during training. To increase the agent's robustness, noise is added to the measurements used as observations in the simulations. R is a uniformly distributed random variable with a maximum amplitude of 1.

Table 3. Agent and task parameters.

Parameter	Value
Initial exploration factor	0.990.99
Minimum exploration factor	0.15
Exploration factor decay	7.5×10-67.5×10-6
Target smooth factor	2.0×10-32.0×10-3
Learning rate	7.5×10-47.5×10-4
Mini-batch size	128128
Look ahead steps	44
Experience buffer length	4.0×1044.0×104
Agents sample time [s]	0.800.80
Discount factor	0.990.99
Gradient decay factor	0.90.90
Maximum number of episodes	1.8×1041.8×104
Load Cases and Task	**Value**

Parameter	Value
Initial position [m]	0.10+0.02*R*0.10+0.02
Final position [m]	0.40+0.05*R*0.40+0.05
External load [kg]	[40 80 120 160 200 240][40 80 120 160 200 240]
Load 1 load sensing pressure *pLS*,1Load 1 load sensing pressure,1 [bar]	[60 100] ± 3*R*[60 100] ± 3
Load 1 flow rate *Q*1Load 1 flow rate 1 [lpm]	10

In a real system, the load pressure from the conventional actuator (*pLS*,1,1) varies according to the load. However, a simplification is made by setting it to constant values to allow for easier interpretation of the results both in the simulation and in the experiments. In the experiments, this signal is emulated with an additional hydraulic circuit. This means that the agent is exposed to 12 different scenarios of external load and pressure on the conventional actuator, and the task is to lift the load from an initial position to a final position while maximising the reward function.

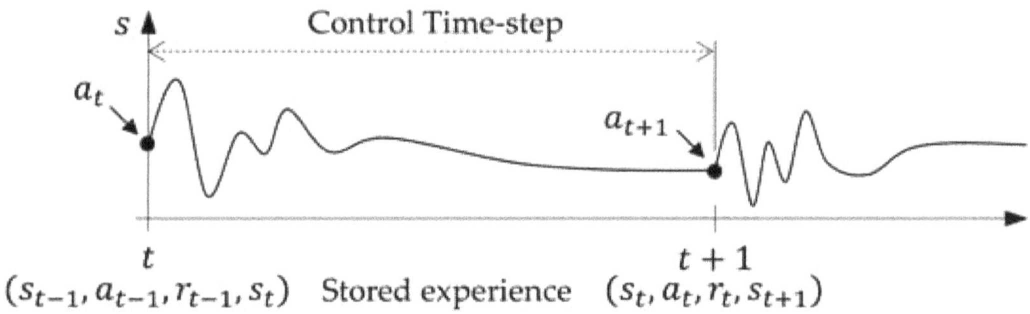

Figure 4. Agent time step versus system dynamics.

Dynamic systems have oscillations, and these must be taken into consideration when setting the agent-control time step, defining the reward function, and considering what type of information is stored as experience. This is represented in **Figure 4**, by showing a dynamic response of the system state (*s*) due to an action *at*.

If the time step is too small, the stored experience will contain dynamic effects rather than steady-state conditions. This might affect the agent's ability to learn because the reward function could give misinformation. However, the reward from the action at *t* is only observed at *t*+1+1 because it is usually a function of the final state and the transition between the states. Therefore, the size of the control time step can be adjusted to avoid these effects. For the current system, the mode selection does not need to occur frequently to avoid unnecessary switching. Thus, the agent time step is set to avoid these oscillations.

4. RESULTS

The agent was trained to lift a load from a certain initial position to a final position, with 12 load cases that including 2 pressure levels on pLS,1,1 and 6 external loads. **Figure 5** presents the agent's training progress, where $Q0$ is the estimated value.

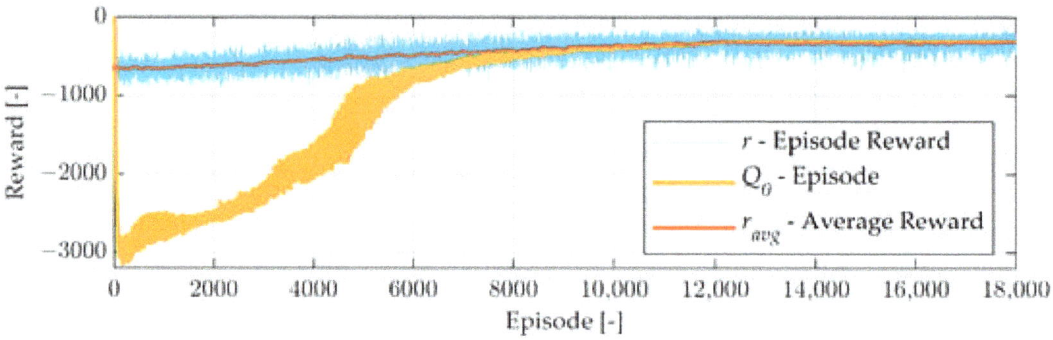

Figure 5. Agent's training progress.

After training, the agent is tested in simulation for all 12 load cases, and its capability to complete the task in an efficient way is evaluated. The energy loss ($Eloss$) is calculated with Equations (6) and (7), with the variables extracted from the simulation.

$$P_{loss} = P_{loss,1} + P_{loss,2} = |Q_1(p_s - p_{LS,1})| + |Q_2(p_s - p_{LS,2})| \quad (6)$$

$$E_{loss} = E_{loss,1} + E_{loss,2} = \int_{t_i}^{t_f} (P_{loss,1} + P_{loss,2})dt \quad (7)$$

4.1. Simulation Results

An estimation of what mode results in the lowest power losses is made by evaluating Equation (6) for each mode and the steady-state conditions of speed (v) and position (x) of the multi-chamber actuator. pLS,2 is calculated based on the external load and the position of the actuator, given the kinematics of the boom arm. The mode that results in the lowest power loss is selected as the best. This gives an indication of what mode selection to expect from the agent's learning. The solution for the two load cases, with 120 kg external load and the two pressure levels on pLS,1 are shown in **Figure 6**. Moreover, the safety boundary condition limiting the number of modes that can be applied to the system is shown, which was implemented as a rule. The resultant pressure (pLS,2) for the different modes is also shown, where it is seen that, to some extent and as expected, the best mode leads to a pressure level close to pLS,1.

What **Figure 6**a,b show is that modes that can exert lower forces are better when the actuator is retracted, while modes that can exert higher forces are better when it is extended. This results in a closer match of pLS,1,1 and pLS,2, as shown in

Figure 6c,d. Due to the geometry of the boom, the further the actuator is extended, the higher the load force on it. However, the mode that results in the lowest power losses is also dependent on the flow rate, Equation (6), and this is the reason why weaker modes might be better (**Figure 6**b) at higher speeds.

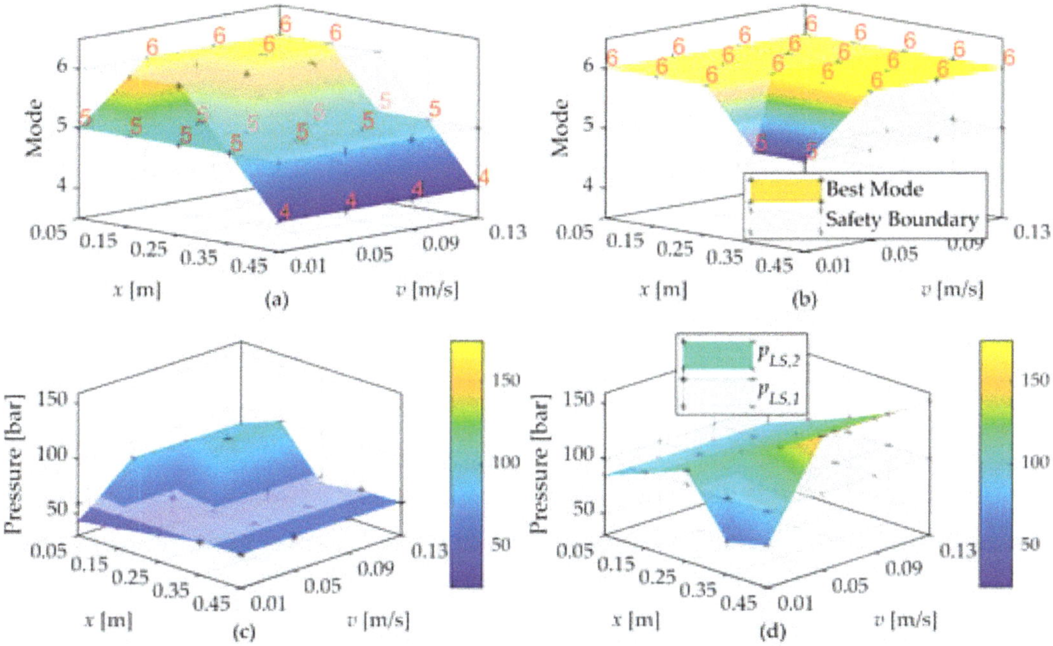

Figure 6. Best modes for 120 kg load. (a) For pLS**,1=60 bar; (b) for** pLS**,1=100 bar; (c) resultant** pLS**,2 for** pLS**,1=60 bar; (d) resultant** pLS**,2 for** pLS**,1=100 bar.**

Figure 6 indicates what to expect from the agent's learning when it comes to the minimisation of energy losses. However, as shown in Equation (2), this is not the only optimisation objective. **Figure 7** and **Figure 8** present simulation results for the trained agent performing the mode selection, for the two load cases. What the figures also show is the performance on the same tests when a conventional actuator with areas equivalent to mode 2 was used instead of the multi-chamber actuator. The abbreviations Conv and RL are used in the figures for the conventional actuator case and the multi-chamber actuator with mode selection, respectively. Power results are normalised with respect to the maximum power of each load case.

Comparing **Figure 7**b with **Figure 7**d shows that the agent does learn to select modes that reduce the energy losses. This results in a better match of pressures than the conventional case, as shown in **Figure 7**c. By choosing a mode with smaller areas and resulting in a better match of pLS,1 and pLS,2, the power losses are reduced significantly. A similar learning and decision-making capacity is observed for the other test case, as shown in **Figure 8**.

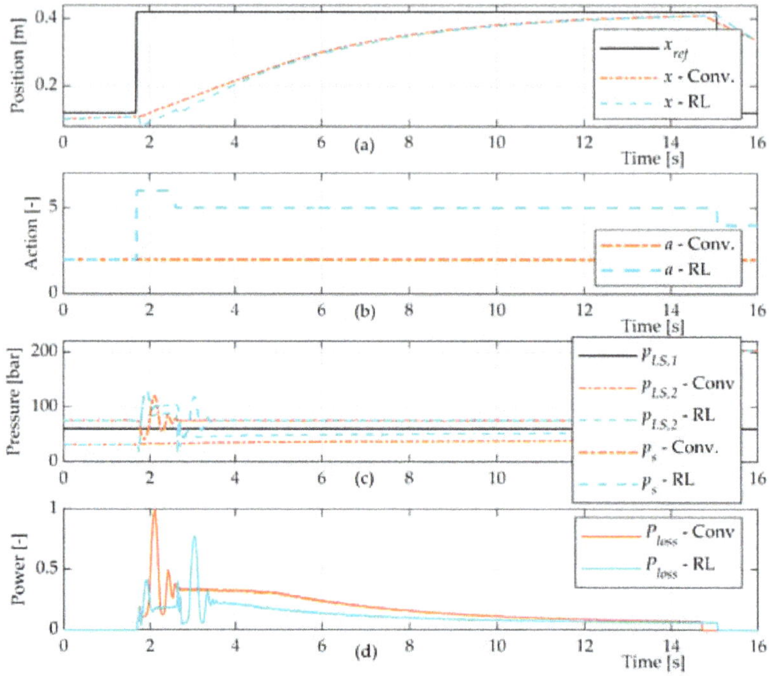

Figure 7. Simulation result, 40 kg load and *pLS*,1=60 bar. (a) Position; (b) action; (c) system and load-sensing pressures; (d) power loss.

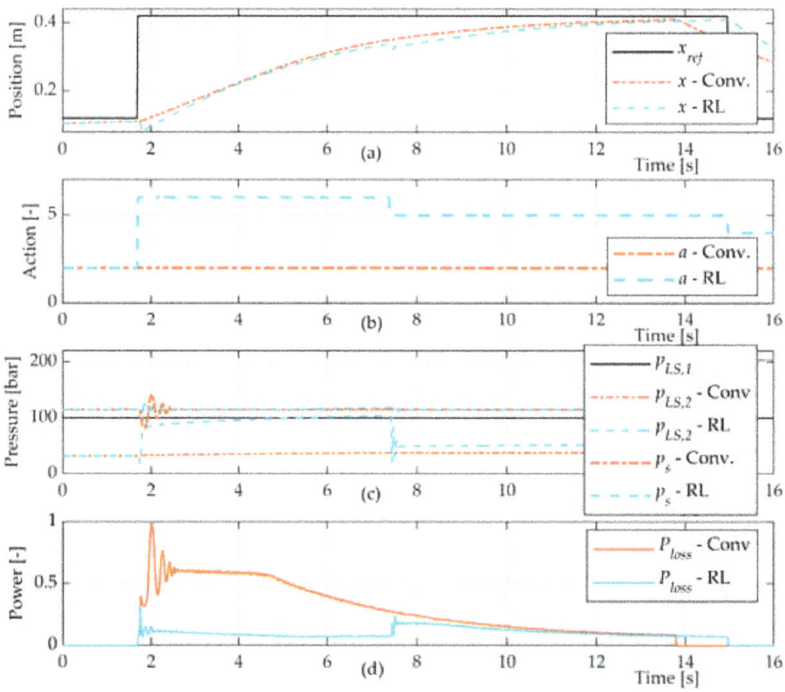

Figure 8. Simulation result, 40 kg load and *pLS*,1=100 bar. (a) Position; (b) action; (c) system and load-sensing pressures; (d) power loss.

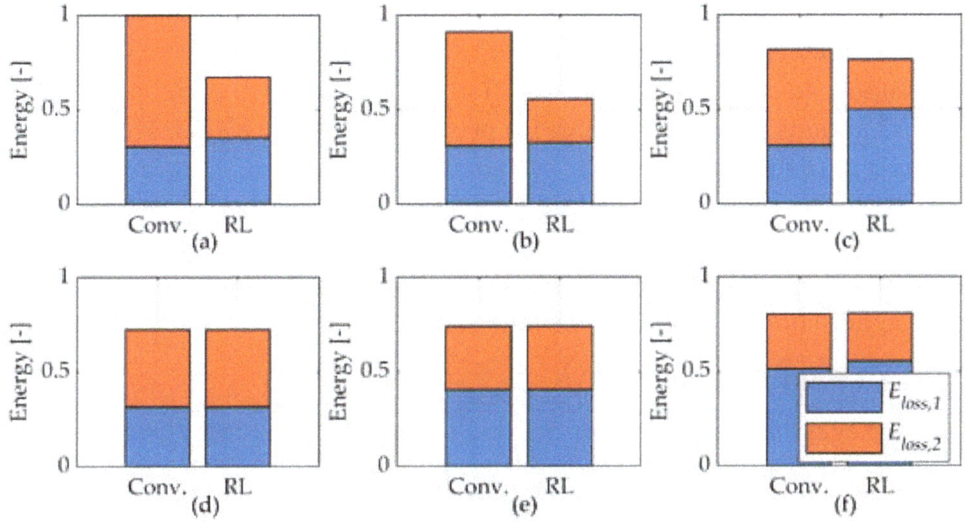

Figure 9. Energy comparison, *pLS*,1=60 bar: (a) 40 kg; (b) 80 kg; (c) 120 kg; (d) 160 kg; (e) 200 kg; (f) 240 kg.

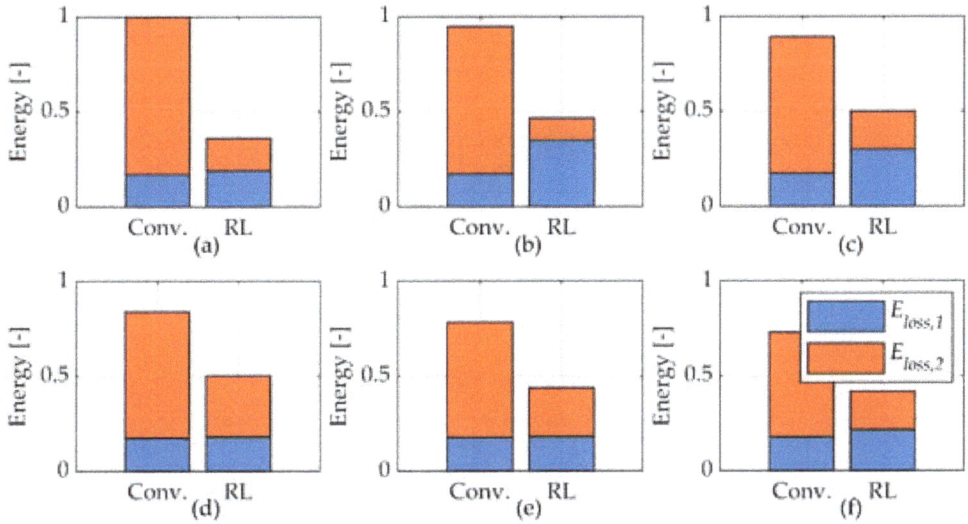

Figure 10. Energy comparison, *pLS*,1=100 bar: (a) 40 kg; (b) 80 kg; (c) 120 kg; (d) 160 kg; (e) 200 kg; (f) 240 kg.

Comparing **Figure 8**b and **Figure 8**d shows that the agent learns to choose the better mode based on the similar principles of reducing the mismatch of pressure (**Figure 8**c) and choosing the modes with a smaller area.

All load scenarios that the agent was trained for were simulated, and Equation (7) is used to compare the energy losses. The results, normalised by the highest total energy losses for all test cases of each level of *pLS*,1, are presented in **Figure 9** and **Figure 10.**

Results from the energy analysis show that the agent learned to select modes that result in lower energy losses for almost all the load scenarios it was trained for. For some load cases, the chosen solution is close to or equal to the conventional case. This happens in the load scenarios presented in **Figure 9**d–f. This means that in the worst case the agent runs the system like a conventional actuator, but it does not choose something worse than that, for example, a mode that cannot drive the load. Although it must be remembered that there is a safety function preventing a weak mode from being applied for high loads, the results showed that most of the time the agent chooses modes within the safety boundary.

4.2. Experimental Tests

The trained agent was deployed to the real system and evaluated on the same load scenarios used in the training. The experimental test results, for the same load cases of **Figure 7** and **Figure 8**, are shown in **Figure 11** and **Figure 12**. Simulation results are plotted in the same figures. However, in the experiments, the agent observes different conditions than in simulation, which can lead to different decisions.

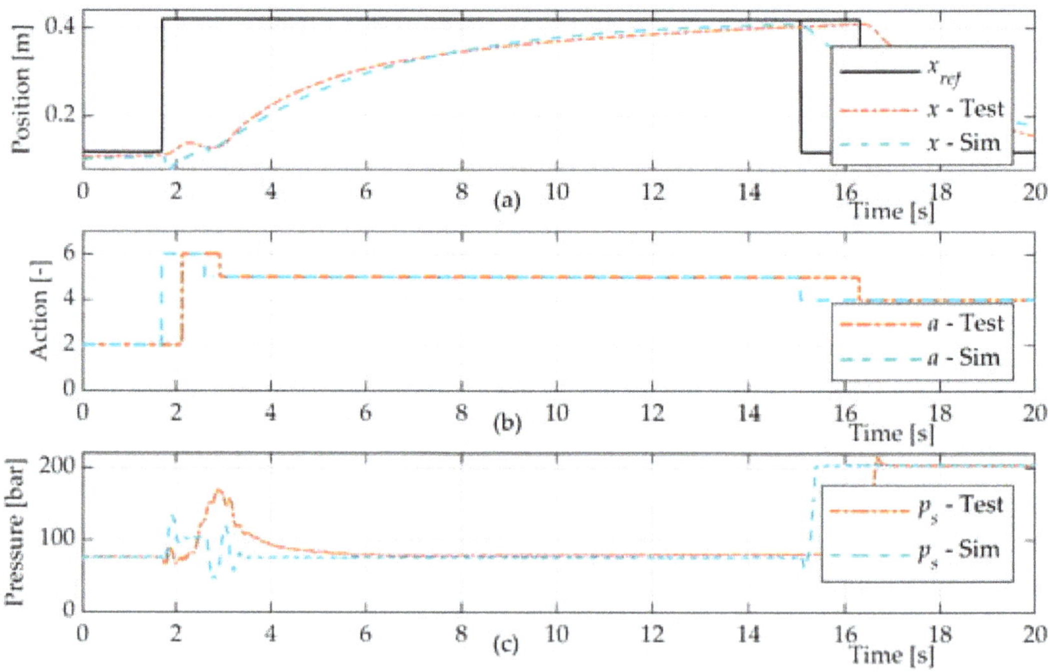

Figure 11. Experimental (Test) and simulation (Sim) results, 40 kg load and ***pLS*,1=60 bar. (a) position and reference; (b) action; (c) system pressure.**

The agent is able to apply in practice the same modes it learned from the optimisation (**Figure 11**b), even though there are differences, for example, in the actuator position (**Figure 11**a) and system pressure (**Figure 11**c).

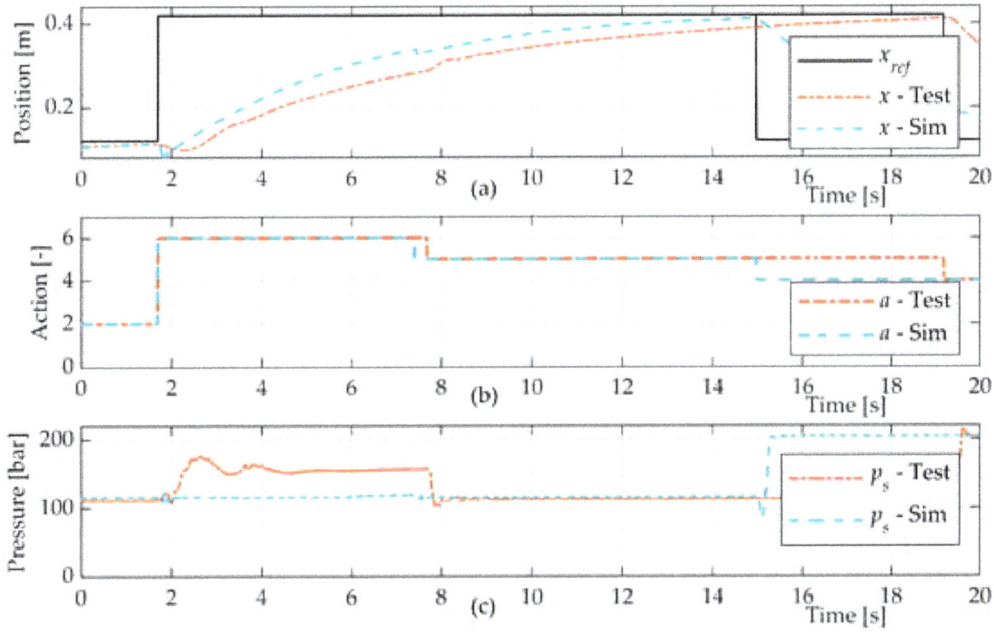

Figure 12. Experimental (Test) and simulation (Sim) results, 40 kg load and *pLS*,1=100 bar. (a) Position and reference; (b) action; (c) system pressure.

As for the previous load case, the agent is able to apply in practice the same modes it learned from the optimisation.

As said, all load cases were tested experimentally, and comparisons of selected modes are shown in **Figure 13** and **Figure 14**, where Sim and Test refer to the simulation and experiment, respectively.

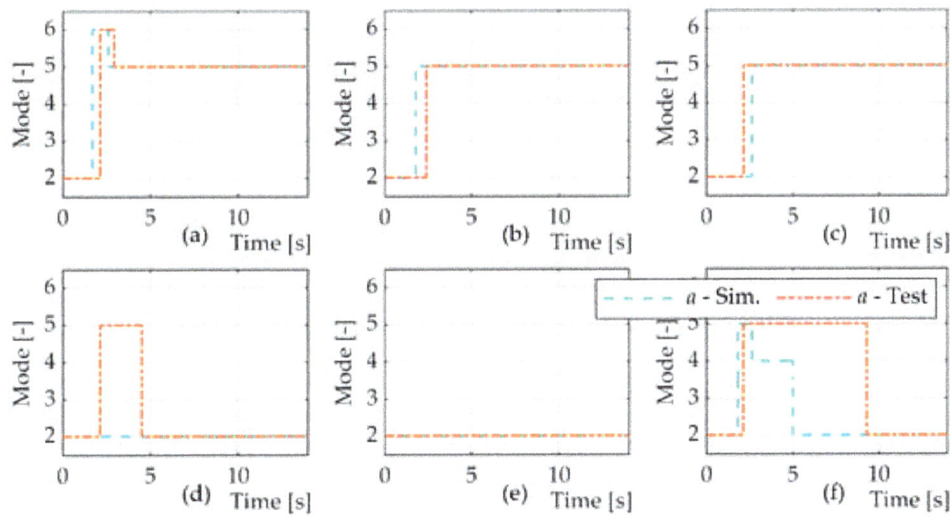

Figure 13. All test cases, *pLS*,1=60 bar: (a) 40 kg; (b) 80 kg; (c) 120 kg; (d) 160 kg; (e) 200 kg; (f) 240 kg.

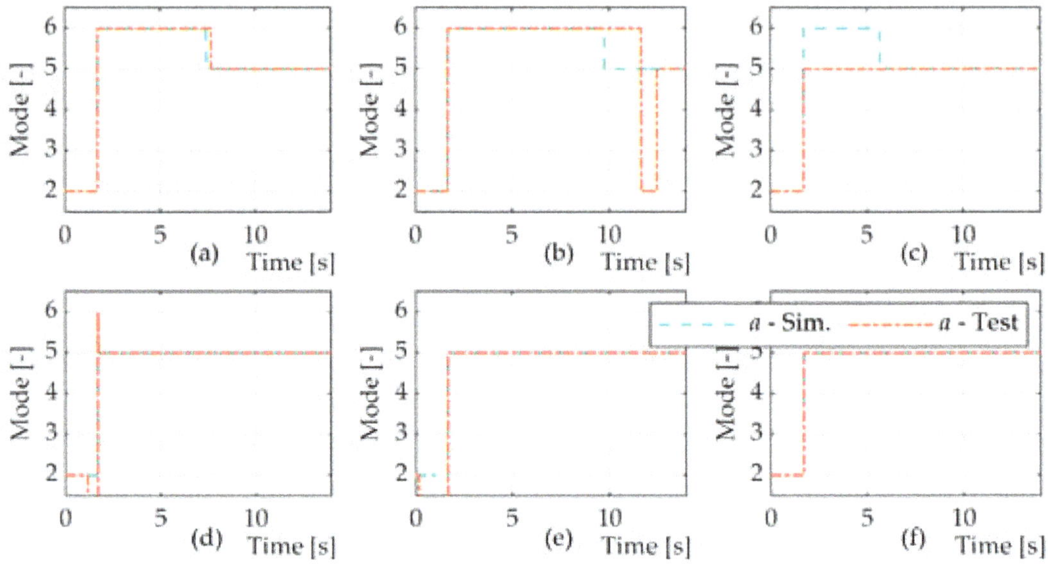

Figure 14. All test cases, *pLS*,1=100 bar: (a) 40 kg; (b) 80 kg; (c) 120 kg; (d) 160 kg; (e) 200 kg; (f) 240 kg.

The experimental results show that in most of the cases the agent was able to implement in the real system the decisions that were also taken in the simulation. The load cases that failed are shown in **Figure 13**d,g and **Figure 14**b,c. Even though these cases show some deviation from what was achieved in the simulation, they are not wrong for the complete test. This means that only a few decisions were wrong in these tests. However, they still followed the logic of choosing weaker modes at the start of the motion and stronger modes at the end of the motion, and, therefore, the agent was still able to complete the motion.

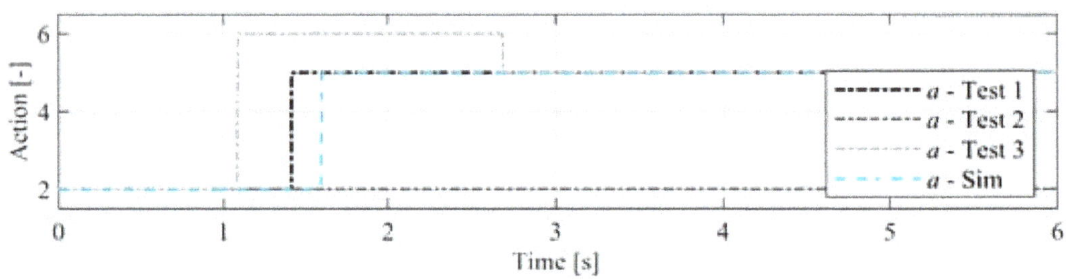

Figure 15. Issues with repeatability for 120 kg load and *pLS*,1=60 bar.

Relatively high sensitivity to the variation in measured variables was observed, and this resulted in low repeatability in some of the load cases. The worst situation was for the load case of 120 kg load and *pLS*,1 bar, an example of which is shown in **Figure 15.**

Figure 15 shows that such controllers, as expected, might underperform when operating under different conditions than they were trained for. However, for the trained agent presented in this study, the decisions, despite being different, still allowed the agent to complete the task in an acceptable way.

5. DISCUSSION

The energy analysis, derived from simulation data, demonstrated a decrease in energy losses due to the agent's selection of appropriate modes. The agent's training did not converge to the global optimal solution that minimizes energy losses. Other factors contribute to the optimization target, therefore the results in Figure 6 are merely approximations of the optimal modes. Future research should focus on identifying the most effective solution and contrasting it with the outcomes achieved by the agent during the training process. The system model was deemed accurate enough to depict the primary properties and behaviors of the system. However, there were discrepancies from the system's actual behavior. While pre-training the agent is adequate, enhancing the agent's performance could be achieved by allowing it to safely interact with the actual system and learn from those encounters to rectify model discrepancies. Machine learning-based controllers are a sort of controller that operates as a black-box. Deploying them to operate real systems carries an inherent risk, sometimes due to the controller acting beyond the training domain. Thus, it is essential to guarantee that the control measures used in the system do not result in any unsafe or unstable outcomes. This work-maintained safety by implementing rules that would override the agent's decisions when needed.

One weakness of this study was that the agent was assessed under similar load levels and tasks to those it was trained for. Random noise was added to all system measurements in the simulation to enhance its robustness during experimental tests. Nevertheless, there remain unanswered questions and an assessment needed to determine the agent's resilience and generalization potential. This is particularly true for more practical load scenarios, such excavation and fluctuating pLS,1 pressure.

6. CONCISE SUMMARY

The chapter showcases simulation and experimental outcomes illustrating the effectiveness of a controller created within a reinforcement learning framework in learning and managing the mode selection of a hydraulic system. The controller underwent simulated training before being utilized to operate the actual system. The primary goal was to optimize the selection of modes for the multi-chamber actuator to minimize resistive control losses in a valve-controlled load-sensing system. The controller autonomously identified optimized control decisions and implemented them in the actual system. The controller demonstrated robustness to discrepancies between the training domain and the application domain, allowing for the extrapolation of knowledge to unfamiliar scenarios. To ensure the safe operation of the system, safety regulations must accompany the trained agent notwithstanding its ability to control the real system. Future research should focus on measuring the proximity of the solution to the global optimum and testing its robustness in practical applications, as several test scenarios showed a high sensitivity to changes in input features.

7. AT A GLANCE

This chapter discusses the creation and application of a reinforcement learning agent to operate as the mode selection for a multi-chamber actuator within a load-sensing architecture. The agent chooses the actuator mode that minimizes energy losses in the system. The agent underwent training in a simulated environment before being put into operation in the actual system. The simulation results demonstrated that the agent was able to decrease energy usage without compromising actuation performance. The experimental results demonstrated the agent's ability to learn through simulation and to control the actual system.

REFERENCES

1. Vukovic, M.; Leifeld, R.; Murrenhoff, H. Reducing fuel consumption in hydraulic excavators—A comprehensive analysis. Energies **2017**, 10, 687. **[Google Scholar]** **[CrossRef]** **[Green Version]**
2. Ketonen, M.; Linjama, M. Simulation study of a digital hydraulic independent metering valve system for an excavator. In Proceedings of the 15th Scandinavian International Conference on Fluid Power, SICFP'17, Linköping, Sweden, 7–9 June 2017. **[Google Scholar]**
3. Raduenz, H.; Ericson, L.; Heybroek, K.; de Negri, V.J.; Krus, P. Extended analysis of a valve-controlled system with multi-chamber actuator. Int. J. Fluid Power **2021**, 23, 79–108. **[Google Scholar]** **[CrossRef]**
4. Linjama, M.; Vihtanen, H.; Sipola, A.; Vilenius, M. Secondary controlled multi-chamber hydraulic actuator. In Proceedings of the 11th Scandinavian International Conference on Fluid Power, SICFP09, Linköping, Sweden, 2–4 June 2009. **[Google Scholar]**
5. Belan, H.C.; Locateli, C.C.; Lantto, B.; Krus, P.; de Negri, V.J. Digital secondary control architecture for aircraft application. In Proceedings of the Seventh Workshop on Digital Fluid Power, Linz, Austria, 26–27 February 2015. **[Google Scholar]**

6. Dell'Amico, A.; Carlsson, M.; Norlin, E.; Sethson, M. Investigation of a digital hydraulic actuation system on an excavator arm. In Proceedings of the 13th Scandinavian International Conference on Fluid Power SICFP2013, Linköping, Sweden, 3–5 June 2013; pp. 505–511. **[Google Scholar] [CrossRef] [Green Version]**

7. Huova, M.; Laamanen, A.; Linjama, M. Energy efficiency of three-chamber cylinder with digital valve system. Int. J. Fluid Power **2010**, 11, 15–22. **[Google Scholar] [CrossRef]**

8. Heemskerk, E.; Eisengießer, Z. Control of a semi-binary hydraulic four-chamber cylinder. In Proceedings of the Fourteenth Scandinavian International Conference on Fluid Power, Tampere, Finland, 20–22 May 2015. **[Google Scholar]**

9. Heybroek, K.; Sjöberg, J. Model predictive control of a hydraulic multichamber Actuator: A feasibility study. IEEE/ASME Trans. Mechatron. **2018**, 23, 1393–1403. **[Google Scholar] [CrossRef]**

10. Donkov, V.H.; Andersen, T.O.; Pedersen, H.C.; Ebbesen, M.K. Application of model predictive control in discrete displacement cylinders to drive a knuckle boom crane. In Proceedings of the 2018 Global Fluid Power Society PhD Symposium (GFPS), Samara, Russia, 18–20 July2018; pp. 408–413. **[Google Scholar] [CrossRef]**

11. Yuan, H.; Shang, Y.; Vukovic, M.; Wu, S.; Murrenhoff, H.; Jiao, Z. Characteristics of energy efficient switched hydraulic systems. JFPS Int. J. Fluid Power Syst. **2014**, 8, 90–98. **[Google Scholar] [CrossRef] [Green Version]**

12. Vukovic, M.; Murrenhoff, H. Single edge meter out control for mobile machinery. In Proceedings of the ASME/Bath Symposium on Fluid Power & Motion Control, FPMC2014, Bath, UK, 10–12 September 2014. **[Google Scholar]**

13. Mnih, V.; Kavukcuoglu, K.; Silver, D.; Graves, A.; Antonoglou, I.; Wiersta, D.; Riedmiller, M. Playing Atari with deep reinforcement learning. arXiv **2013**, arXiv:1312.5602v1. **[Google Scholar]**

14. Zhu, Q.; Wang, Q. Real-time energy management controller design for a hybrid excavator using reinforcement learning. J. Zheijang Univ.-Sci. A (Appl. Phys. Eng.) **2017**, 18, 855–870. **[Google Scholar] [CrossRef]**

15. Zhang, W.; Wang, J.; Liu, Y.; Gao, G.; Liang, S.; Ma, H. Reinforcement learning-based intelligent energy management architecture for hybrid construction machinery. Appl. Energy **2020**, 275, 115401. **[Google Scholar] [CrossRef]**

16. Backman, S.; Lindmark, D.; Bodin, K.; Servin, M.; Mörk, J.; Löfgren, H. Continuous control of an underground loader using deep reinforcement learning. Machines **2021**, 9, 216. [**Google Scholar**] [**CrossRef**]

17. Andersson, J.; Bodin, K.; Lindmark, D.; Servin, M.; Wallin, E. Reinforcement learning control of a forestry crane manipulator. In Proceedings of the 2021 IEEE/RSJ International Conference on Intelligent Robots and Systems (IROS), Prague, Czech Republic, 27 September–1 October 2021. [**Google Scholar**]

18. Egli, P.; Hutter, M. A general approach for the automation of hydraulic excavator arms using reinforcement learning. IEEE Robot. Autom. Lett. **2022**, 7, 5679–5686. [**Google Scholar**] [**CrossRef**]

19. Sutton, R.S.; Barto, A.G. Reinforcement Learning—An Introduction, 2nd ed.; MIT Press: Cambridge, MA, USA, 2018; ISBN 9780262039246. [**Google Scholar**]

20. Berglund, D.; Larsson, N. Controlling a Hydraulic System Using Reinforcement Learning: Implementation and Validation of a DQN-Agent on a Hydraulic Multi-Chamber Cylinder System. Master's Thesis, Linköping University, Linköping, Sweden, 2021. [**Google Scholar**]

21. HOPSAN Multi-Domain System Simulation Tool, Division of Fluid and Mechatronic System, Department of Management and Engineering, Linköping University, Linköping, Sweden. Available online: **https://liu.se/en/research/hopsan** (accessed on 1 June 2022).

22. DQN Agent, Mathworks Reinforcement Learning Toolbox. Available online: **https://mathworks.com/help/reinforcement-learning/ug/dqn-agents.html** (accessed on 1 June 2022).

07 DESIGNING A VARIABLE PRESSURE MULTI-PRESSURE RAIL SYSTEM FOR AGRICULTURAL USE

1. INTRODUCTION: AN OVERVIEW

Global warming and climate change have received significant attention in recent years, leading to numerous technological developments. A crucial factor is the utilization of fossil fuels resulting in the release of greenhouse gases. Off-road mobile machinery, such as agriculture and construction equipment, are significant in this context [1]. The advanced hydraulic systems used in agriculture have low efficiency, averaging 20%, and are not suitable for future efficiency needs. This study will concentrate on agricultural equipment systems. An agricultural tractor is commonly used to carry out specific tasks in a field, such as planting, seeding, baling, etc. [2]). These tools have actuators that are usually powered by the high-pressure hydraulic system of the tractor via hydraulic remote connections. Figure 1 below shows the most prevalent configuration of the hydraulic circuit. The tractor features a load-sensing hydraulic system consisting of a variable displacement pump with a differential flow limiter and a load-sensing valve known as an electro-hydraulic remote valve. This circuit is highly energy-efficient for managing many actuators [3]. The pump delivery pressure stabilizes at the maximum actuator pressure, together with a pump margin $pmar$, to reduce unnecessary pressurization of the system.

The hydraulic control system consists of actuators that carry out agricultural tasks and a local flow control valve at each actuator to adjust its speed. This setup enables the installation of several actuators on a single remote line in the tractor, as depicted in Figure 1 (left), and ensures that the control performance of the implement is not affected by the tractor's brand or technological level.

The LS system's principle of operation is compromised by this setup as the LS pump is required to match the demand of the implement valves rather than the commanded flow rate at the distant valve. This is accomplished by regulating the highest flow rate at the Electronic Health Record (EHR) and allowing the pressure compensators (PCs) of the equipment to reach the maximum pressure of the pump *p**. Thus, the pump operates under pressure saturation regardless of the load on the actuators. The hydraulic control system's low energy efficiency is seen in Figure 1's power map, with the orange area indicating energy loss from throttling. If the system operated as an LS system, the pump pressure would be at *pLS*, the maximum actuator load *pActmax*, plus the pump margin *pmar* to guarantee proper functionality of the control valve, rather than *pmax*.

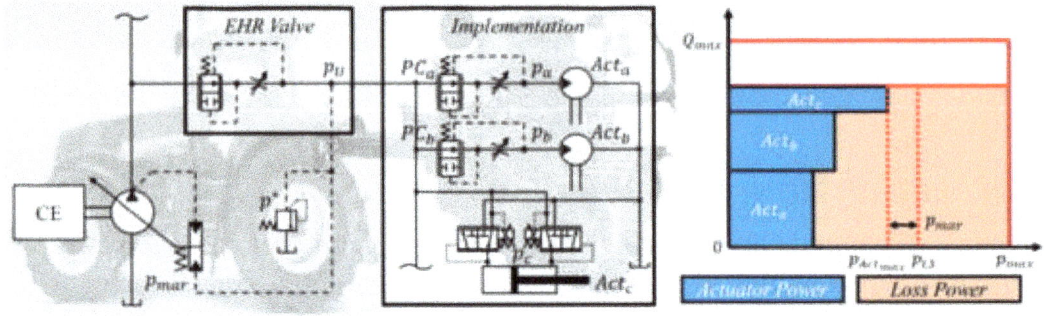

Figure 1. Hydraulic circuit for tractor and planter system with redundant flow control valve (left) and power loss (Right).

Given the widespread use of LS systems, various techniques have been documented in literature to enhance their energy efficiency. Bedotti et al. [4] aimed to decrease throttling losses by minimizing the LS margin of the system. Decreasing the margin may impact system performance, although reference [4] suggests a safe reduction of 23%, resulting in power savings of around 4%. Tian et al. [5] suggested a way to adjust the LS margin of a pump dynamically to decrease throttling losses without compromising system performance, but did not include losses caused by highly changing loads. Siebert et al. proposed a method to decrease throttling losses by increasing the outlet pressure of actuators, hence lowering losses while keeping a consistent pressure drop across the actuator. The approach experienced a 44% decrease in throttling losses during a simulation. Another important technique is independent metering, where the intake and output control apertures are separated and controlled individually, enabling more flexible control and power optimization with accurate dynamics.

The previously mentioned solutions do not address the control conflict among flow control valves, which causes inefficiency in tractor-implement operation due to pressure saturation of the supply pump. To prevent pressure saturation, it is recommended to measure the load pressure at the actuator (*pa*, *pb*, or *pc*, Figure 1 right) instead than at the EHR output (*pU*). Yet, this technique is typically avoided due to its inability to provide enough dynamic behavior, as the pump flow variation is not rapid enough to handle the actuator load fluctuations. Additionally, it would create interdependence between the tractor and planter technology.

Thus, in order to improve energy economy in hydraulic control systems, various layout layouts should be explored. Commonly, research highlights displacement control actuation as a very effective way for controlling a function, as it completely eliminates throttling losses [3,9]. This technique necessitates a separate flow supply (such as a hydraulic pump) for each function, making it expensive and unfeasible for a tractor-implement system that needs to accommodate numerous functions. Displacement control technology was successfully applied in construction equipment [10,11,12,13,14]. Busquets et al. [15] suggested a solution to decrease the number of pumps by creating a control sharing protocol for pump flow based on typical excavator cycles. This design allows for the control of all excavator functions with only four pumps. Implementing such a method in agricultural systems is challenging due to their connection to implements with uncertain motor cycles. Pressure-controlled flow supplies are an uncommon hydraulic actuation method in off-road vehicles. In these systems, the supply pump is not adjusted to correspond to a specified flow rate, as shown in Figure 1 (left), but instead to set a specified pressure. The basic application of this idea is the constant pressure (CP) system. In this system, the supply pump establishes the pressure in a shared pressure rail that links the pump to the actuator, with control valves regulating the flow from this rail.

The CP system separates the actuators from the prime mover, enabling power management measures to be applied at the prime mover. The CP system attains a greater control bandwidth by utilizing the actuator response that operates on the control valve receiving flow from the rail and is not reliant on the supply pump. A basic CP system with one pressure rail experiences high throttling losses due to the need for the rail pressure to be higher than the maximum load pressure, resulting in slowing down the flow to other actuators at a lower pressure. Advanced CP systems, as outlined in reference [17], tackle this issue through two methods: directly managing actuator displacement to manage speed and torque, or use hydraulic transformers to lower rail pressure without throttling. Volvo successfully applied the first strategy on a 30-ton excavator. This design had an innovative variable-area cylinder and demonstrated a 34–50% improvement in fuel efficiency. Various hydraulic transformer designs proposed in references [19,20] have not been widely used due to their complexity, high cost, and untested reliability.

The multi-pressure rail (MPR) system discussed in this study does not fit within the previously defined categories for advanced CP systems. Instead, it suggests utilizing several pressure rails. Lumkes et al. [21] initially proposed the concept of the MPR system and conducted feasibility testing on a backhoe test rig. The results demonstrated functionality, but the investigation did not focus on specific controller design or possibilities for power-saving. In their study, Dengler et al. suggested incorporating a fixed pressure medium pressure rail to achieve a possible 20% efficiency improvement compared to a state-of-the-art LS system in a simulation. The MPR system discussed in this research is most closely related to the STEAM excavator created by the IFAS team at RWTH Aachen and the hybrid hydraulic architecture (HHEA) developed by Li et al. Both systems utilized different versions of MPR designs for building machinery.

The STEAM excavator system utilizes engine management algorithms and recovers energy from overrunning loads wherever feasible. Published data indicates a 30% decrease in gasoline usage. The HHEA system proposes using extra hydraulic and electric machines, one for each actuator, to mimic hydraulic transformers and minimize throttle loss, even though it has not been tested in a real car. The system is expected to decrease energy usage by 40% based on the simulation. Both topologies do not change the immediate pressure level in the rails because of the high dynamic characteristics of the motor cycles commonly found in construction equipment. Opgenoorth et al. (2017) introduced a Multi-Pressure Rail (MPR) design for an electrically powered excavator, featuring adjustable rail pressures. Their simulation demonstrated a 29% increase in efficiency compared to a conventional Load-Sensing (LS) system. No controller design or experimental validation was provided. Bertolin and Vacca conducted a parametric research in reference [28] to analyze the impact of varying cylinder areas on the efficiency of the proposed system.

The analysis also takes into account varying numbers of rails in the system and how to optimize their pressure levels to enhance system efficiency. The authors of this work previously proposed many alternatives for the design and management of a variable pressure rail MPR system in a prior publication [29]. This study utilized a physical test rig to demonstrate the fundamental controllability of the system and to assess the comparative efficiency of several configurations including pumps and accumulators. The study found that the two-pump system without accumulators showed superior dynamic behaviors compared to the variable pressure system controlled by both supply systems, as it lacked the additional capacitance provided by the accumulators.

This work aims to develop and test a variable pressure MPR system tailored for an agricultural-implement system. These systems typically have numerous actuators with low dynamic needs and minimal opportunities for energy recuperation. The chapter aims to discuss the design of the hydraulic system and measure the energy savings possible in a real-world scenario. The paper introduces the MPR system design, taking into account the distinctive system characteristics in Section 2. Section 3 provides a detailed introduction to the multi-level controller. A test rig is created, modeled, constructed, and tested to confirm the stability and effectiveness of the proposed MPR controller in Section 4. Section 5 displays two often used reference vehicles in the field: a tractor and an implement. A comprehensive system model, which incorporates the reference tractor, planter, and the proposed MPR components, is developed, verified, and utilized to forecast the power conservation and efficiency enhancement under the machine's typical operating conditions in Section 6. The data indicates a maximum of 59% decrease in power at the pump shaft, resulting in an 89.8% increase in system efficiency.

2. MPR SYSTEM DESCRIPTION

Figure 2 shows the design of the MPR system with a generic number of pressure rails as considered in this work. It includes a supply system, pressure rails, pressure select and control valve sets (PSCV), and actuators. The supply system controls the pressure level in each rail. The rails connect the supply system to each PSCV. Then, the PSCVs connect the most appropriate rails to the inlet and outlet of each actuator to minimize throttling loss. The PSCVs also control the flow rate to match the commanded velocity.

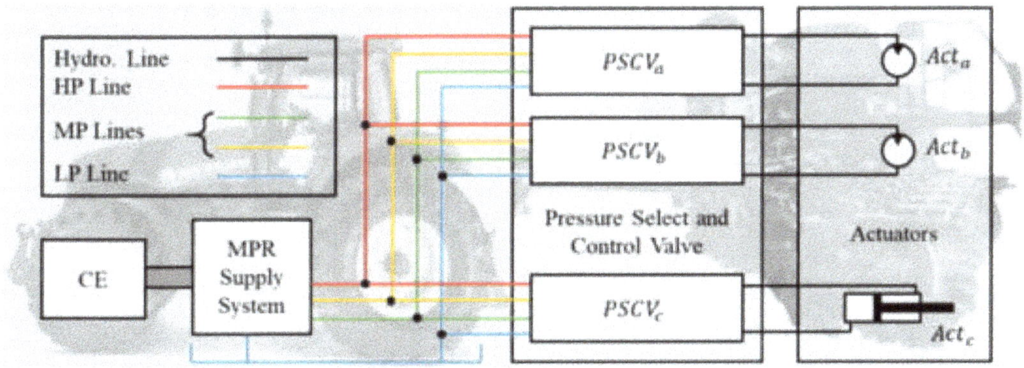

Figure 2. Generic design of the multi-pressure rail system with four pressure rails.

Figure 3 and **Figure 4** show an example of how the three-rail MPR system works for both hydraulic motors and cylinders with PSCV. First, the PSCV pressure selection stage (V1–V6) will be opened or closed to connect the line $A1$ and $B1$ to the pressure rails according to the command from the supervisory controller.

Then, the metering stage control valve (VA and VB) will control the actuator speed tracking with the actuator controller. Using different combinations of inlet/outlet pressure levels, the MPR makes available to each actuator multiple output torque and force ranges. For example, considering an MPR system with three rails, a hydraulic motor can operate under six possible modes: high pressure at the inlet with low pressure at the outlet (HP-LP), HP-MP, MP-LP, LP-MP, LP-HP, and MP-HP. There are also three idle modes: LP-LP, MP-MP, and HP-HP. Due to the pressure losses in the lines and selection and metering valves, the pressure margin *pmar* needs to be considered to leave enough pressure drop for proper functioning of the actuator. As a result, the torque range for each working mode is shown in **Table 1.**

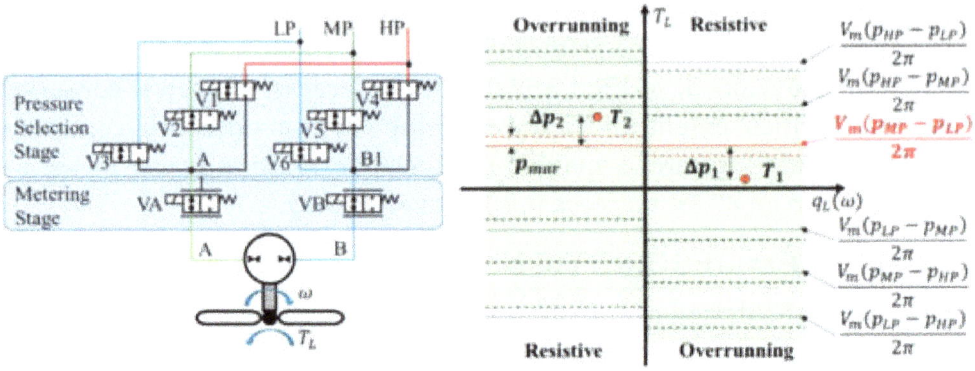

Figure 3. Working principle for MPR system with motor actuator.

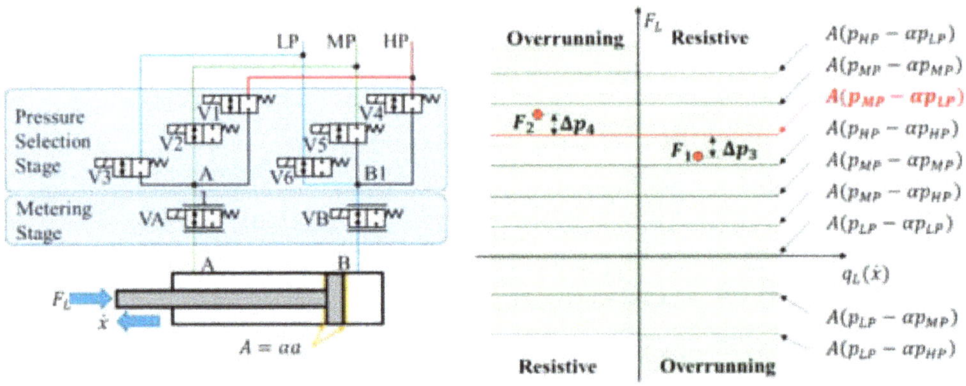

Figure 4. Working principle for MPR system with cylinder actuator.

If the motor load is at the value T1 as shown in **Figure 3**, the system should work under mode MP-LP (the red line in the figure) to minimize the throttling loss to Δ*p*1. Clearly, this is a great advantage in comparison with the baseline system of **Figure 1** (left) which puts all the actuators under the maximum pressure differential. In case of an overrunning load, such as *T*2, the MPR has the potential to recover energy by connecting the motor inlet to LP and the outlet to MP. This allows the motor to work in pumping mode, thus inputting energy back to the system, under a minimal throttling loss Δ*p*2.

Table 1. Working Mode Torque Range for Three-Rail MPR System.

Load Condition	Mode	Lower Limit (Nm)	Upper Limit (Nm)
Resistive	**MP-LP**	0	$\frac{V_m(p_{MP}-p_{LP}-p_{mar})}{2\pi}$
	HP-MP	$\frac{V_m(p_{MP}-p_{LP}-p_{mar})}{2\pi}$	$\frac{V_m(p_{HP}-p_{MP}-p_{mar})}{2\pi}$
	HP-LP	$\frac{V_m(p_{HP}-p_{MP}-p_{mar})}{2\pi}$	$\frac{V_m(p_{HP}-p_{MP}-p_{mar})}{2\pi}$

Load Condition	Mode	Lower Limit (Nm)	Upper Limit (Nm)
Overrunning	LP-LP/MP-MP/HP-HP	$\frac{V_m(p_{LP}-p_{MP}+p_{mar})}{2\pi}$	0
	LP-MP	$\frac{V_m(p_{MP}-p_{HP}+p_{mar})}{2\pi}$	$\frac{V_m(p_{LP}-p_{MP}+p_{mar})}{2\pi}$
	MP-HP	$\frac{V_m(p_{LP}-p_{HP}+p_{mar})}{2\pi}$	$\frac{V_m(p_{MP}-p_{HP}+p_{mar})}{2\pi}$
	LP-HP	Maximum System Torque	$\frac{V_m(p_{LP}-p_{HP}+p_{mar})}{2\pi}$

The hydraulic cylinder, illustrated in **Figure 4** (pressure margin lines are omitted), works under a similar principle. The only difference is for cylinders with differential areas, the idle modes for hydraulic motors can also output force, such as force levels $(pHP-\alpha pHP)$, $A(pMP-\alpha pMP)$, and $A(pLP-\alpha pLP)$. As a result, there are nine force lines in total compared to the six torque lines of a symmetric actuator. When the cylinder has a resistive load, such as $F1$ in **Figure 4** (right), the best working mode is MP-LP (the red line in the figure) with the throttling loss $\Delta p3$. With overrunning load $F2$, the system prefers working in the same mode to recover energy to the MP rail with throttling loss $\Delta p4$.

The case of unidirectional motors with no overrunning loads is further considered in this section to present general design considerations for the MPR system. This case is particularly relevant for agricultural tractor–implement systems, and in particular for the reference tractor–planter as it will be described in **Section 4.**

2.1. Number of Rails

The number of rails is a critical parameter in an MPR system, as it directly determines the number of modes available for the actuator connections. A two-rail system has only one mode, as each actuator can be connected only with the HP line and the LP line. In a three-pressure MPR, there are three modes: supply from medium pressure (MP) to LP rail (MP-LP), from HP to MP rail (HP-MP), and from HP to LP rail (HP-LP). The number of modes (nmd) is determined by the number of rails (nR) as seen in Equation (1) with a unidirectional motor as the actuator.

$$n_{md} = \frac{n_R(n_R - 1)}{2} \quad (1)$$

Intuitively, the more rails there are in the system, the less throttling losses will occur. However, a high number of rails can be impractical, as it implies a high number of components (detrimental for costs and spatial requirements) and added control complexity. Therefore, the optimal number of rails comes from a compromise between energy loss and technical-economic feasibility. Considering all the benefits and costs of adding rails, the three-rail MPR system is the most suited choice for the reference system which will be introduced in **Section 4**. As a result, the MPR system design in this study will continue with three pressure rails. The simulation results that drove this decision can be seen in **Section 6**.

2.2. Supply System

Based on the analysis and experimental work in [**29**], to achieve the target of unidirectional, resistive load conditions present in most agricultural implements, the supply system shown in **Figure 5** is most suitable due to its fast response in rail pressure control. This system uses one pump per pressure rail, allowing for a fast control response for both pressures independently. Most of the time this is the operating mode in which the system will run. Should a user require a higher flow than one pump can provide, however, a merging valve is included which connects both pumps to one rail. This setup allows for the pumps to be downsized somewhat, in turn causing them to typically run at higher displacements and efficiencies. Finally, the MP pump must be able to go over center in order to handle any cases in which flow coming from the HP rail to the MP rail exceeds the flow from the MP to LP rails.

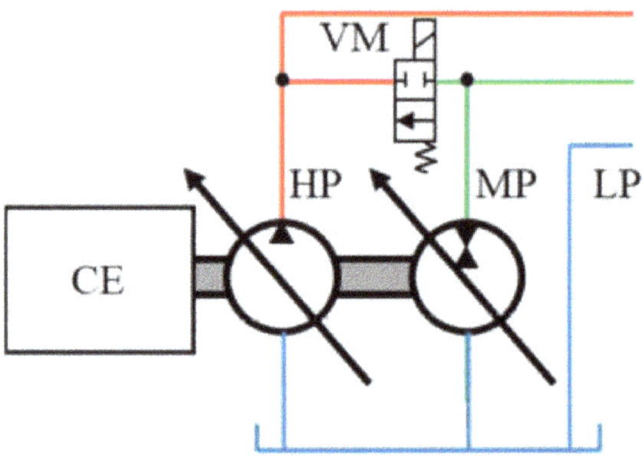

Figure 5. MPR supply system design.

To size the pumps, there are two options. For the first option, both pumps should be sized large enough to supply the full flow required by the users to eliminate any possibility of pump flow saturation for both rails. This sizing option will ensure the system always remains in normal MPR mode and no merging of flow is needed. However, the pumps will suffer from lower efficiency due to working at a lower displacement. For the second option, owing to the merging valve, the total flow from both pumps needs to satisfy the maximum required flow rate as shown in Equation (2). This provides design freedom on how to size HP and MP pumps and an optimization method is proposed in this work to determine the best size for both pumps.

$$V_{P_{HP}} + V_{P_{MP}} \geq \frac{q_{max}}{\omega_p} \quad (2)$$

To find the best size for both pumps, a cost function J is defined in Equation (3). This cost function evaluates only the average power loss during the reference machine's normal working condition, which represents the bulk of the machine's operating time throughout its lifetime. The first half of the cost function represents the summation of volumetric $qspk$ and hydromechanical losses $Tspk$ for the pump k at corresponding working pressure $ppk-pLP$, and rotational speed ωpk. The second half calculates the throttling loss in the control valves for all $nact$ actuators at their flow rate qj, and pressure drop across their control valve $pin-\Delta pi-pout-$. pin and $pout$ are the inlet and outlet rail pressures that are selected by the pressure selection stage valve in **Figure 4**, and they are either HP, MP, or LP. The actuator load pressure Δpi is the pressure drop across the actuator suction side and deliver side. With the cost Function (3) and Constrain (2), the best pump size that could satisfy the flow requirements and minimize the system loss could be found for the selected target machine.

$$J = \sum_{k=1}^{2} \left(q_{s_{pk}} \left(p_{pk} - p_{LP} \right) + T_{s\,pk} \omega_{p_k} \right) + \sum_{j=1}^{n_{act}} q_j \left(p_{in} - \Delta p_i - p_{out} \right) \quad (3)$$

Due to limited resources, this study will use the same size of units as the reference machine (45 cc), which follows the first sizing option, and all the controller design and simulation results are with respect to that size.

2.3. Pressure Select and Control Valve Set

There are many ways to achieve these two functions. The most generic design is shown in **Figure 6** (left). It includes six on/off valves (V1–V6) as the pressure select valves and two proportional valves (VA and VB) as the actuator control valves.

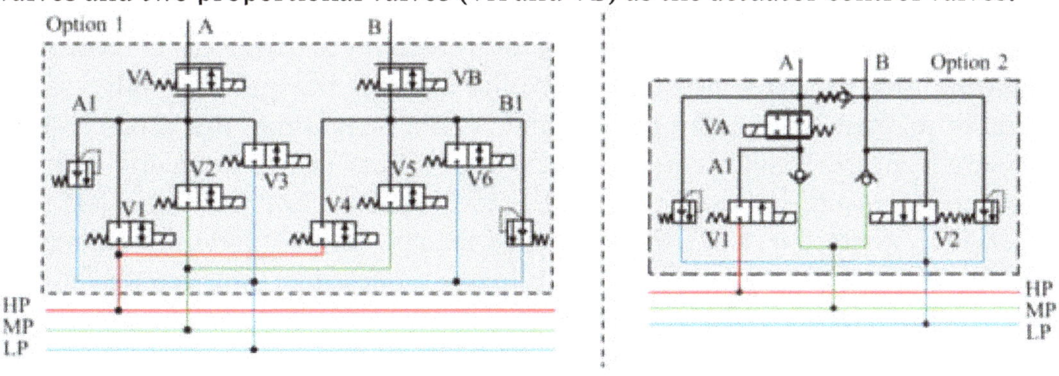

Figure 6. MPR pressure select and control valve set designs.

Two pressure relief valves and check valve sets are used to protect the circuit from overpressure or cavitation. In this setup, the actuator can operate in both directions (A to B, or B to A) in both resistive and overrunning loading conditions, as both the inlet and return of the actuator could be connected to any rail. Moreover, both metering-in and metering-out control strategies are possible. This design is very flexible but also expensive, as it uses many solenoid-operated valves.

To better fit the valve set design to the actuator, the load characteristic (mentioned at the end of the state-of-the-art section) should be considered. Based on these characteristics, the architecture of **Figure 6** (right) would be a more suitable design for this study. It only has two on/off valves (V1 and V2) with two check valves at the pressure select stage. The inlet side can connect to HP and MP pressure rails and the outlet can connect to MP and LP pressure rails. On top of that, it has only one metering valve (VA) on the A side for meter-in control. An anti-cavitation check valve is placed between the actuator work ports and two pressure relief and check valves are used to protect the circuit. This design can no longer achieve all working conditions as option 1 can, but is adequate for the previously discussed unidirectional actuators, without overrunning loads, and greatly reduces the number of solenoids, and thus cost and control complexity.

3. MPR CONTROLLER DESIGN

The control structure for the proposed MPR system has two levels: the actuator level controllers and the supervisory level controller. The actuator level controllers include the actuator speed controller, the rail pressure controller, and the pump displacement controller. The supervisory controller is the online optimization controller that sets the HP and MP rail pressure to minimize total power loss due to fluid throttling.

The basic control structure is shown in **Figure 7.** The mechanical and hydraulic components of the tractor and its implement are in red and orange boxes, and the controllers are in green boxes. When the actuator command comes in, the supervisory controller records it and passes it on to the actuator speed controller. Following this, the actuator speed controller controls the metering valve of the actuator to track the actuator command. In the meantime, the actuator load pressure is measured and sent back to the supervisory controller, along with the actuator command, for use in the rail pressure online optimization. Next, the supervisory controller sends the optimized HP and MP rail pressure commands and the optimal working mode for each actuator to the rail pressure controller and the actuator selection valves. In addition, a feedforward displacement command is generated from the actuator speed and rail pressure commands and is then sent to the pump displacement controller. The rail pressure controller causes the rail pressure to follow the optimized command by sending the pressure feedback displacement command to the pump displacement controller. Finally, the pump displacement controller maneuvers the displacement of the HP and MP pumps.

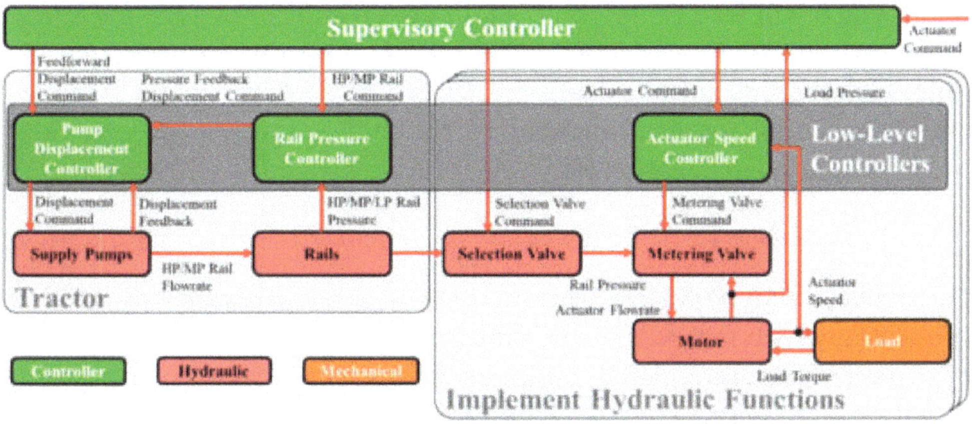

Figure 7. MPR control structure overview.

The detailed controller design is shown in the following sections.

3.1. Supervisory Controller

The supervisory controller will give the best HP and MP rail pressures and the corresponding mode for each actuator to minimize the throttling loss in the MPR system.

$$p_{HP} = \max(\Delta p_1, \Delta p_2, \ldots, \Delta p_n) + p_{mar} \tag{4}$$

To determine the HP rail pressure Equation (4) is used. The HP rail pressure is equal to the highest load pressure in the system plus a pressure margin of *pmar* to overcome any losses in the line and the losses associated with the pressure drop necessary for the control valve to properly function.

As the HP rail pressure is easily determined, the only independent variable is the MP rail pressure (*pMP*). To this end, an optimization method is used based on the cost function defined in Equation (5). This variable is constrained by Equations (7) and (8) which will be introduced later to maintain the pressure margin for proper actuator functionality. The three summations in the cost function represent the throttling loss for each actuator at all three modes. For example, when actuator *i* is working in mode MP-LP, its throttling loss is the product between its flow rate *qi* and the pressure drops across the control valve, *pMP-Δpi-pLP*. The last part of the cost function accounts for the power loss of the HP and MP supply pumps.

$$\begin{aligned} P(p_{MP}) = & \sum_{i=1}^{MP-LP} q_i \left(p_{MP} - \Delta p_i - p_{LP}\right) + \sum_{j=1}^{HP-MP} q_j \left(p_{HP} - \Delta p_j - p_{MP}\right) \\ & + \sum_{k=1}^{HP-LP} q_k \left(p_{HP} - \Delta p_k - p_{LP}\right) + q_{p_{HP}}(p_{HP} - p_{LP})(1 - \eta_{p_{HP}}) \\ & + q_{p_{MP}}(p_{MP} - p_{LP})(1 - \eta_{p_{MP}}) \end{aligned} \tag{5}$$

This cost function is discontinuous. When actuators work in a different combination of modes, Equation (5) has to be rearranged with different constraints for pMP, which will lead to different local optimal choices of pMP for difference combinations. In order to find the global minimum of this cost function, all the possible combinations must be evaluated. With a system that has nR working rails and $nact$ actuators, the total number of possible working mode combinations $ncomb$ is calculated in Equation (6).

$$n_{comb} = \left(\frac{n_R(n_R - 1)}{2}\right)^{n_{act}} \tag{6}$$

For a five-actuator system working with a three-pressure rail system, the $ncomb$ is 243, which means the cost function needs to be evaluated 243 times at every instant to insure a global minimum for the cost function. This makes the optimization quite impractical for an online implementation. An incomplete list of possible combinations is shown in **Table 2.**

Table 2. Incomplete list of possible combinations for three-rail five-actuator MPR system.

Combi Nation	Actuators 1	2	3	4	5	Combi Nation	Actuators 1	2	3	4	5
1	MP-LP	MP-LP	MP-LP	MP-LP	MP-LP	82	HP-MP	MP-LP	MP-LP	MP-LP	MP-LP
2	MP-LP	MP-LP	MP-LP	MP-LP	HP-MP	83	HP-MP	MP-LP	MP-LP	MP-LP	HP-MP
3	MP-LP	MP-LP	MP-LP	MP-LP	HP-LP	84	HP-MP	MP-LP	MP-LP	MP-LP	HP-LP
4	MP-LP	MP-LP	MP-LP	HP-MP	MP-LP	85	HP-MP	MP-LP	MP-LP	HP-MP	MP-LP
...						...					

Fortunately, there are ways to reduce the list of combinations to be considered. First, organize actuators 1–5 with respect to their load pressure in decreasing order into actuators a–e. Following this, apply the rules below.

Rule 1: an actuator a should always work in mode HP-LP. After organizing the actuators, it is important to note that a-e is not necessarily in the same order as 1–5. From the organized list of load pressures, it is clear that actuator a, as the highest load pressure actuator, should always be working in mode HP-LP because the HP rail pressure is set based on this pressure. As a result, 162 combinations can be eliminated.

Rule 2: when an actuator moves to a lower mode, the actuators with a lower load pressure should never move to a higher mode. This is because there will always be a better combination to take its place. **Figure 8** graphically describes this concept. In this setup, assuming the minimum pressure drop is guaranteed, the best combination is 229 (HP-LP, HP-LP, HP-MP, HP-MP, and MP-LP) with the optimized HP and MP pressure. The throttling loss for actuators d and e are $\Delta pdHP{-}MP$ and $\Delta peMP{-}LP$, respectively. If actuators d or e were working in mode HP-LP instead of their current mode, this would result in combinations 231 (HP-LP, HP-LP, HP-

MP, HP-MP, and HP-LP), 232 (HP-LP, HP-LP, HP-MP, HP-LP, and MP-LP), or 234 (HP-LP, HP-LP, HP-MP, HP-LP, and HP-LP) It is clear that from the plot, none of those combinations will result in lower throttling loss than combination 231, as the throttling losses of $\Delta pdHP{-}LP$ and $\Delta peHP{-}LP$ are greater than $\Delta pdHP{-}MP$ and $\Delta peMP{-}LP$. As a result, an additional 60 combinations are eliminated.

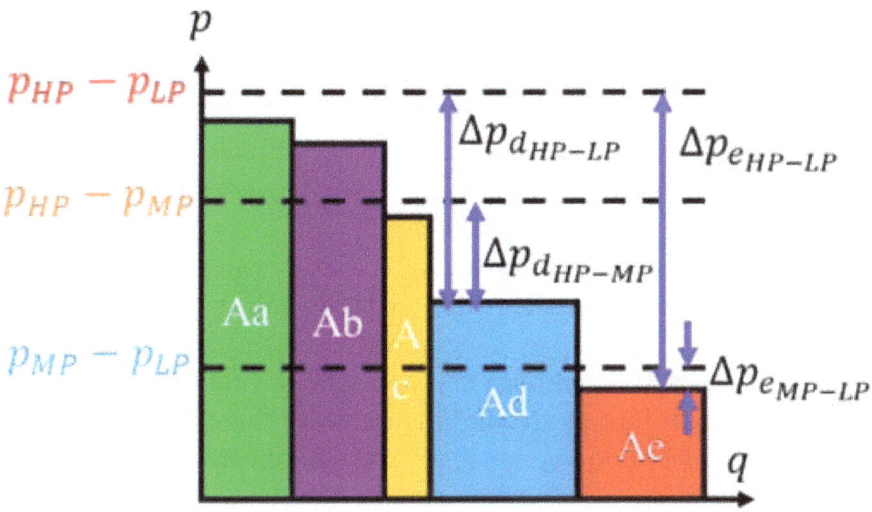

Figure 8. Pressure and flow rate diagram for MPR actuators after sorting.

Rule 3: at least one actuator should work in mode MP-LP. This rule is based on the recirculated excess flow in the MP rail. If no actuator is working in mode MP-LP, such as combination 203 (HP-LP, HP-MP, HP-MP, HP-MP, and HP-MP), there is no actuator taking flow from the MP rail, and all the flow coming from actuators working in mode HP-MP will need to be recirculated through the MP pump back to the shaft; this process generates losses. Therefore, it is certain that if the actuators working in mode HP-MP switch to mode MP-LP with correct MP pressure, such as combination 163 (HP-LP, MP-LP, MP-LP, MP-LP, and MP-LP), it will result in less losses for the system, as the pump loss due to recirculation is eliminated. As a result, five more combinations are eliminated.

To summarize, after applying the above rules, the final list of 16 available combinations is listed in **Table 3.**

Table 3. List of final combinations.

Combi	Actuators					Combi	Actuators				
Nation	a	b	c	d	e	Nation	a	b	c	d	e
163	HP-LP	MP-LP	MP-LP	MP-LP	MP-LP	218	HP-LP	HP-LP	MP-LP	MP-LP	HP-MP
164	HP-LP	MP-LP	MP-LP	MP-LP	HP-MP	221	HP-LP	HP-LP	MP-LP	HP-MP	HP-MP
167	HP-LP	MP-LP	MP-LP	HP-MP	HP-MP	226	HP-LP	HP-LP	HP-MP	MP-LP	MP-LP
176	HP-LP	MP-LP	HP-MP	HP-MP	HP-MP	229	HP-LP	HP-LP	HP-MP	HP-MP	MP-LP

Combi	Actuators					Combi	Actuators				
Nation	a	b	c	d	e	Nation	a	b	c	d	e
190	HP-LP	HP-MP	MP-LP	MP-LP	MP-LP	235	HP-LP	HP-LP	HP-LP	MP-LP	MP-LP
199	HP-LP	HP-MP	HP-MP	MP-LP	MP-LP	236	HP-LP	HP-LP	HP-LP	MP-LP	HP-MP
202	HP-LP	HP-MP	HP-MP	HP-MP	MP-LP	238	HP-LP	HP-LP	HP-LP	HP-MP	MP-LP
217	HP-LP	HP-LP	MP-LP	MP-LP	MP-LP	241	HP-LP	HP-LP	HP-LP	HP-LP	MP-LP

For each combination, the MP rail pressure must fall within the upper and lower constraints to ensure the combination works properly. The lower constraint is determined by Equation (7). The total pressure difference from the MP rail to the LP rail needs to be higher than the highest load working in mode MP-LP plus the pressure margin to ensure proper functioning of the control valve. The upper constraint is determined by Equation (8). The total pressure difference from the HP rail to the MP rail needs to be higher than the highest load pressure that is working in mode HP-MP plus the pressure margin. If the constraints for one combination are incompatible, then the combination is eliminated for that calculation. After rearranging the constraints, as shown in Equations (7) and (8), **Table 4** shows the complete list of all lower and upper constraints for all combinations.

$$p_{MP} - p_{LP} \geq \max(\Delta p_{i_{MP-LP}}) + p_{mar} => p_{MP} \geq \max(\Delta p_{i_{MP-LP}}) + p_{mar} + p_{LP} \quad (7)$$

$$p_{HP} - p_{MP} \geq \max(\Delta p_{i_{HP-MP}}) + p_{mar} => p_{MP} \leq -\max(\Delta p_{i_{HP-MP}}) + p_{HP} - p_{mar} \quad (8)$$

Table 4. List of constrains for the final 16 combinations.

Combi	Constraint *		Combi	Constraint *		Combi	Constraint *		Combi	Constraint *	
Nation	Lower	Upper	Nation	Lower	Upper	Nation	Lower	Upper	Nation	Lower	Upper
163	Δ*pb*	*inf*	190	Δ*pc*	-Δ*pb*	218	Δ*pc*	-Δ*pe*	235	Δ*pd*	*inf*
164	Δ*pb*	-Δ*pe*	199	Δ*pd*	-Δ*pb*	221	Δ*pc*	-Δ*pd*	236	Δ*pd*	-Δ*pe*
167	Δ*pb*	-Δ*pd*	202	Δ*pe*	-Δ*pb*	226	Δ*pd*	-Δ*pc*	238	Δ*pe*	-Δ*pd*
176	Δ*pb*	-Δ*pc*	217	Δ*pc*	*inf*	229	Δ*pe*	-Δ*pc*	241	Δ*pe*	*inf*
* All lower constraints need to add *pmar+pLP*, and all upper constraints need to add *pHP-pmar*.											

Finally, with a closer look at the cost function (5), the value of the cost function is a linear function with the independent variable pMP. This means if the derivative of the cost function with respect to pMP is positive, the minimum pMP should be used to achieve the lowest value of the cost function, which is at the lower constraint of pMP, and vice versa. This further simplifies the optimization to just one evaluation of the sign of the derivative with respect to pMP and one evaluation of the cost function at the corresponding constraint for each combination.

3.2. Real-Time Implementation of Supervisory Controller

The supervisory control logic shown above is operated at a constant frequency on a central ECU, checking continuously to ensure that the system stays in the optimal operating condition. The speed at which the supervisory controller runs can be far lower than that at which the actuator level controls run, 10 Hz and 100 Hz, respectively, for example. This real-time implementation does, however, introduce two control challenges that need to be addressed to ensure the system can operate stably.

First, the control logic proposed above assumed that the system reached a steady state immediately after switching modes and stayed there until the next switch. However, during real drive cycles, small fluctuations of actuator pressures may lead to switches between optimal modes and cause excessive switching between combinations if the value of the cost function for multiple combinations get very close. This could potentially make the system unstable. To overcome this issue, a power loss margin *Pmar* is added to the cost function comparison. The supervisory controller will allow a switch only if the new combination is outperforming the current combination in cost by *Pmar*. There is a possibility that the current combination will not work in the new time step. For example, the current combination at the new time step may have contradictory *pMP* upper and lower constraints, which lead to no possible *pMP* existing for the current combination. In this case, to ensure that the system functions correctly, the supervisory controller will force a switch to the new combination, regardless of *Pmar*. This way, switches will happen much less frequently at the cost to savings of *Pmar*, which is small compared to the total savings, and the benefits of system stability.

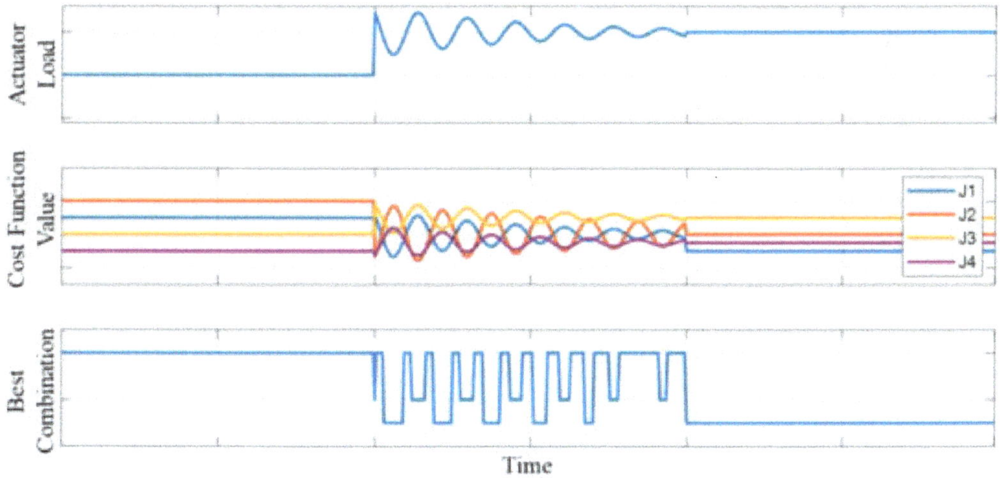

Figure 9. Conceptual example of actuator mode switching dynamic reaction.

Second, during transient condition, such as multi-actuator switching, the system pressure experiences a discontinuity that will impose actuator load pressure oscillation for a short time. An example is shown in **Figure 9.** A simple step change is demonstrated in a single actuator load (top), along with a sample of four of the cost functions (middle), and the actuator's connection mode (bottom). The simulation used accounts for other actuators and cost functions, but only the relevant ones are shown. When the load change occurs, there are transient oscillations in the system that cause corresponding oscillations in the cost functions. The magnitude of these oscillations is instantaneously greater than *Pmar*, and if left unchecked, could lead to uncontrolled switching as seen in the figure. To resolve this, a brief cooldown time, *tcd*, was added to the controller, forcing a minimum time (two seconds for the reference controller) between switches. As with the *Pmar* logic, an exception was implemented allowing the system to switch modes immediately if its current operating condition is unable to satisfy the pressure margin constraints.

3.3. Actuator Level Controllers

The actuator controller is a simple closed-loop PI (proportional integral) controller for actuator speed (**Figure 10**). The actuator command comes from the supervisory controller and the actuator speed is fed back from a speed sensor to close the control loop.

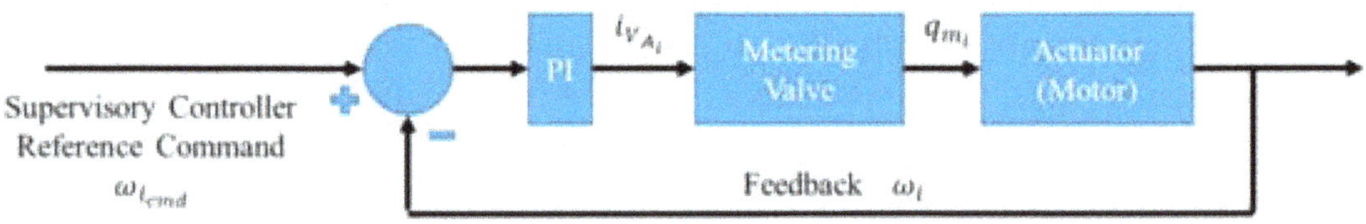

Figure 10. Actuator speed controller.

The rail pressure controller and pump displacement controller form a cascaded PI controller (**Figure 11**). The outer loop is the rail pressure controller that takes rail pressure commands from the supervisory controller and outputs the reference pump displacement command to the inner loop. The rail pressure measurements are used to close the control loop. The inner loop is a feedback–feedforward controller. In addition to the command from the outer loop, the feedforward pump displacement command is sent from the supervisory controller, which is a displacement estimation from the actuator modes and flow needed. The measured pump displacement is used to close the control loop.

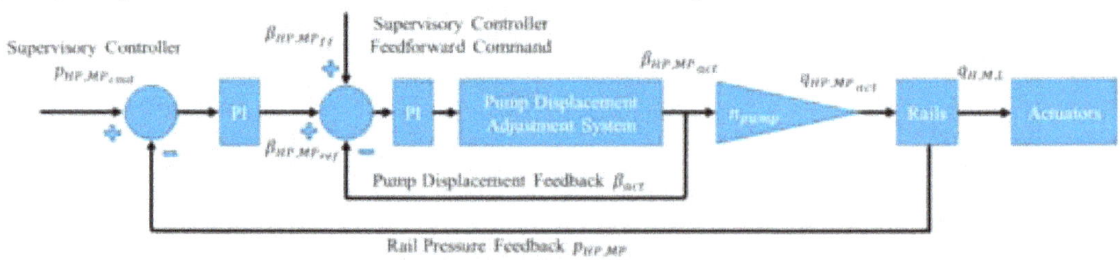

Figure 11. Rail pressure and pump displacement controller.

4. MPR TEST RIG AND REFERENCE MACHINE

4.1. MPR Test Rig Design

A standalone test rig was designed to validate the effectiveness of the control scheme being used. The system was designed as shown in **Figure 12.** Key features in the design were variable load units, allowing the simulation of a wide variety of loading conditions, an unloading valve, enabling the switch between one and two pump supply systems, and a switching valve and accumulator system, enabling the function of the one pump system. The system also used the more general design for the valve set outlined above. This was done to ensure flexibility in future work with the system.

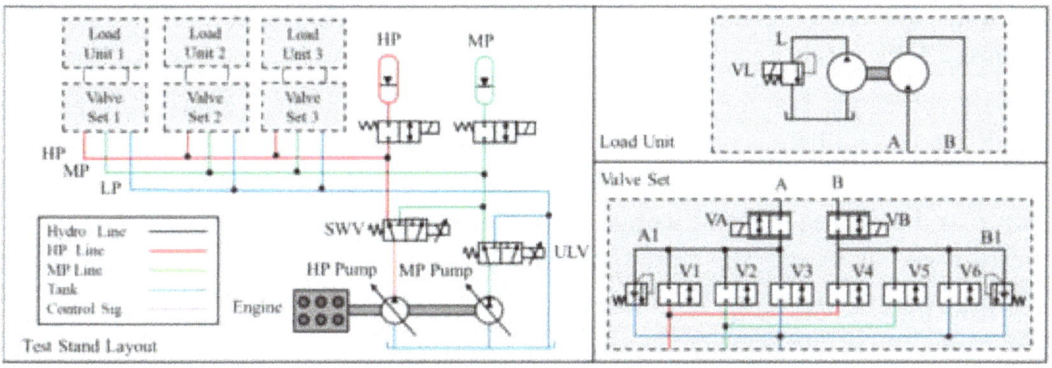

Figure 12. Hydraulic circuit for the MPR test rig.

The test rig design shown above was designed to be able to operate as both a one pump and two pump MPR system, by using the unloading valve (ULV) to isolate the MP pump and using the switching valve (SWV) to switch the HP pump between the rails, as it was initially designed for use on [**29**]. For the purposes of this paper, however, the system is used in the two-pump configuration.

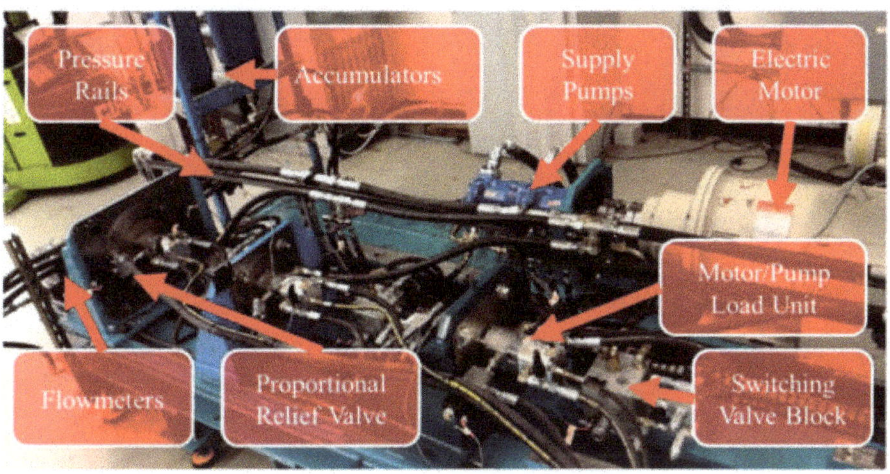

Figure 13. Completed test stand.

The system is controlled using an NI CRio controller, which in turn uses Bosch Rexroth's BODAS pump controllers and pressure transducers to control the hydraulic components of the system. **Figure 13** shows the completed test stand implementation.

4.2. Reference Machine

To show the power savings introduced by the proposed MPR system and controller design, two reference machines (a tractor and an implement) are selected and modeled. The reference machines have been instrumented and tested in the field under normal working conditions to gather baseline data for building and validating the hydraulic model. Finally, the actual power consumption by the hydraulic system on the reference machines for the baseline machine and the proposed MPR machine are compared.

The reference tractor is a T8.390 New Holland tractor. The tractor and its simplified hydraulic circuit are shown in **Figure 14.** It has two sets of centralized load-sensing hydraulic systems. The first system has the first HP pump as a flow source and steering system, using the power beyond function and remote valves one to three as users. The flow first goes through the steering priority valve to satisfy the steering needs, then continues to supply the rest of the functions. The second system is powered by the second HP pump with the hitch function and EHR valves four to six. The implement is connected to the hydraulic system through the tractor EHR. An LP pump is used to power all low-pressure functions, such as the brake, clutch, differential lock, etc. The LP system, which is also a low power system, is not included in this study.

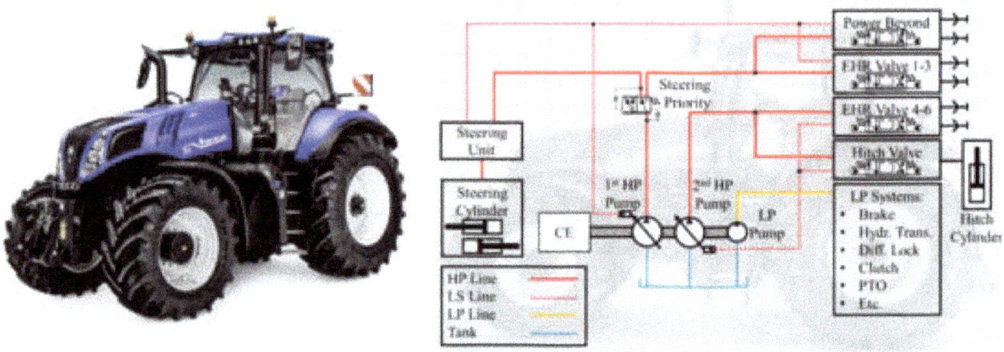

Figure 14. Reference tractor and its hydraulic circuit.

The reference implement is a Case Early Riser 2150 16-row planter. The planter and its simplified hydraulic circuit are shown in **Figure 15.** The planter has two hydraulic circuits. Circuit one contains a bulk fill fan system, fertilizer system, and compressor system. Circuit two contains an alternator system, vacuum system, weight management cylinder system (WMC), and hydraulic down pressure cylinder system (HDPC). These functions, except the compressor system, are continuously working during normal planter working conditions. In addition to the motor actuators, the two-cylinder systems also appear in the system. The cylinder systems follow the terrain and maintain a commanded downward force to ensure planting quality.

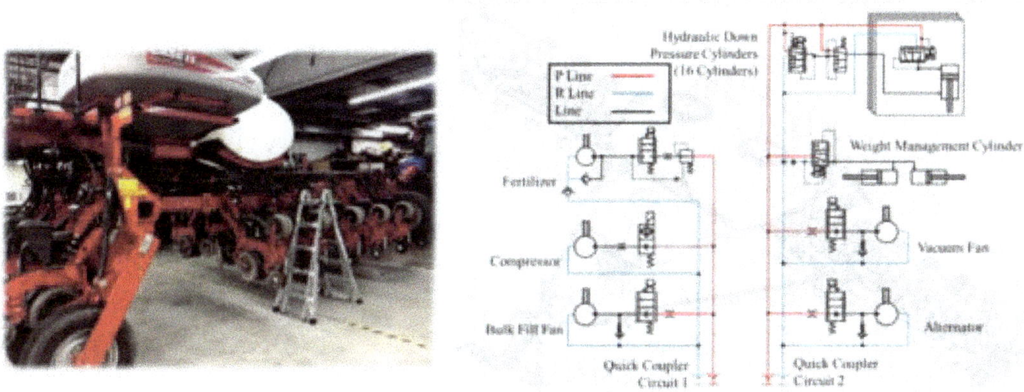

Figure 15. Reference planter and its hydraulic circuit.

5. System Modeling Methodology

This section presents the lumped parameter models used to estimate the effectiveness of the MPR solution in terms of controllability and energy savings. Specifically, the test rig model is used to validate the proposed MPR control performance. The tractor–planter model is instead used to predict the total power saving in realistic working conditions.

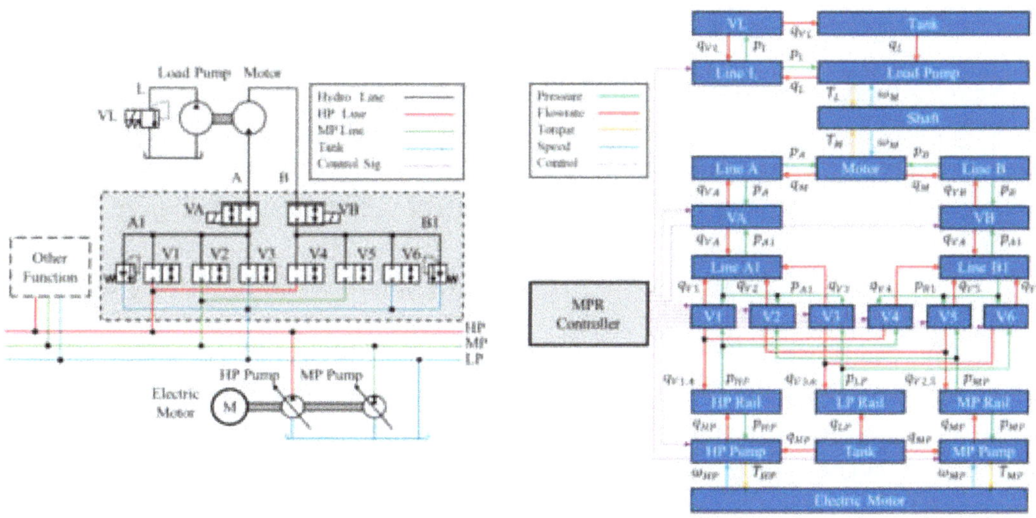

Figure 16. Block diagram for test rig modeling (left: test rig hydraulic circuit, right: modeling block diagram).

5.1. Test Rig Model

For easier understanding of the test rig modeling, the system is decomposed into individual components that could be modeled with a governing equation, such as hydraulic motor, pumps, valves, lines, etc. The exchange of information between the blocks is indicated in the block diagram in **Figure 16**. The inputs for the simulation are the actuator commanded speeds, commanded loads, and electric motor speed. They are directly fed into the MPR controller through the simulation. All other variables such as pressure, flow rate, torque, speed, and power are calculated internally in the simulation.

For simplicity, the electric motor in this model is assumed to be an ideal rotational speed source. The motor shaft speed ωm is simulated using Equation (9). $Ishaft$ is motor shaft inertia. TL is the load torque on the motor shaft. Tm are the torque outputs by the motor.

$$I_{shaft}\dot{\omega_m} = T_m - T_L \tag{9}$$

Following this, the hydraulic unit (HP and MP pumps, motor, and load pump), effective flowrate $qeff$, and torque $Teff$ are modeled using Equations (10) and (11). V is the unit maximum displacement. The volumetric loss qs and hydromechanical loss Ts are determined using empirical lookup tables. The losses are related to unit fractional displacement β, pressure drop across the unit Δp, and unit speed ω. A more detailed description on the setup of the hydraulic unit model can be found in [**30**].

$$q_{eff} = \pm\beta V\omega - q_s(\beta, \Delta p, \omega) \tag{10}$$

$$T_{eff} = \pm\frac{\beta V\Delta p}{2\pi} - T_s(\beta, \Delta p, \omega) \tag{11}$$

Next, the pressure dynamics in each control volume (HP and MP rail, line A1, B1, A, B, and L) are modeled by the pressure build up Equation (12), where the K is the bulk modulus, Vol is the volume, qin/out is the flowrate that flows into or out of the control volume, and the $dVdt$ is the control volume rate of change.

$$\dot{p} = \frac{K}{Vol}\left(q_{in} - q_{out} - \frac{dV}{dt}\right) \tag{12}$$

Finally, the flowrate qV through valves (V1–6, A, B, and L) are modeled using the orifice Equation (13). The maximum orifice area A for each valve is different and it is calculated from the valve datasheet. The normalized opening of the valve yV depends on the type of valve. For a proportional valve, the control input, which is normalized between 0 to 1, serves as the valve opening. For check valves and electrical on/off valves, the opening is either 0 or 1 depending on the working logic. For valves with hydraulic pilot pressure, a force balance between solenoid force, hydraulic force, and spring force is applied to calculate the valve opening.

$$q_V = \alpha_D A y_V \sqrt{\frac{2\Delta p}{\rho}} \quad (13)$$

A summary of the components and parameters used in this simulation model is shown in **Table 5** below.

Table 5. Components and parameters used for test rig modeling.

Oil Density	850 kg/m3 850 kg/m3
Discharge Coef.	0.7
Bulk Modulus	1.7e9 Pa 1.7e9 Pa
El. Motor Speed	1500 rpm 1500 rpm
Supply Pumps	Bosch Rexroth A10VO (EOC-P), max. displacement 45 cc/rev Bosch Rexroth A10VO (EOC-P), max. displacement 45 cc/rev
Load Pump	Bosch Rexroth AZPV, 8 cc/rev, custom made
Load Motor	Bosch Rexroth AZMV, 8 cc/rev, custom made
Load Inertia	0.001·kg·m2 0.001·kg·m2
On/Off Valve	Bosch Rexroth VEI-16-10A-NC (5 bar Δp Bosch Rexroth VEI-16-10A-NC (5 bar for 30 lpm q for 30 lpm q)
Proportional Valve	Bosch Rexroth KSVSR2 (14 bar Δp Bosch Rexroth KSVSR2 (14 bar for 75 lpm q for 75 lpm q)
Load Valve	Bosch Rexroth VEP-5B-2S-P (max. cracking pressure 220 bar Bosch Rexroth VEP-5B-2S-P (max. cracking pressure 220 bar, flow rate pressure gradient 7.5 L/min/bar flow rate pressure gradient 7.5 L/min/bar

5.2. Complete System Model

The complete tractor–planter model was built in the Simcenter Amesim environment. The model includes the tractor model and the planter model, in both the commercial solution and the proposed MPR component configuration. The simplified hydraulic circuit is shown in **Figure 17**. It is a 3-rail MPR system with 5 actuators. The 5 actuators are alternator motor, vacuum fan motor, bulk fill fan motor, fertilizer motor, and the cylinder systems.

The models for the tractor and planter were built following the same lumped parameter approach as the test rig model. Previous studies conducted by this team have presented the model development and its validation for the commercial tractor system [**31,32**] and tractor–implement system [**33**]. The baseline simulation results that will be presented in the next section are the straight output of these previously presented models. The relevant portions of the tractor model (engine, pumps, and control valves) and of the validated planter model (motors, cylinders, control valves, and corresponding load) will serve as the baseline for comparison with the MPR supply system and the MPR working actuator, respectively. The test rig model (HP and MP rails, PSCV, and MPR controllers) will serve as the MPR component models which transfer power in between the tractor and planter models.

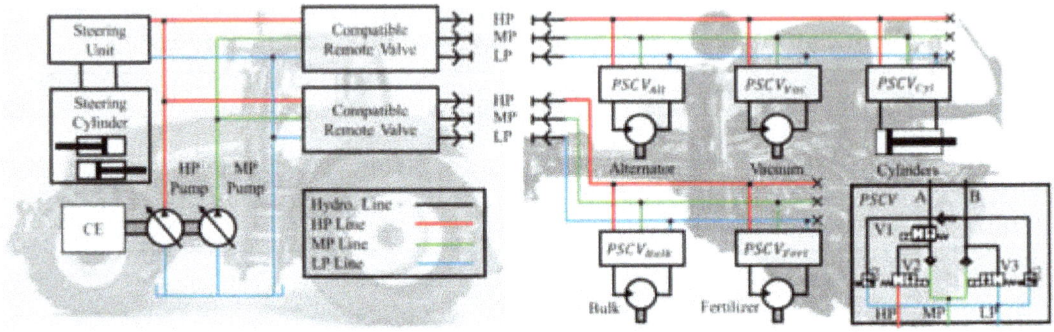

Figure 17. Simplified hydraulic circuit for the MPR system for the reference machines.

One important note regarding the model used is to consider the downforce cylinder system of the planter. An accurate modeling of the cylinder system presents complexity in determining the correct force inputs from the ground, as well as including a proper dynamic model for the bodies connected to the cylinder. To simplify this portion of the model, and ensure a realistic simulation, a simplified load module was built to recreate the cylinder flow/pressure dynamic using experimental data from real measurements as shown in **Figure 18.** This module will enforce the flow rate needed from the cylinder systems (which was measured from experiments) and use the same pilot-operated, electric-controlled, 3/2 proportional pressure-reducing valves to track the measured working pressure (also measured from experiments) in the cylinders (CH). Therefore, this load module consumes the flow rate and outputs the controlled cylinder pressure as measured from the experiments.

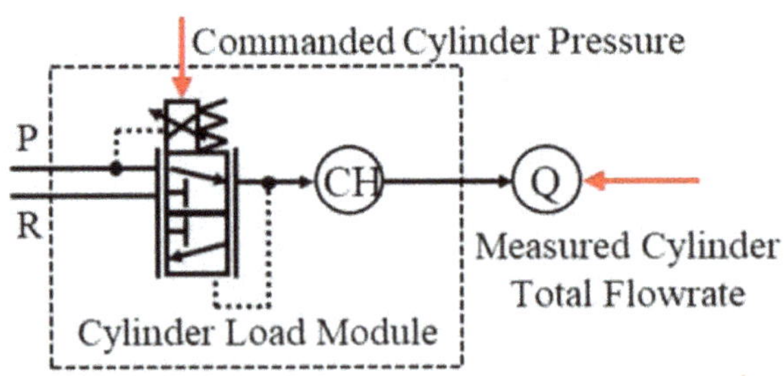

Figure 18. Cylinder systems model diagram.

6. RESULTS AND DISCUSSION

This section firstly presents the simulation used to select a three-rail system, then discusses the validation of the test rig model, the planter model, and the power consumption prediction of the complete MPR tractor–implement model compared to the baseline solution.

6.1. Rail Number Analysis

A representative drive cycle was acquired for the selected reference machine, seen in **Table 6**, which displays typical usage. It was run through a streamlined version of the model discussed above, with most transient dynamics removed to focus on steady state performance.

Table 6. Representative drive cycle for the rail number analysis.

Normalized Cmd	Time	Alternator	Bulk Fill Fan	Fertilizer	Vacuum Fan	Cylinders
Normal	5–38%	60%	60%	~40%	~63%	Dynamic

Averaged over time, the simulation results are shown in **Figure 19**, with each additional rail improving loss reduction; by far the most remarkable improvement occurred following the addition of two to three pressure rails. These results motivated the selection of a three-rail system for this application, as the relatively minor reduction in losses for rails above three was judged not to be worth the additional control complexity.

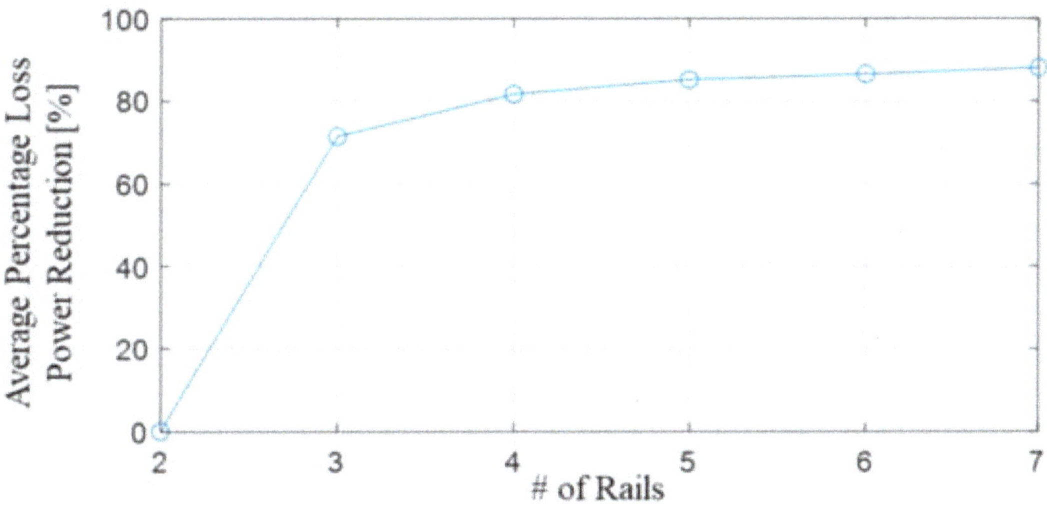

Figure 19. Loss reduction for number of rails.

6.2. Test Rig Model Validation Result

A drive cycle was generated for the test rig using the output of the supervisory controller to select loading conditions and command speeds for the three actuators to force the system into a sequence of mode switches. These switches were selected specifically to test the controllability and stability of the system. This drive cycle is outlined in **Table 7**.

Table 7. Validation drive cycle for MPR test rig.

Time	(A) 0–10 s		(B) 10–30 s		(C) 30–50 s		(D) 50–70 s	
	Command	Mode	Command	Mode	Command		Command	Mode
$\omega 1cmd$ [rpm]	1200	3	1200	3	1200	3	600	2
$p1L$ [bar]	160		160		160		20	
$\omega 2cmd$ [rpm]	800	1	800	1	800	3	800	1
$p2L$ [bar]	20		100		140		100	
$\omega 3cmd$ [rpm]	600	1	600	2	600	1	1200	3
$p3L$ [bar] bar	20		20		20		160	

The results from the test rig model are shown in **Figure 20**. The first row shows the switch from working condition B to C (started at 30 s) in the drive cycle of **Table 7**. The second row shows the last switch from working condition C to D (started at 50 s), which is the hardest switch. The plot on the left shows the actuators speed tracking, the middle plot shows the actuators load pressure and corresponding optimized rail pressures, and the right plot shows the actuators operating mode as determined by the supervisory controller.

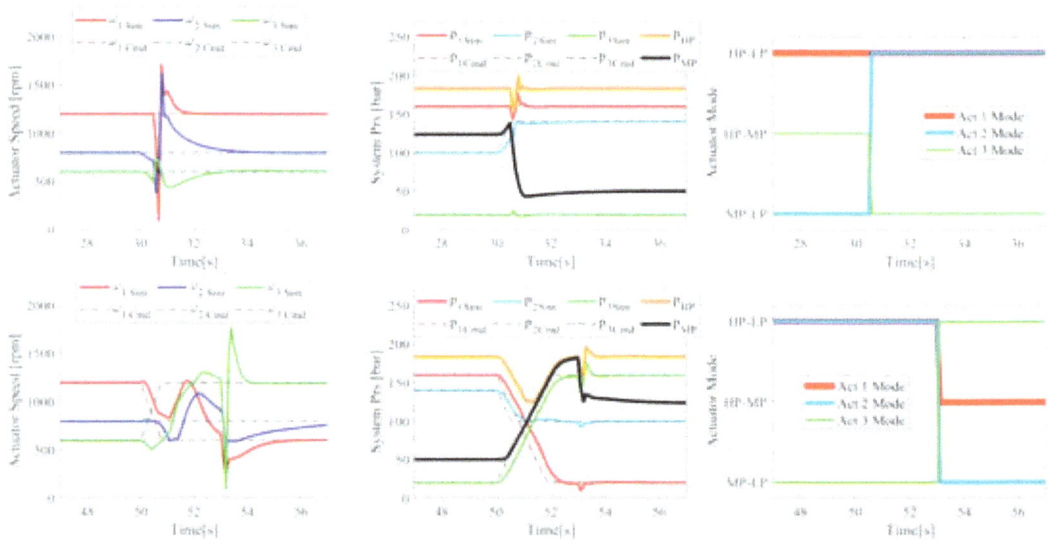

Figure 20. Test rig simulation results. Top row: mild switch; bottom row: hard switch.

Overall, the system behaves as expected and both the actuator level controller and supervisory controller perform correctly. The steady state command tracking is excellent, and the system is stable with a minimum number of mode switches. From the results, a speed peak for all actuators at the moment of the mode switch can be noticed. This occurs because when the switch happens, the local actuator speed controller, not being informed of the change in supply pressure, must react from its previous state. As a result, the sudden pressure change will create a flow spike and change the pressure drop across the motor. This aspect is accentuated in the test rig actuators because they have an extremely low mechanical inertia, being simply a pump directly connected to a hydraulic motor. Therefore, a short torque spike is reflected in a corresponding speed peak. The second reason is because the on/off valves have different opening and closing times. The on/off valves on the test rig have a closing time approximately three times longer than the opening time. When a switch happens, there will be a very short time during which the two rails are directly connected, causing the higher pressure rail flow to rush to the lower pressure rail, resulting in flow loss to the actuators and a subsequent drop followed by an overshoot in speed. A more sophisticated actuator-level control method could help alleviate some of these issues. The corresponding systems on the reference machine, however, are largely fan drives with very large inertias which will be much more robust against fluctuations in supply flow and pressure; as a result, the current level of control complexity is suitable for progressing to on-machine implementation.

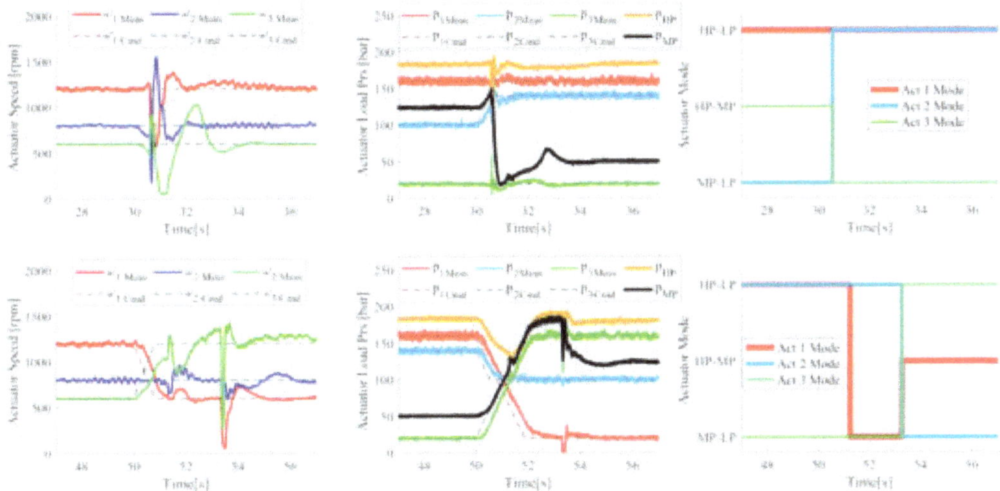

Figure 21. Test rig measurement results.

6.3. Complete System Performance and Power Consumption Prediction Result

This section presents the simulation result for the proposed MPR system applied to the tractor and the planter. All controllers are running at a fixed speed of 10 Hz. First, the drive cycle will be introduced. Then, system tracking and mode switching results will be presented to demonstrate system functionality. Finally, the pump shaft power consumption comparison will be introduced to show the power saving and efficiency gains.

The drive cycle includes two working conditions: normal and high-speed conditions. The normal and high-speed conditions are representative of two commonly used operating modes for this implement, according to the manufacturer, and are run back-to-back for the simulation. In reality, they are not run together, but this is a simplification only to show the operating conditions of both systems, and to provide a command switch to display transient behavior. According to the machine manufacturer, most users operate the planter in normal working conditions when planting. Therefore, power saving is most important in normal conditions. The drive cycle is shown in **Table 8**. For confidentiality, all parameters are normalized. This drive cycle was executed in the field during the baseline tests. The command for the alternator and the bulk fill fan is a constant motor speed. Fertilizer and vacuum fan systems do not have a direct motor speed command, so the measured flow rate of these systems is used to back-calculate the actual motor speed as the modeled system speed command. The cylinder system uses measured pressure values and flow rates from actual system operation for the load module to recreate the load condition, as discussed in **Section 5**.

Table 8. Representative drive cycle for the complete system model.

Normalized Cmd	Time	Alternator	Bulk Fill Fan	Fertilizer	Vacuum Fan	Cylinders
Normal	5–38%	60%	60%	~40%	~63%	Dynamic
High-Speed	60–90%	80%	80%	~60%	~70%	Dynamic

Figure 22 shows the results from the proposed three-rail MPR system for the reference machines. The vacuum and cylinder system has been chosen to show the system performance and controller activity. The left plot shows the results for the vacuum fan and the right plot shows the results for the cylinder systems from the same simulation. The top plot shows the actuator command tracking (left y-axis) and system mode (right y-axis), while the bottom plot shows the actuator load pressure with optimized rail pressures.

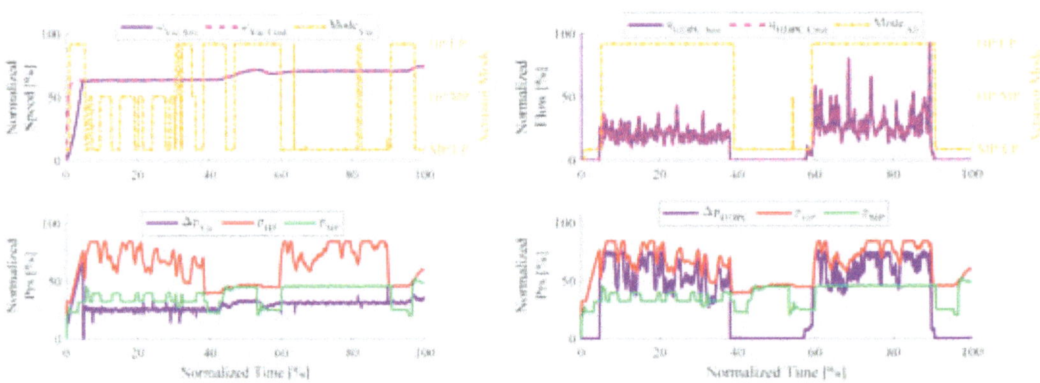

Figure 22. Complete system model result on MPR system performance.

From the results, all sub-systems track their command closely, as demonstrated by the vacuum. At mode switching, very little speed disturbance could be observed in the motor actuators. The tracking results prove that the proposed MPR system achieves at least the same level of performance as the baseline systems. In addition, from the bottom plot, the HP rail pressure closely follows the highest actuator pressure (which is the cylinder system pressure), the MP rail pressure is continuously adjusting with respect to the actuators load pressures, and the corresponding mode for each actuator is switching to ensure the power loss due to throttling is minimized. This shows that the supervisory controller works as expected to minimize system power loss.

In this drive cycle, when active, the cylinder system is always the highest load actuator, dominating the HP rail pressure. The high dynamic pressure variation leads to rapid mode switching in other actuators. This is more severe in normal working conditions. At high-speed working conditions, the cylinder system load is less dynamic, and the system becomes more stable as fewer switches occur. The designed controller is robust enough to ensure all systems stay stable and maintain good tracking, despite this fast dynamic.

Finally, **Figure 23** shows the pump shaft power consumption comparison between the baseline and the MPR system. There is no doubt that the MPR system generates great power savings compared to the baseline simulation. The MPR system demonstrates that some of the mode switches reduce the system power. This is due to the pressure transient in chambers A_1 and B (**Figure 6** right) at the moment of actuator mode switch. When it happens, for example when MP-LP switches to HP-LP, the pressure difference between the HP rail and the inlet chamber (which is currently at MP) creates a very short but large flow surge to charge the chamber pressure to the HP pressure. This flow peak, although it is short, shows its effect on the power consumption.

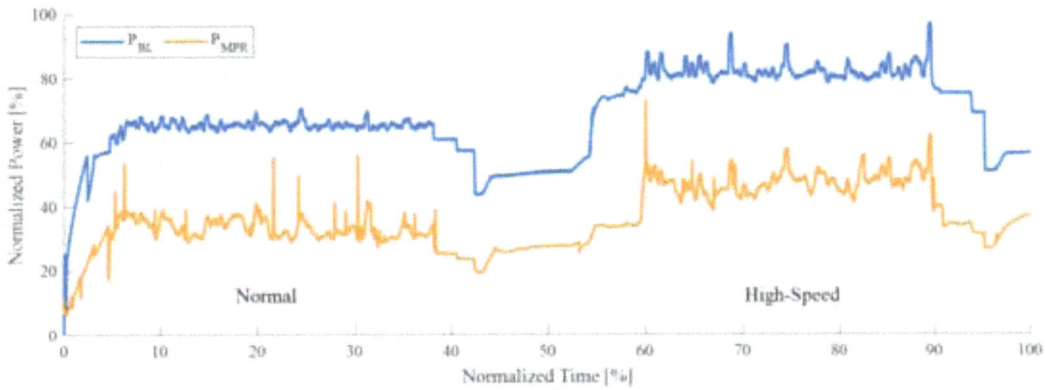

Figure 23. Complete system model result on power comparison between baseline, option one and option two.

Table 9 summarizes the normalized total power, the efficiency for each simulation, and the corresponding power reduction and system efficiency gain as a percentage. As stated, the MPR system results in great power saving and efficiency gain.

Table 9. Complete system model result on average power saving and efficiency gain for baseline and MPR system.

	Baseline		MPR System	
	Normal	High-Speed	Normal	High-Speed
Total Power [%]	65.41	81.90	34.46	47.27
Reduction [%]	-	-	47.32	42.28
Efficiency [%]	20.19	27.89	38.32	48.32
Efficiency Gain [%]	-	-	89.80	73.25

7. CONCISE SUMMARY

This study utilized the multi-pressure rail (MPR) hydraulic system in agricultural machinery like tractors and towed implements, which typically operate under prolonged steady-state circumstances with unidirectional hydraulic motors as the primary actuator. The system design and controller designs were offered with careful consideration of the application's characteristics. The suggested MPR system aims to reduce throttling loss at the actuator control valves by intelligently selecting pressures in the HP and MP rails. This system provides a superior option to the conventional centralized load sensing system for off-road agricultural machinery.

The study included a comprehensive MPR system design, covering the rail quantity, supply system, and pressure select and control valve set (PSCV). The ideal number of rails was initially established at three, considering the balance between power efficiency, system expenses, and intricacy for the selected device. Two pumps were used for the supply system to independently provide each train with water. This design eliminated the need for accumulators on the rails and facilitated precise rail pressure management to reduce throttling loss efficiently. Subsequently, two distinct PSCV designs were introduced. A multi-level controller was created to regulate the intended MPR system. The supervisory controller manages rail pressure and actuator operation mode to reduce throttling loss, while actuator controllers maintain subsystem stability and command tracking. A comprehensive design was provided. The suggested supervisory controller architecture aligns with the first sizing choice for the supply system, ensuring that neither pump flow saturates. The statement is accurate regarding the pump size and the flow needs of the reference cars. If the rails experience flow saturation due to insufficient pump flow, the supervisory controller's case elimination procedure and case selection logic may require modification, presenting a promising area for future improvement.

A test rig was constructed to assess the stability and command tracking capability of the proposed MPR system. Once the test rig was devised, a lumped parameter model was created. A drive cycle was then developed to test various crucial mode switchings and forecast the behavior of the test rig, assisting in controller tuning. The test rig was constructed and operated using the identical drive cycle to confirm the effectiveness of the controller design. The results indicated that the suggested MPR architecture and controller are stable and exhibit outstanding performance in both steady state and transient settings.

A high-fidelity lumped parameter model was developed for the full system, which includes the reference tractor, planter, and the proposed MPR system components, to forecast the power saving and efficiency gain of the proposed MPR system in comparison to the baseline system. The model was thoroughly validated, and a measured baseline drive cycle was utilized as a command input for each actuator in the entire system model. The results demonstrate that the suggested MPR system offers equivalent performance to the baseline system while also delivering significant power savings and efficiency improvements. The MPR system attained a maximum of 59.4% power reduction at the pump shaft and 89.8% increase in system efficiency under typical working conditions.

8. AT A GLANCE

This chapter offers a way to minimize energy dissipation in the hydraulic control system of agricultural tractors and their attachments. The approach, known as a multi-pressure rail (MPR), delivers power to hydraulic functions based on pressure control logic, rather than the conventional flow control logic found in hydraulic systems used in off-road vehicles. The suggested hydraulic control system eliminates redundant flow control valves in the current system, which result in unnecessary throttling losses and reduce overall energy efficiency. Previous research on MPR technology focuses on construction vehicles, aiming to enable energy recovery during overrunning loads and improve engine control. This research focuses on agricultural applications whose functions mostly run under resistive load situations with sluggish dynamics, providing a chance to solve throttle losses. The paper presents a variable pressure control approach to manage the instantaneous pressure at each rail. A stationary test rig is designed and utilized to validate a numerical simulation model of the MPR system and its control strategy for the development of both the controller and hydraulic system architecture. The system's dynamic behavior during function changes between different pressure levels is closely examined to minimize flow oscillations to each hydraulic function. The validated simulation model is utilized to forecast the energy conservation of the MPR solution in a real-world scenario including a 435 hp hydraulic tractor operating a 16-row planter. The authors have access to the operating characteristics throughout typical drive cycles. The data indicates a maximum of 59% decrease in power at the pump shaft, resulting in an 89.8% increase in system efficiency.

REFERENCES

1. Kean, A.J.; Sawyer, R.F.; Harley, R.A. A Fuel-Based Assessment of Off-Road Diesel Engine Emissions. J. Air Waste Manag. Assoc. **2000**, 50, 1929–1939. **[Google Scholar] [CrossRef] [PubMed]**
2. Love, L.J.; Lanke, E.; Alles, P. Estimating the Impact (Energy, Emissions and Economics) of the US Fluid Power Industry; Oak Ridge National Laboratory: Oak Ridge, TN, USA, 2012. **[Google Scholar]**
3. Vacca, A.; Franzoni, G. Hydraulic Fluid Power: Fundamentals, Applications, and Circuit Design; Wiley: Hoboken, NJ, USA, 2021. **[Google Scholar]**
4. Bedotti, A.; Campanini, F.; Pastori, M.; Riccò, L.; Casoli, P. Energy saving solutions for a hydraulic excavator. Energy Procedia **2017**, 126, 1099–1106. **[Google Scholar] [CrossRef]**
5. Tian, X.; Stump, P.; Vacca, A.; Fiorati, S.; Pintore, F. Power-Saving Solutions for Pre-Conpensated Load-Sensing Systems on Mobile Machines. Trans. ASABE **2021**, 64, 1435–1448. **[Google Scholar] [CrossRef]**
6. Siebert, J.; Wydra, M.; Geimer, M. Efficiency Improved Load Sensing System—Reduction of System Inherent Pressure Losses. Energies **2017**, 10, 941. **[Google Scholar] [CrossRef]**
7. Eriksson, B.; Palmberg, J.-O. Individual metering fluid power systems: Challenges and opportunities. Proc. Inst. Mech. Eng. Part I J. Syst. Control. Eng. **2011**, 225, 196–211. **[Google Scholar] [CrossRef]**
8. Lyu, Z.; Chen, Z.; Yao, B. Energy Saving Motion Control of Independent Metering Valves and Pump Combined Hydraulic System. IEEE ASME Trans. Mechatron. **2019**, 24, 1909–1920. **[Google Scholar] [CrossRef]**
9. Murrenhoff, H.; Sgro, S.; Vukovic, M. An overview of energy saving architectures for mobile applications. In Proceedings of the 9th International Fluid Power Conference, Aachen, Germany, 24–26 March 2014. **[Google Scholar]**
10. Williamson, C.; Zimmerman, J.; Ivantysynova, M. Efficiency Study of an Excavator Hydraulic System Based on Displacement-Controlled Actuators. In Proceedings of the ASME/BATH Symposium on Fluid Power and Motion Control, Bath, UK, 14–16 September 2008. **[Google Scholar]**

11. Zimmerman, J. Toward Optimal Multi-Actuator Displacement Controlled Mobile Hydraulic Systems, West Lafayette. Ph.D. Thesis, Purdue University, West Lafayette, IN, USA, 2012. **[Google Scholar]**

12. Daher, N.; Ivantysynova, M. Yaw stability control of articulated frame off-highway vehicles via displacement controlled steer-by-wire. Control Eng. Pract. **2015**, 45, 46–53. **[Google Scholar]** **[CrossRef]**

13. Zhang, S.; Li, S.; Minav, T. Control and Performance Analysis of Variable Speed Pump-Controlled Asymmetric Cylinder System uder Four-Quadrant Operation. Actuators **2020**, 9, 123. **[Google Scholar]** **[CrossRef]**

14. Fassbender, D.; Brach, C.; Minav, T. Using Displacement Control for Single Cylinders on an Electric Mobile Machine—Improved Efficiency Versus Increased Component Costs. In Proceedings of the 13th International Fluid Power Conference, 13. IFK, Aachen, Germany, 13–15 June 2022. **[Google Scholar]**

15. Busquets, E. Advanced Control Algorithms for Compact and Highly Efficient Displacement-Controlled Multi-Actuator and Hydraulic Hybrid Systems, West Lafayette. Ph.D. Thesis, Purdue University, West Lafayette, IN, USA, 2016. **[Google Scholar]**

16. Murrenhoff, H. Regelung von Verstellbaren Verdrängereinheiten am Konstant-Drucknetz; RWTH: Aschen, Germany, 1983. **[Google Scholar]**

17. Dreher, T. The Capability of Hydraulic Constant Pressure System with a Focus on Mobile Machines. In Proceeding of the 6th FPNI-PhD Symp., West Lafayette, IN, USA, 15–19 June 2010. **[Google Scholar]**

18. Heybroek, K.; Sahlman, M. A hydraulic hybrid excavator based on multi-chamber cylinders and secondary control—Design and experimental validation. Int. J. Fluid Power **2018**, 19, 91–105. **[Google Scholar]** **[CrossRef]**

19. Vael, G.E.M.; Achten, P.A.J.; Fu, Z. The Innas Hydraulic Transformer the Key to the Hydrostatic Common Pressure Rail; International: Warrendale, PA, USA, 2000. **[Google Scholar]** **[CrossRef]** **[Green Version]**

20. Shen, W.; Huang, H.; Pang, Y.; Su, X. Review of the Energy Saving Hydraulic System Based on Common Pressure Rail. IEEE Access **2017**, 5, 655–669. **[Google Scholar]** **[CrossRef]**

21. Lumkes, J.; Andruch, J. Hydraulic Circuit for Reconfigurable and Efficient Fluid Power Systems. In Proceedings of the 12th Scandinavian International Conference on Fluid Power, Tampere, Finland, 18–20 May 2011. **[Google Scholar]**

22. Dengler, P.; Groh, J.; Geimer, M. Valve control concepts in a constant pressure system with an intermediate pressure line. In Proceedings of the 21st International Conference on Hydraulics and Pneumatics, Ostrava, Czech Republic, 1–3 June 2011. [**Google Scholar**]

23. Dengler, P.; Geimer, M.; Baum, H.; Schuster, G.; Wessing, C. Efficiency improvement of a constant pressure system using an intermediate pressure line. In Proceedings of the 8th International Fluid Power Conference, Dresden, Germany, 26–28 March 2012. [**Google Scholar**]

24. Vukovic, M.; Leifeld, R.; Murrenhoff, H. STEAM—A Hydraulic Hybrid Architecture for Excavators. In Proceedings of the 10th International Fluid Power Conference, Dresden, Germany, 8–10 March 2016. [**Google Scholar**]

25. Vukovic, M.; Leifeld, R.; Murrenhoff, H. Reducing Fuel Consumption in Hydraulic Excavators—A Comprehensive Analysis. Energies **2017**, 10, 687. [**Google Scholar**] [**CrossRef**]

26. Siefert, J.; Li, P.Y. Optimal Control and Energy-Saving Analysis of Common Pressure Rail Architectures: HHEA and STEAM. In Proceedings of the BATH/ASME 2020 Symposium on Fluid Power and Motion Control, Virtual, 9–11 September 2020. [**Google Scholar**]

27. Opgenoorth, A.; Hass, C.; Frischkorn, K.; Schmitz, K. Hydraulische Mehrdrucksysteme für mobile Arbeitsmaschinen mit elektrischen Antrieben. In Proceedings of the Hybride und Energieeffiziente Antriebe für Mobile Arbeitsmaschinen: 8. Fachtagung, Karlsruhe, Germany, 23 February 2021. [**Google Scholar**]

28. Bertolin, M.; Vacca, A. A Parametric Study on Architectures Using Common-Pressure Rail Systems and Multi-Chamber Cylinders. In Proceedings of the IEEE Global Fluid Power Society PhD Symposium, Napoli, Italy, 12–14 October 2022. [**Google Scholar**]

29. Guo, X.; Madau, R.; Lengacher, J.J.; Vacca, A.; Cardoso, R. Multi-Pressure Rail System Design with Variable Pressure Control Strategy. In Proceedings of the 13th International Fluid Power Conference, 13. IFK, Aachen, Germany, 13–15 June 2022. [**Google Scholar**]

30. Guo, X.; Vacca, A. Advanced Design and Optimal Sizing of Hydrostatic Transmission Systems. Actuators **2021**, 10, 243. [**Google Scholar**] [**CrossRef**]

31. Tian, X.; Gomez, J.C.; Vacca, A.; Fiorati, S.; Pintore, F. Analysis of Power Distribution in the Hydraulic Remote System of Agricultural Tractors Through Modelling and Simulations. In Proceedings of the ASME/BATH 2019 Symposium on Fluid Power and Motion Control, Longboat Key, FL, USA, 7–9 October 2019. [**Google Scholar**]

32. Tian, X.; Vacca, A.; Fiorati, S.; Pintore, F. An Analysis of the Energy Consumption in the High-Pressure System of an Agricultural Tractor through Modeling and Experiment. In Proceedings of the 77th International Conference on Agricultural Engineering, Hannover, Germany, 9 November 2019; pp. 9–18. **[Google Scholar]** **[CrossRef]**

33. Tian, X.; Guo, X.; Stump, P.; Vacca, A.; Fiorati, S.; Pintore, F. New Hydraulic Control Technologies for Improving the Energy Efficiency of the Hydraulic System of Agricultural Tractors and Their Implements. In Proceedings of the IEEE Global Fluid Power Society PhD Symposium, Napoli, Italy, 12–14 October 2022. **[Google Scholar]**

08 RISK PRIORITIZATION CONCEPT IN FAILURE MODE AND EFFECTS ANALYSIS (FMEA) OF FLUID POWER COMPONENTS

1. INTRODUCTION: AN OVERVIEW

Although not the most recent advancement in engineering, fluid power systems are still extensively utilized in industrial drive systems. At times, their distinctive features make them irreplaceable by other systems. Extensive use, particularly in critical applications prioritizing safety, necessitates enhancing reliability. Fluid power component failures are intricate because of the interplay of high-pressure fluid, solid and chemical impurities, and structural parts. The reasons indicated and the movement of the related elements increase the probability of failure in the fluid power component. Research on the failures and reliability of fluid power systems or their components is conducted using a variety of instruments and approaches.

Reference [1] discusses the monitoring and modeling of gradual failure in a standard fluid power system. Y. Lee et al. [2] researched a hydraulic system malfunction that caused a fire on the wind turbine. References [3,4,5] discuss hydraulic pump failures. Studies are carried out on failures, their analysis, and their impact on system operation for fluid power components [6,7,8]. Watton J. compiled a detailed analysis of fluid power system failures, while another study focused on the reliability of mechanical elements, including components of the fluid power system, and determined the base failure rate.

Fluid power components' reliability can be enhanced by methods such as failure modes and effects analysis (FMEA), fault tree analysis (FTA), root cause analysis (RCA), and their variations. FMEA is a widely used tool for enhancing system quality, although its risk assessment component has been questioned and identified as a method weakness. The conventional FMEA employs a Risk Priority Number (RPN) calculated by multiplying Severity (S), Occurrence (O), and Detection (D). Equal weighting of those factors can lead to similar outcomes for various combinations and may cause confusion regarding safety.

Table 1. The AP classification [19].

S 9–10											S 7–8											S 4–6										
D\O	10	9	8	7	6	5	4	3	2	1	D\O	10	9	8	7	6	5	4	3	2	1	D\O	10	9	8	7	6	5	4	3	2	1
10	H	H	H	H	H	H	H	H	H	L	10	H	H	H	H	H	H	H	M	M	L	10	H	H	H	M	M	M	M	L	L	L
9	H	H	H	H	H	H	H	H	H	L	9	H	H	H	H	H	H	H	M	M	L	9	H	H	H	M	M	M	M	L	L	L
8	H	H	H	H	H	H	H	H	H	L	8	H	H	H	H	H	H	H	M	M	L	8	H	H	H	M	M	M	M	L	L	L
7	H	H	H	H	H	H	H	H	H	L	7	H	H	H	H	H	H	H	M	M	L	7	H	H	H	M	M	M	M	L	L	L
6	H	H	H	H	H	H	H	M	M	L	6	H	H	H	H	H	M	M	M	M	L	6	H	H	H	M	M	L	L	L	L	L
5	H	H	H	H	H	H	H	M	M	L	5	H	H	H	H	H	M	M	M	M	L	5	H	H	H	M	M	L	L	L	L	L
4	H	H	H	H	H	H	H	L	L	L	4	H	H	H	H	H	M	M	L	L	L	4	M	M	M	M	M	L	L	L	L	L
3	H	H	H	H	H	H	H	L	L	L	3	H	H	H	H	H	M	M	L	L	L	3	M	M	M	M	M	L	L	L	L	L
2	H	H	H	H	H	H	H	L	L	L	2	H	H	H	H	H	M	M	L	L	L	2	M	M	M	M	M	L	L	L	L	L
1	H	H	H	H	H	M	M	L	L	L	1	H	H	H	M	M	M	M	L	L	L	1	M	M	M	L	L	L	L	L	L	L
S 2–3											S 1																					
D\O	10	9	8	7	6	5	4	3	2	1	D\O	10	9	8	7	6	5	4	3	2	1											
10	M	M	M	L	L	L	L	L	L	L	10	L	L	L	L	L	L	L	L	L	L											
9	M	M	M	L	L	L	L	L	L	L	9	L	L	L	L	L	L	L	L	L	L											
8	M	M	M	L	L	L	L	L	L	L	8	L	L	L	L	L	L	L	L	L	L											
7	M	M	M	L	L	L	L	L	L	L	7	L	L	L	L	L	L	L	L	L	L											
6	M	M	M	L	L	L	L	L	L	L	6	L	L	L	L	L	L	L	L	L	L											
5	M	M	M	L	L	L	L	L	L	L	5	L	L	L	L	L	L	L	L	L	L											
4	L	L	L	L	L	L	L	L	L	L	4	L	L	L	L	L	L	L	L	L	L											
3	L	L	L	L	L	L	L	L	L	L	3	L	L	L	L	L	L	L	L	L	L											
2	L	L	L	L	L	L	L	L	L	L	2	L	L	L	L	L	L	L	L	L	L											
1	L	L	L	L	L	L	L	L	L	L	1	L	L	L	L	L	L	L	L	L	L											

The classic Failure Mode and Effects Analysis (FMEA) was officially established in industrial standards in the early 1960s. Subsequently, the industry identified shortcomings in the risk assessment process. The 2008 FMEA manual states that the RPN should no longer be considered a recommended approach for determining the requirement for action. Furthermore, it states that failure modes rated with severity numbers 9 or 10 require appropriate mitigation. The most recent update [19] included an action priority (AP) ranking table where the AP is determined by a combination of S, O, and D, with severity being the most influential factor. system 1 displays the AP classification system. The failure modes are categorized into three levels of risk assessment: high risk (H), medium risk (M), and low risk (L) for design or process FMEA.

This release appears to be an advancement from the previous one since it eliminates the subjective nature of risk evaluation. For severity values of 9-10, the risk is classified as medium for detection and occurrence levels of 5-6 and 2-3 respectively, or low for detection and occurrence levels of 1-3.

Additional limitations include the transformation of verbal concepts into numerical data and uncertainty associated with subjective opinions, expertise, and knowledge of specialists assessing systems. Various strategies and technologies are utilized to address the above listed flaws. Additional criteria or relevant weights can be incorporated into the typical risk evaluation. Another method involves including different viewpoints on the risk, like those of the consumer or maintenance.

Various techniques are employed to reduce the uncertainties in the FMEA process. The prevalent methodologies include the gray theory set [24], linguistic theory [25], fuzzy sets [26], and reasoning theory [27].

Liu et al. conducted a comprehensive literature analysis on the methods and tools utilized in Failure Mode and Effects Analysis (FMEA) for risk assessment and uncertainty reduction. Recent studies have concentrated on addressing the limitations of traditional Failure Mode and Effects Analysis (FMEA) by introducing new techniques or integrating existing ones. Reference [29] outlines an FMEA technique that categorizes failure modes by pairs of risk factors: S and O, S and D, and O and D. The risk factors pair findings were examined using grey relation analysis. Yu et al. [30] applied the cloud model theory to reduce linguistic uncertainties and utilized the VIKOR model to establish risk priority. The authors in reference [31] adapted the FMEA framework for IT based on recommendations from the literature review. Reference [32] applied the cloud model to enhance the FMEA. Yelda et al. [33] introduced a three-stage fuzzy risk assessment method using FMEA. The authors in reference [34] suggested a modified approach to the FMEA method by incorporating fuzzy rule base (FRB) and grey relations theory (GRT) to address the limitations of standard FMEA approaches. Uncertainty stemming from varying experts' knowledge backgrounds is discussed in reference [35], which utilizes prospect theory. Another method for transforming language phrases into numerical data is discussed in reference [36]. Fuzzy numbers and traditional measurement of alternatives are utilized in ranking through the Compromise Solution (MARCOS) approaches. Shi et al. [37] have suggested combining hesitant linguistic preference relations (HLPRs) and an extended dynamic consensus model in FMEA. Another way to modify FMEA is by considering the impact of failure modes and how these impacts are reduced in the system.

Scholars continue to focus on and address the unresolved limitations of FMEA. The risk assessment in FMEA remains challenging despite the variety of approaches and instruments available. Existing approaches in the literature are either overly complex or too computationally costly for practical implementation in industrial standards. The recent practical application of Failure Mode and Effects Analysis (FMEA) by AIAG and VDA helps eliminate subjective judgments in risk evaluation. The current risk ranking system may result in inaccurately assessed failure modes for safety. The ambiguity associated with the subjective nature of experts' expertise and background remains significant. We aimed to adapt FMEA to resemble traditional analysis while avoiding the need of intricate instruments and procedures, and at the same time reducing common uncertainties. Our method is based on the premise that every system is designed to carry out specific tasks, and any potential faults have the potential to disrupt this process. We classify failures of individual components into two categories: those that prevent the system from accomplishing a task and those that allow the main task to be done but with large or minor faults. We suggest evaluating the system's risk by considering the mix of failures and associated components in terms of its capacity to carry out the specified task(s). After categorizing the failures of associated components, they are ranked based on the product of occurrence (O) and detection (D) factors. The proposed approach defines the occurrence (O) as a function of subfactors that influence the probability of failure events, including the manufacturing process, duty time, and functions carried out by specific components. The detection (D) remains consistent with traditional FMEA.

This methodology builds upon our prior work [39], which introduced a similar method. This approach's originality lies in the risk categorization and the mechanism used to assess the occurrence (O) component. Unlike our prior study, where a base failure rate was utilized for assessment, in this method, it was determined according to the functions executed by components. We suggest using additional modification subfactors to differentiate failures caused by reasons other than regular functioning. The occurrence (O) and detection (D) variables hold equal importance. The suggested approach was utilized in the risk evaluation for the Failure Mode and Effects Analysis (FMEA) of the flow control valve.

2. METHODOLOGY

The methodology given is fundamentally akin to the conventional FMEA paradigm. Most steps are analogous. As indicated in our prior research [40], possible failures are common in fluid power components analyzed in the traditional FMEA.

2.1. Assumptions

This study aimed to qualitatively analyze failures and their consequences on components within fluid power systems. Complex failures may occur in systems, when an initial failure can lead to subsequent failures. Thus, we include just basic types of failure, and the study was conducted solely for primary mechanical failures. We presume that just a limited amount of crucial data is accessible for the component under investigation. Specific technical data and manufacturing/ assembly methods are not available.

Another assumption is that problems can be discovered without the need for an advanced diagnostic system or disassembly. Only fundamental measurements like pressure and flow rate are accessible.

2.2. Method

One step in the FMEA process involves conducting a functional study of the system to identify its fundamental components, potential failures, and underlying causes. Formal guidelines for this process are not clearly established and may vary based on the specific scenario. The proposed method operates under the assumption that every system is designed to accomplish one or more tasks. If a failure happens, the system may not be able to complete the planned task or may only partially do it with some reduction in performance.

We assume that system S consists of a set of components $ci \in C$, $i=\{1...n\}$, which can be connected to each other in any way (parallel, series, or series-parallel). Each component ci performs a specific function or several functions $fj \in F$,=$\{1,2\}$=1,2 in the system S. The set of component functions fi can be determined as follows:

- $f1$: main component functions;
- $f2$: auxiliary functions.

Each components ci may have a certain state determined by the failure $fa \in Fa$,$a=\{fc,fm,fn\}$, which can be determined as follows:

- fc: failures, which stop s component from executing its function;
- fm: failures, which allow a component to execute its function with a major degradation;
- fn: failures, which allow a component to execute its function with a minor degradation.

We introduce the failure importance factor fi as:

$$f_i = \begin{cases} 3, & \text{for } f_c \\ 2, & \text{for } f_m \\ 1, & \text{for } f_n \end{cases} \quad (1)$$

Among components ci, when creating a system we can distinguish those:

- That are essential to perform the intended system's task: main components ce;
- That are supplementing main components and ensure the proper system operation: major components cm;
- Whose failure has little effect on the main task of the system: minor components cn;
- Additional components whose failure does not affect the main task of the system ca.

We introduce the component importance factor cim as:

$$c_{im} = \begin{cases} 4, & \text{for } c_e \\ 3, & \text{for } c_m \\ 2, & \text{for } c_n \\ 1, & \text{for } c_a \end{cases} \quad (2)$$

Additionally, we assume that failures that may appear for some components may influence other components' failures. For that reason, we have implemented a component failure factor cin which can be described as:

$$c_{in} = \frac{k}{n} \quad (3)$$

where:

k is the number of components that a given component may influence;

n is the total number of components.

The system severity Se we can define as the ability to perform the intended task. We assume that risk assessment is evaluated primarily by severity Se, which can be expressed as:

$$S_e = (c_{im} + c_{in}) \cdot f_i \quad (4)$$

The severity Se is valued with numbers 1–15.

The occurrence O we define as an estimator of failure for corresponding components. We assumed that the main contributor to the likelihood of component failure is the number of functions (ff factor) the component realizes with modification factors λi:

$$O = \alpha\left(f_f + \sum \lambda_i\right) \quad (5)$$

where:

$$f_f = \frac{\sum f_1}{\sum f_1 + \sum f_2} \quad (6)$$

$$\lambda_i = \lambda_{mo} + \lambda_{ma} + \lambda_{dt} + \lambda_{ev} \quad (7)$$

The factors above arise from the analysis of fluid power components [**10**], which states that the reliability of fluid power components depends on:

- Relative motion λmo;
- Tolerances and surface quality which requires more sophisticated manufacturing processes λma;
- Duty time λdt;
- Environmental conditions λev.

Each modification factor λi has equal weight and may reach a maximal value equal to 0.25. The function factor ff maximal value is 4. The evaluation of all mentioned factors can be set arbitrarily and then renormalized to the required range.

Since the mentioned study defines failure rate, it was necessary to introduce a corrective factor α, which allows distinguishing failures which occur during regular operation or in other circumstances. The coefficient is valued in the following way:

- Normal operation: 2;
- Operation outside specified specification: 1.5;
- Maintenance: 1;
- Assembly: 0.5.

The detection D has been defined and evaluated in the following way:

- Detection of failure is almost certain: 1;
- Detection is very high: 2;
- Detection is high: 3;
- Detection is moderately high: 4;
- Detection is moderate: 5;
- Detection is low: 6;
- Detection is very low: 7;
- Detection is remote: 8;
- Detection is very remote: 9;
- Detection is nearly not possible: 10.

According to our primary hypothesis, the risk assessment is realized based on severity Se, which defines whether the system is able to perform intended tasks. The prioritization of failures inside task failures is performed by product occurrence O and detection D.

3. CASE STUDY

The investigated fluid power component is a two-way flow control valve presented in **Figure 1**. The valve's purpose is to maintain a constant flow rate regardless of the pressure difference between ports A and B. When the pressure line is connected to port A, the flow rate is controlled by the throttling gap between piston C21 and sleeve C3. The solenoid sets the gap by setting the position of the piston C21. The constant flow rate is provided by a pressure compensator (piston C22 and nozzle C4).

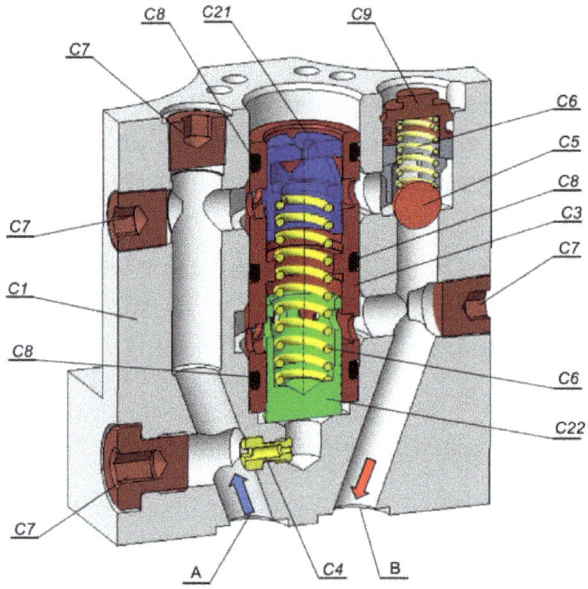

Figure 1. Flow control valve: 2-way flow control valve type UDRD6 by Ponar Wadowice: C1—body, C21—piston 1, C22—piston 2, C3—sleeve, C4—nozzle, C5—ball, C6—spring, C7—plug, C8—sealing ring, C9—check valve seat, A and B—valve port.

We can distinguish the following components whose failure will disturb the main task of the valve:

- Body ($c1$);
- Piston 1 ($c21$);
- Piston 2 ($c22$);
- Sleeve ($c3$);
- Nozzle ($c4$);
- Spring ($c6$);
- Plug ($c7$);
- sealing ring ($c8$).

The following components, when failures occur, may still complete the primary task but with some limitations:

- Ball ($c5$);
- Check valve seat ($c9$);
- Oil ($c10$), it should be understood as a component that transfers pressure energy.

Failures that leave the valve unable to perform the task are:

- Fracture ($fc1$);
- Yielding ($fc2$);
- Seizing ($fc3$);
- Aging ($fc4$);
- Hardening ($fc5$);
- Thread stripping ($fc6$);
- Clogging ($fc7$);
- Extreme contamination ($fc8$).

Failures that still allow the valve to perform the task but with a major degradation are:

- Erosion ($fm1$);
- Wear ($fm2$);
- Aeration ($fm3$);
- Contamination ($fm4$);
- Chemical action ($fm5$).

Failures that still allow the valve to perform the task but with a minor degradation are the same as the aforementioned. The factor which defines whether the failure belongs to major or minor degradation is its intensity. For simplicity, we will not consider failures with minor degradation. Functions that are performed by individual components are presented in **Table 2**.

Table 2. Functions performed by valve components.

Component ci	Primary Function $f1$	Secondary Function $f2$
body ($c1$	connecting with the system fixing piston/sleeve position seals flow management	fixing check valve mounting seat for spring seat for nozzle
piston 1 ($c21$)	flow rate setting sealing	-
piston 2 ($c22$)	flow rate control sealing	-

Component c_i	Primary Function $f1$	Secondary Function $f2$
sleeve ($c3$)	guiding pistons sealing	-
nozzle ($c4$)	pressure control	-
ball ($c5$)	-	closing/opening flow
spring ($c6$)	positioning	-
plug ($c7$)	closing flow ducts	-
sealing ring ($c8$)	sealing	-
check valve seat ($c9$)	-	guiding ball
	-	guiding spring
oil ($c10$)	energy transfer	greasing
	-	cooling

Task failures for the corresponding components are presented in **Table 3.**

Table 3. Task failures.

Valve Function	Component	Failure
The primary task is not realized	body $c1$	fracture $fc1$ yielding $fc2$ thread stripping $fc6$
	piston 1 $c21$	fracture $fc1$ yielding $fc2$ seizing $fc3$
	piston 2 $c22$	fracture $fc1$ yielding $fc2$ seizing $fc3$
	sleeve $c3$	fracture $fc1$ yielding $fc2$ seizing $fc3$
	nozzle $c4$	fracture $fc1$ clogging $fc7$
	spring $c6$	fracture $fc1$ yielding $fc2$
	plug $c7$	fracture $fc1$ thread stripping $fc6$
	sealing ring $c8$	fracture $fc1$ aging $fc4$
	oil $c10$	hardening $fc5$ extreme contamination $fc8$

Valve Function	Component	Failure
The major degradation of the primary task	body $c1$	erosion $fm1$
	piston 1 $c21$	wear $fm2$
	piston 2 $c22$	chemical action $fm5$
	sleeve $c3$	
	plug $c7$	
	ball $c5$	fracture $fc1$
		seizing $fc3$
		erosion $fm1$
		wear $fm2$
		chemical action $fm5$
	spring $c6$	chemical action $fm5$
	sealing ring $c8$	wear $fm2$
		chemical action $fm5$
	check valve seat $c9$	fracture $fc1$
		seizing $fc3$
		yielding $fc2$
		thread stripping $fc6$
		erosion $fm1$
		wear $fm2$
		chemical action $fm5$
	oil $c10$	aeration $fm3$
		contamination $fm4$
		chemical action $fm5$

4. RESULTS

Table 4 and **Table 5** present the risk assessment of failures and corresponding components. **Table 4** presents failures and corresponding component ratings when the primary task of the valve is not realized. The rank is created based on the severity number Se. For the failures with equal Se, the rank is evaluated based on a product of occurrence (O) and detection (D). **Table 5** presents failures and corresponding component ratings when the primary task of the valve is maintained with major deficiencies.

Table 4. The rank of failures when the primary task is not completed.

Rank		Severity					Occurrence						Detection	
		cim	cin	fi	Se	ff	λmo	λma	λdt	λev	α	O	D	$O{\cdot}D$
1	$c10fc8$	4	0.81818	3	14.4545	1.2	0	0	0	0.125	1.5	1.9875	4	7.95
2	$c1fc2$	4	0.54545	3	13.6363	4	0.0625	0.125	0.25	0.25	1.5	7.0313	6	42.1875
3	$c3fc3$	4	0.54545	3	13.6363	1.6	0.0625	0.25	0.25	0.0625	2	4.45	9	40.05
4	$c21fc1$	4	0.54545	3	13.6363	1.6	0.125	0.25	0.1875	0.0625	1.5	3.3375	9	30.0375
	$c21fc2$	4	0.54545	3	13.6363	1.6	0.125	0.25	0.1875	0.0625	1.5	3.3375	9	30.0375
	$c3fc1$	4	0.54545	3	13.6363	1.6	0.0625	0.25	0.25	0.0625	1.5	3.3375	9	30.0375
	$c3fc2$	4	0.54545	3	13.6363	1.6	0.0625	0.25	0.25	0.0625	1.5	3.3375	9	30.0375
5	$c1fc1$	4	0.54545	3	13.6363	4	0.0625	0.125	0.25	0.25	1.5	7.0313	3	21.0938
6	$c21fc3$	4	0.54545	3	13.6363	1.6	0.125	0.25	0.25	0.0625	2	4.575	6	27.45
7	$c1fc6$	4	0.54545	3	13.6363	4	0.0625	0.125	0.25	0.25	0.5	2.3438	2	4.6875
8	$c22fc1$	4	0.45454	3	13.3636	1.6	0.125	0.25	0.25	0.0625	1.5	3.4313	9	30.8813
9	$c22fc2$	4	0.45454	3	13.3636	1.6	0.125	0.25	0.25	0.0625	1.5	3.4313	9	30.8813
	$c22fc3$	4	0.45454	3	13.3636	1.6	0.125	0.25	0.25	0.0625	2	4.575	6	27.45
10	$c6fc2$	4	0.36363	3	13.0909	0.8	0.0625	0.0625	0.25	0.0625	1.5	1.8563	9	16.7063
11	$c6fc1$	4	0.36363	3	13.0909	0.8	0.0625	0.0625	0.25	0.0625	1.5	1.8563	7	12.9938
12	$c4fc7$	4	0.18181	3	12.5454	0.8	0.0625	0.125	0.1875	0.0625	2	2.475	9	22.275
13	$c4fc1$	4	0.18181	3	12.5454	0.8	0.0625	0.125	0.1875	0.0625	1.5	1.8563	6	11.138
14	$c8fc4$	4	0.09090	3	12.2727	0.8	0.0625	0.0625	0.25	0.0625	2	2.475	9	22.275
15	$c8fc5$	4	0.09090	3	12.2727	0.8	0.0625	0.0625	0.25	0.0625	2	2.475	9	22.275
	$c8fc1$	4	0.09090	3	12.2727	0.8	0.0625	0.0625	0.25	0.0625	2	2.475	6	14.85
16	$c7fc1$	4	0.09090	3	12.2727	0.8	0.0625	0.0625	0.0625	0.25	0.5	0.6183	1	0.6188
	$c7fc6$	4	0.09090	3	12.2727	0.8	0.0625	0.0625	0.0625	0.25	0.5	0.6183	1	0.6188

Table 5. The ranks of functional failures with major degradation.

Rank		Severity				Occurrence							Detection	
		cim	cin	fi	Se	ff	λmo	λma	λdt	λev	α	O	D	$O{\cdot}D$
1	$c8fm2$	4	0.0909	3	12.2727	0.8	0.0625	0.0625	0.25	0.0625	2	2.4750	10	24.75
	$c8fm5$	4	0.0909	3	12.2727	0.8	0.0625	0.0625	0.25	0.0625	2	2.4750	10	24.75
2	$c10fm3$	4	0.8182	2	9.6364	1.2	0.0625	0.0625	0.25	0.25	2	3.65	4	14.6
	$c10fm4$	4	0.8182	2	9.6364	1.2	0.0625	0.0625	0.25	0.25	2	3.65	4	14.6
	$c10fm5$	4	0.8182	2	9.6364	1.2	0.0625	0.0625	0.25	0.25	2	3.65	4	14.6
3	$c1fm1$	4	0.5455	2	9.0909	4	0.0625	0.1250	0.25	0.25	2	9.3750	10	93.75
	$c1fm2$	4	0.5455	2	9.0909	4	0.0625	0.1250	0.25	0.25	2	9.3750	10	93.75
	$c1fm5$	4	0.5455	2	9.0909	4	0.0625	0.1250	0.25	0.25	2	9.3750	10	93.75
4	$c21fm5$	4	0.5455	2	9.0909	1.6	0.1250	0.25	0.25	0.0625	2	4.5750	10	45.75
5	$c21fm1$	4	0.5455	2	9.0909	1.6	0.1250	0.25	0.1875	0.0625	2	4.45	10	44.5
	$c21fm2$	4	0.5455	2	9.0909	1.6	0.1250	0.25	0.1875	0.0625	2	4.45	10	44.5
	$c3fm1$	4	0.5455	2	9.0909	1.6	0.0625	0.25	0.25	0.0625	2	4.45	10	44.5
	$c3fm2$	4	0.5455	2	9.0909	1.6	0.0625	0.25	0.25	0.0625	2	4.45	10	44.5
	$c3fm5$	4	0.5455	2	9.0909	1.6	0.0625	0.25	0.25	0.0625	2	4.45	10	44.5
6	$c22fm1$	4	0.4545	2	8.9091	1.6	0.1250	0.25	0.25	0.0625	2	4.575	10	45.75
	$c22fm2$	4	0.4545	2	8.9091	1.6	0.1250	0.25	0.25	0.0625	2	4.575	10	45.75
	$c22fm5$	4	0.4545	2	8.9091	1.6	0.1250	0.25	0.25	0.0625	2	4.575	10	45.75
7	$c6fm5$	4	0.3636	2	8.7273	0.8	0.0625	0.0625	0.125	0.0625	2	2.225	10	22.25
8	$c4fm1$	4	0.1818	2	8.3636	0.8	0.0625	0.125	0.1875	0.0625	2	2.475	10	24.75
	$c4fm2$	4	0.1818	2	8.3636	0.8	0.0625	0.125	0.1875	0.0625	2	2.475	10	24.75
	$c4fm5$	4	0.1818	2	8.3636	0.8	0.0625	0.125	0.1875	0.0625	2	2.475	10	24.75
9	$c9fc1$	2	0.2727	3	6.8182	0.4	0.0625	0.125	0.125	0.0625	1.5	1.1625	10	11.625

Rank			Severity						Occurrence				Detection	
	$c9fc2$	2	0.2727	3	6.8182	0.4	0.0625	0.125	0.125	0.0625	1.5	1.1625	10	11.625
10	$c9fc6$	2	0.2727	3	6.8182	0.4	0.0625	0.125	0.125	0.0625	0.5	0.3875	10	3.875
11	$c5fc3$	2	0.1818	3	6.5455	0.8	0.1250	0.125	0.125	0.0625	2	2.475	10	24.75
12	$c5fc1$	2	0.1818	3	6.5455	0.8	0.1250	0.125	0.125	0.0625	1.5	1.8563	10	18.5625
13	$c7fm1$	3	0.0909	2	6.1818	0.8	0.0625	0.0625	0.0625	0.25	2	2.475	10	24.75
	$c7fm2$	3	0.0909	2	6.1818	0.8	0.0625	0.0625	0.0625	0.25	2	2.475	10	24.75
	$c7fm5$	3	0.0909	2	6.1818	0.8	0.0625	0.0625	0.0625	0.0625	2	2.1	10	21
14	$c9fm1$	2	0.2727	2	4.5455	0.4	0.0625	0.125	0.125	0.0625	2	1.55	10	15.5
	$c9fm2$	2	0.2727	2	4.5455	0.4	0.0625	0.125	0.125	0.0625	2	1.55	10	15.5
	$c9fm5$	2	0.2727	2	4.5455	0.4	0.0625	0.125	0.125	0.0625	2	1.55	10	15.5
15	$c5fm1$	2	0.1818	2	4.3636	0.8	0.1250	0.125	0.125	0.0625	2	2.475	10	24.75
	$c5fm2$	2	0.1818	2	4.3636	0.8	0.1250	0.125	0.125	0.0625	2	2.475	10	24.75
	$c5fm5$	2	0.1818	2	4.3636	0.8	0.1250	0.125	0.125	0.0625	2	2.475	10	24.75
	$c5fm5$	2	0.1818	2	4.3636	0.8	0.1250	0.125	0.125	0.0625	0.5	0.6188	10	6.1875

The results of failures with minor degradation are essentially the same as those presented in **Table 4**. The only difference is failure intensity.

5. DISCUSSION

The ranking mechanism in this study enabled the identification of potential faults that could render the valve component incapable of performing its intended function. Some failures within the assumed categories have equal severity values (Se). Introducing the product of occurrence (O) and detection (D) as an extra evaluation criterion did not enable differentiation between their rankings. It suggests that an additional element not covered in this study would need to be utilized. One possible remedy could involve adjusting the detection factor (D), which has not been altered, while challenges with uncertainties linked to converting linguistic data into quantitative data persist at a rather high level. This method utilizes factors derived from failure history and study on things that are similar but not identical. It is most appropriate for usage in the design phase or the early stages of failure analysis. Furthermore, it can be integrated with other tools and techniques like root cause analysis (RCA) or fault tree analysis (FTA). The grade for failures with mild degradation of the valve's intended function is equivalent to that for substantial degradation. The only variation lies in the degree of failure.

As the majority of failures in **Table 5** are caused during the normal operational condition, it is certain the degradation of the valve's main task will depend on the intensity of wear or erosion of valve components. Another feature that can be observed is the source of the failures which stop the valve from performing its task. All of those failures are mainly caused during assembling, maintenance work, or operation outside specifications. Analogously, those failures which cause major or minor degradation of the valve are mainly produced during normal operation.

The presented method can be easily extended to any fluid power components, however, we are unable to perform FMEA for electrical/electronic components. It can be used only for purely mechanical or hydromechanical components. Proposed subfactors in occurrence (O) were defined only as typical for mechanical or hydromechanical failures. The severity values *Se* can be set arbitrary with any values; the proposed methods only prioritize failures for related components inside categorized failures on those which stop the valve from performing its task and those which cause a major or minor malfunction.

Comparison with the Traditional FMEA Risk Assessment

The **Table 6** presents a risk analysis using the traditional approach for selected failures and related components by invited experts. The risk has been expressed as the RPN (product of severity (S), occurrence (O), and detection (D)) and by using action priority (AP).

Table 6. The RPN and AP.

Failure/Component	S	O	D	RPN	AP
$c1fc1$	10	3	3	90	L
$c7fc6$	10	1	1	10	L
$c1fc2$	10	3	6	180	H
$c211fc1$	10	3	9	270	H
$c21fc2$	10	3	9	270	H
$c21fc3$	10	5	6	300	H
$c22fc1$	10	3	9	270	H
$c22fc2$	10	3	9	270	H
$c22fc3$	10	5	6	300	H
$c21fm1$	8	7	10	560	H
$c21fm2$	8	7	10	560	H
$c22fm1$	8	7	10	560	H
$c22fm2$	8	7	10	560	H

According to the table, the highest RPN value was reached for failure wear $fm2$ and erosion $fm1$ for the components pistons 1 $c211$ and piston 2 $c22$. For both, the action priority (AP) is high (H). This is overestimating those failures, which can cause a major or minor malfunction in the valve. In contrast, the failure fracture $fc11$ for the body $c1$ and thread stripping $fc6$ for plug $c7$ are underestimated. Their RPN values are lower than other failures, and action priority (AP) is low (L). Those failures, when they occur, stop the valve's operation.

6. CONCISE SUMMARY

Although FMEA and its new standards have been continuously enhanced, certain issues persist without resolution. The risk assessment concept in this research, along with the evaluation of potential flaws' likelihood, helps reduce ambiguity in the collected data. The suggested method classifies failures and important components based on their impact on the ability to complete the desired task. The strategy outlined in this study is versatile and applicable to all components of the fluid power system. The proposed method can detect faults that may lead to critical system failure, which could be overlooked by both the old FMEA method and the most recent one used in the industry.

The proposed methods do not necessitate computationally intensive tools or simulation techniques. Essentially, it closely resembles the conventional FMEA process. The proposed method can be customized to meet specific demands and assigned variable weights due to the failures and component categorization. The method demonstrated its excellence compared to the recent AIAG and VDA FMEA standard and the conventional FMEA framework. Some failures on primary valve components may not be identified as critical to the system. An expansion of the method proposed could involve redefining the detection factor to reduce the remaining high uncertainty and incorporating new risk factors like manufacturing or maintenance costs.

7. AT A GLANCE

FMEA is a widely used tool for decades and is also used as an industrial standard. However, there are two main drawbacks of this analysis that have been specified from the beginning. The first one is risk prioritization, which is expressed by a risk priority number (RPN). The RPN is a product of three factors with equal weight: severity (S), occurrence (O), and detection (D), which may produce equal risk priority for different combinations of S, O, and D. The second is the uncertainties caused by converting linguistic terms into quantitative data. The essential data used in the FMEA strongly depend on subjective experts' opinions, knowledge, and experience. For decades, various attempts of overcoming these weaknesses have been made, not only by academics but also by industry. The Automotive Industry Action Group (AIAG) and Verband der Automobilindustrie (VDA) have created an FMEA handbook that defines action priority (AP) depending on the combination of severity, occurrence, and detection numbers. This study presents an alternative to risk prioritization in FMEA based on failures of the tasks which analyzed systems perform. The fundamental factors S, O, and D have been redefined in a way to minimize uncertainties. The proposed method has been implemented in the flow control valve and can be easily applied in mechanical engineering applications.

REFERENCES

1. Zhao, X.; Zhou, C.; Zhao, L.; Zhang, S.; Wen, P.; Jiang, J. Condition evaluation model of fluid power system in gradual failure based on data envelopment analysis. Comput. Fluids **2015**, 110, 219–226. **[Google Scholar] [CrossRef]**
2. Lee, Y.; Lee, G.; Yang, J.; Park, J.; Baek, D. Failure analysis of a hydraulic power system in the wind turbine. Eng. Fail. Anal. **2020**, 107, 104218. **[Google Scholar] [CrossRef]**
3. Ma, Z.; Wang, S.; Shi, J.; Li, T.; Wang, X. Fault diagnosis of an intelligent hydraulic pump based on a nonlinear unknown input observer. Chin. J. Aeronaut. **2018**, 31, 385–394. **[Google Scholar] [CrossRef]**
4. Li, T.; Wang, S.; Zio, E.; Shi, J.; Ma, Z. A numerical approach for predicting the remaining useful life of an aviation hydraulic pump based on monitoring abrasive debris generation. Mech. Syst. Signal Process. **2018**, 136, 106519. **[Google Scholar] [CrossRef]**

5. Hast, D.; Findeisen, R.; Streif, S. Detection and isolation of parametric faults in hydraulic pumps using a set-based approach and quantitative–qualitative fault specifications. Control Eng. Pract. **2015**, 40, 61–70. [**Google Scholar**] [**CrossRef**]
6. Gianni, N.; Tito, M. Failure of a heavy-duty hydraulic cylinder and its fatigue re-design. Eng. Fail. Anal. **2011**, 18, 1030–1036. [**Google Scholar**]
7. Roquet, P.; Gamez-Montero, P.J.; Castilla, R.; Raush, G.; Codina, E. A Simplified Methodology to Evaluate the Design Specifications of Hydraulic Components. Appl. Sci. **2018**, 8, 1612. [**Google Scholar**] [**CrossRef**]
8. Guo, Y.; Xiong, G.; Zeng, L.; Li, Q. Modeling and Predictive Analysis of Small Internal Leakage of Hydraulic Cylinder Based on Neural Network. Energies **2021**, 14, 2456. [**Google Scholar**] [**CrossRef**]
9. Watton, J. Modelling, Monitoring and Diagnostic Techniques for Fluid Power Systems; Springer: London, UK, 2007. [**Google Scholar**]
10. Handbook of Reliability Prediction for Mechanical Equipment; Naval Surface Warfare Center: West Bethesda, MD, USA, 2011.
11. Stirrup, T.S.; Chamberlin, H.C.; Curran, K.L.F. Scaled Wind Farm Technology Hydraulic System Failure Modes and Effects Analysis; Sandia National Laboratories: Albuquerque, NM, USA; Livermore, CA, USA, 2015. [**Google Scholar**]
12. Fabis-Domagala, J.; Domagala, M.; Momeni, H. A Matrix FMEA Analysis of Variable Delivery Vane Pumps. Energies **2021**, 14, 1741. [**Google Scholar**] [**CrossRef**]
13. Zhang, F.; Cheng, L.; Gao, Y.; Xu, X.; Wang, Y. Fault tree analysis of a hydraulic system based on the interval model using latin hypercube sampling. J. Intell. Fuzzy Syst. **2019**, 37, 8345–8355. [**Google Scholar**] [**CrossRef**]
14. Hemati, A.; Shooshtari, A.A. Gear Pump Root Cause Failure Analysis Using Vibrations Analysis and Signal Processing. J. Fail. Anal. Preven. **2020**, 20, 1815–1818. [**Google Scholar**] [**CrossRef**]
15. Liu, H.C.; Liu, L.; Liu, N. Risk evaluation approaches in failure mode and effects analysis: A literature review. Expert Syst. Appl. **2013**, 40, 828–838. [**Google Scholar**] [**CrossRef**]
16. MIL-P 1629; USA Military Standard, Procedure for Performing a Failure Mode, Effects and Criticality Analysis (MIL-P 1629, USA). Military Specifications and Standards: Washington, DC, USA, 1949.
17. J1739 200901; Potential Failure Mode and Effects Analysis in Design (Design FMEA), Potential Failure Mode and Effects Analysis in Manufacturing and Assembly Processes (Process FMEA). SAE International: Warrendale, PA, USA, 2009.

18. AIAG. Potential Failure Mode and Effect Analysis (FMAE), 4th ed.; AIAG: Automotive Industry Action Group, 2008; Available online: **https://www.aiag.org/store/publications/details?ProductCode=FMEA-4** (accessed on 20 July 2022).

19. AIAG & VDA FMEA Handbook. AIAG: Automotive Industry Action Group. 2019. Available online: **https://www.aiag.org/store/publications/details?ProductCode=FMEAAV-1** (accessed on 20 July 2022).

20. Anes, V.; Henriques, E.; Freitas, M.; Reis, L. A new risk prioritization model for failure mode and effects analysis. Qual. Reliab.Eng. Int. **2018**, 34, 516–528. **[Google Scholar] [CrossRef]**

21. Tang, Y.; Zhou, D.; Chan, F.T.S. AMWRPN: Ambiguity Measure Weighted Risk Priority Number Model for Failure Mode and Effects Analysis. IEEE Access **2018**, 6, 27103–27110. **[Google Scholar] [CrossRef]**

22. Koomsap, P.; Charoenchokdilok, T. Improving risk assessment for customer-oriented FMEA. Total Qual. Manag. Bus. Excell. **2018**, 29, 1563–1579. **[Google Scholar] [CrossRef]**

23. Pancholi, N.; Bhatt, M. FMECA-based maintenance planning through COPRAS-G and PSI. J. Qual. Maint. Eng. **2018**, 24, 224–243. **[Google Scholar] [CrossRef]**

24. Li, Z.; Chen, L. A novel evidential FMEA method by integrating fuzzy belief structure and grey relational projection method. Eng. Appl. Artif. Intell. **2019**, 77, 136–147. **[Google Scholar] [CrossRef]**

25. Shi, S.; Fei, H.; Xu, X. Application of a FMEA method combining interval 2-tuple linguistic variables and grey relational analysis in preoperative medical service process. IFAC-PapersOnLine **2019**, 52, 1242–1247. **[Google Scholar] [CrossRef]**

26. Filo, G.; Fabis-Domagala, J.; Domagala, M.; Lisowski, E.; Momeni, H. The idea of fuzzy logic usage in a sheet-based FMEA analysis of mechanical systems. MATEC Web Conf. **2018**, 183, 03009. **[Google Scholar] [CrossRef]**

27. Qin, J.; Xi, Y.; Pedrycz, W. Failure mode and effects analysis (FMEA) for risk assessment based on interval type-2 fuzzy evidential reasoning method. Appl. Soft Comput. **2020**, 89, 10613. **[Google Scholar] [CrossRef]**

28. Liu, H.C.; Chen, X.; Dua, C.; Wang, Y. Failure mode and effect analysis using multi-criteria decision making methods: A systematic literature review. Comput. Ind. Eng. **2019**, 135, 881–897. **[Google Scholar] [CrossRef]**

29. Ouyang, L.; Che, Y.; Yan, L.; Park, C. Multiple perspectives on analyzing risk factors in FMEA. Comput. Ind. **2022**, 141, 103712. **[Google Scholar] [CrossRef]**

30. Yu, J.; Wu, S.; Chen, H.; Yu, Y.; Fan, H.; Liu, J. Risk assessment of submarine pipelines using modified FMEA approach based on cloud model and extended VIKOR method. Process. Saf. Environ. Prot. **2021**, 155, 555–574. **[Google Scholar]**

31. Apol Pribadi, S.; Nina Fadilah, N. The consistency analysis of failure mode and effect analysis (FMEA) in information technology risk assessment. Heliyon **2020**, 6, e03161. **[Google Scholar]** **[CrossRef]**

32. Lipeng, W.; Fang, Y.; Fang, W.; Zijun, L. FMEA-CM based quantitative risk assessment for process industries—A case study of coal-to-methanol plant in China. Process. Saf. Environ. Prot. **2021**, 149, 299–311. **[Google Scholar]** **[CrossRef]**

33. Yelda, Y.; Gülin Feryal, C. A FMEA based novel intuitionistic fuzzy approach proposal: Intuitionistic fuzzy advance MCDM and mathematical modeling integration. Expert Syst. Appl. **2021**, 183, 115413. **[Google Scholar]** **[CrossRef]**

34. Shamsu, H.; Jin, W.; Christos, K.; Musa, B. Modified FMEA hazard identification for cross-country petroleum pipeline using Fuzzy Rule Base and approximate reasoning. J. Loss Prev. Process. Ind. **2022**, 74, 104616. **[Google Scholar]** **[CrossRef]**

35. Yan, Z.; Chuanhao, F.; Hengjie, Z. From diversity to consensus: Impacts of opinion evolution and psychological behaviours in failure mode and effect analysis. Appl. Soft Comput. **2022**, 128, 109399. **[Google Scholar]** **[CrossRef]**

36. Soumava, B.; Shankar, C. Failure analysis of CNC machines due to human errors: An integrated IT2F-MCDM-based FMEA approach. Eng. Fail. Anal. **2021**, 130, 105768. **[Google Scholar]** **[CrossRef]**

37. Shi, H.; Liu, Z.; Liu, H.C. A new linguistic preference relation-based approach for failure mode and effect analysis with dynamic consensus reaching process. Inf. Sci. **2022**, 977–993. **[Google Scholar]** **[CrossRef]**

38. Qun, W.; Guozhu, J.; Yuning, J.; Wenyan, S. A new approach for risk assessment of failure modes considering risk interaction and propagation effects. Reliab. Eng. Syst. Saf. **2021**, 216, 108044. **[Google Scholar]** **[CrossRef]**

39. Fabis-Domagala, J.; Domagala, M.; Momeni, H. A Concept of Risk Prioritization in FMEA Analysis for Fluid Power Systems. Energies **2021**, 14, 6482. **[Google Scholar]** **[CrossRef]**

40. Fabis-Domagala, J.; Momeni, H.; Filo, G.; Bikass, S.; Lempa, P. Matrix FMEA Analysis of the flow conrol valve. QPI **2019**, 1, 590–595. **[Google Scholar]**

09 UTILIZING SUPERVISED LEARNING AND NEURAL NETWORKS FOR DATA-DRIVEN CONDITION MONITORING OF A HYDRAULIC PRESS

1. INTRODUCTION: AN OVERVIEW

Throughout the operational lifespan of an industrial equipment, the deterioration of several components within the system is unavoidable. Machine elements deteriorate over time, particularly in applications that entail transmitting high power in harsh circumstances, such as those found in hydraulics. Maintenance actions are necessary to preserve good machine availability and process quality. Planning maintenance actions can be tough for machine operators as they need to balance between early downtimes from maintenance work and downtimes from lower process quality or machine damage. Condition monitoring technologies can be used to efficiently schedule maintenance periods and identify potential breakdowns early. The goal of condition monitoring is to assess the present health status of a machine by automatically analyzing operational data collected from it [1].

Various methods for condition monitoring have been studied in the realm of hydraulics. The literature explores solutions that vary from rule-based systems to model-based approaches to completely data-driven approaches [2,3]. The current integration of measurement technology and information technology in industrial products has shifted focus to data-driven techniques. These techniques, commonly referred to as machine learning, seek to derive a model from data using algorithms to link patterns in recorded data with machine health states. In supervised learning approaches, the machine's health status can be identified by processing observed sensor and control signal sets algorithmically. Furthermore, using the data labeling, it is possible to trace the source of abnormal machine behavior back to problems in the system components.

Challenges persist in implementing data-driven methods for condition monitoring. A significant problem is the absence of techniques for the broad and universal creation of data-driven solutions. Another obstacle is the need for extensive data. To create strong models, the training data should encompass a wide variety of potential operating scenarios and equipment setups. Machine learning-based systems are considered black-box models, making it difficult for human inspectors to understand the decision-making process of these algorithms.

This work introduces a mechanism for creating data-driven condition monitoring systems for industrial hydraulics. A feature-based supervised learning approach is utilized to extract features from multivariate, cyclic, and non-stationary time series data. Furthermore, the approach involves a step for selecting features, aiding in the identification of pertinent data attributes and providing an understanding of how errors impact observable system metrics. A flexible method is recommended for setting up and training classification models, enabling the modular configuration of models for specific machine and component issues. Research is conducted on a simulation model of a hydraulic press test bench, ensuring that data availability does not hinder the investigation.

2. RELATED WORK AND STATE OF THE ART

This section provides a summary of past research on condition monitoring in hydraulic applications. The text discusses the general notion of neural networks in relation to important design factors for their configuration and application in a fault classification solution.

2.1. Condition Monitoring in Hydraulics

In general, applying machine learning for the condition monitoring of hydraulic applications has proven to yield satisfactory results [**2**,**7**,**8**]. Most publications in this field focus on the use of data-driven techniques for monitoring single hydraulic components such as pumps and valves [**9**,**10**,**11**,**12**]. The algorithms mainly used in these studies are based on variations of neural networks, which are used to process raw time series for the inference of the health states of the components. In [**13**], features are extracted from the raw time series in order to detect anomalies in the operation of a speed-variable pump by the use of methods of unsupervised learning. A similar feature-based approach is pursued in [**14**] in the framework of a systematic investigation of techniques of supervised learning for monitoring an axial piston pump. As typically only a few sensors are applied at the component level, data are scarce in these use cases, such that mostly only one single sensor recording can be considered. On the contrary, applying condition monitoring on a system level enables to include further sensor data available in the system. Hence, a component fault cannot only be inferred directly from data acquired on that component, but can also be inferred from indirect effects on other components and the overall system behavior. This concept is successfully applied in [**15**,**16**] for hydraulic system test benches by also using a feature-based approach. The features mostly used are measures from descriptive statistics, such as the mean, median and variance of a signal, while, for rotating machinery, features from the frequency domain are often considered [**15**]. Alternatively, features can be manually defined, as in [**17**]. In [**18**], the authors have investigated refinement strategies of generic feature generation, where including process knowledge into the feature generation was shown to improve the detection of faults. Furthermore, in [**14**] and [**16**], a feature reduction is performed by applying filter methods and analytical methods of principal component analysis, respectively. While a significant reduction in features can be achieved with a principal component analysis or a linear discriminant analysis, the resulting feature sets are not interpretable by a human inspector after the transformation [**19**].

2.2. Design Considerations for Neural Networks

In contrast to conventional approaches, data-driven methods depend less on domain-specific knowledge but extract the knowledge implicitly or explicitly from data. In the field of machine learning, this process of knowledge extraction is referred to as model training, where the model is a mathematical structure that is fit to reproduce a desired input–output behavior. The type of model mainly considered in this study is feedforward neural networks in the form of multilayer perceptrons, as shown in **Figure 1**.

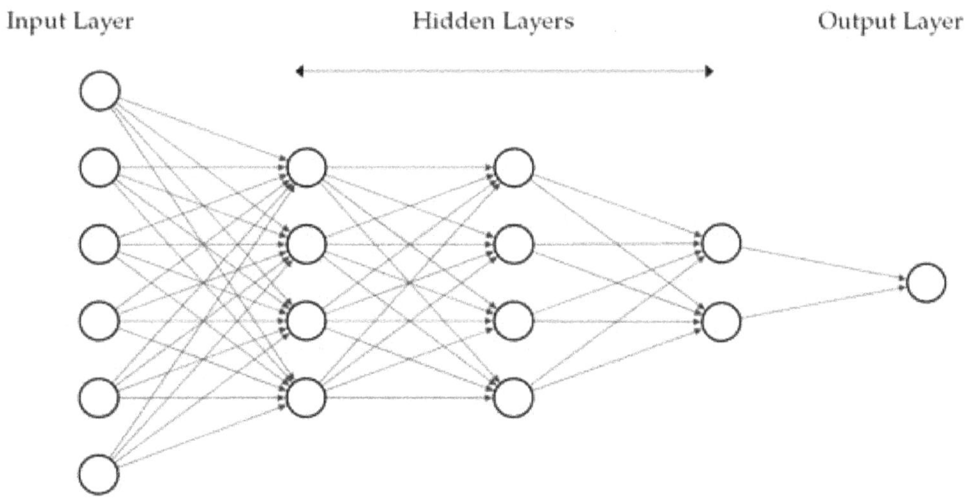

Figure 1. Feedforward neural network with sic nodes in the input layer, three hidden layers and one node in the output layer.

Feedforward neural networks consist of neural units sequentially stacked in layers, where the outputs of a preceding layer form the inputs for the subsequent layer. In regular fully connected neural networks, the outputs of all units of a preceding layer are connected to each neural unit of the next layer, where each connection is attributed a weight parameter. Within the units, the respective input values are multiplied with the corresponding weight parameter, summed up and passed through an activation function. The main purpose of the activation function is applying a transformation to the data between the layers and using nonlinear activation functions to enable nonlinear mappings between input samples and the output of the network [**20**]. Accordingly, typically used activation functions are the sigmoid activation function, tanh activation function or a rectified linear activation function (ReLU).

The layers located between the input layer and the output layer of the network are referred to as hidden layers. Theoretically, using one hidden layer with a nonlinear activation and an arbitrary high number of nodes in the hidden layer allows to approximate any given input–output relationship [**21**]. Accordingly, combining multiple layers linearly, as done in a multilayer perceptron, preserves this characteristic of a universal approximator. Moreover, combining multiple layers adds to the representational capacity of a neural network, which, in turn, can allow to use less nodes in the network in total compared to the network with a single hidden layer [**20**].

In order to obtain a desired input–output mapping from a neural network, the weight parameters of the model are adjusted algorithmically during the model training. For this, each training sample is first passed to the input of the model in a forward pass, where the model outputs a value. After this, the deviation between the model output from the forward pass and the desired output is computed according to a predefined loss function. Subsequently, in the reverse pass, the gradient of the loss function is computed with respect to the model weights, so that it can be inferred which model weights to adjust in order to minimize the loss. While the reverse pass is performed by generic gradient-descent-based algorithms, the selection of the loss function is linked to the format of the model output of the model, i.e., whether continuous values or discrete class labels are predicted [**22**]. In contrast to the weight parameters of the model, the structural parameters (number of layers, number of nodes and activation functions) and fitting parameters (loss function and optimization algorithm) remain fixed during training. Therefore, these parameters are referred to as the hyperparameters of a model. Applying neural networks hence requires to define these hyperparameters for the specific problem at hand. As no general rule exists for setting the hyperparameters of neural networks, especially for the structural hyperparameters, they are often determined based on experience or by implementing and comparing several setups. The latter approach is also referred to as grid search.

Furthermore, methods of supervised learning are based on fitting models to training data, for which the expected output of the model is known. In the case of condition monitoring, this can be any indicator of the machine health status, which, in turn, can be represented by a single value or be composed of multiple values. The former, the single-value case, is referred to as single-target prediction, while the latter case is referred to as multi-target prediction. Furthermore, in the case of classification, where the task is to predict discrete classes for a set of inputs, a distinction can be made based on the number of possible classes that each target can represent. In the most basic case, targets are binary values, but cases of multiple classes are possible as well.

2.3. Feature Representation of Time Series Data

As the number of input nodes of a neural network is fixed, the dimensionality of the input data must be uniform among all data instances, too. However, time series data, as typically acquired from sensors in industrial machines, can have varying lengths, depending on variations in the process or the sampling rates of different sensors. Moreover, the longer the time series, the more input nodes are required, and the number of parameters of the neural network unfavorably increases. Therefore, describing a time series by a fixed number of representative characteristics is one common means to obtain the compatibility of time series and a fixed number of model inputs [**22**]. These characteristics are referred to as features and can be either defined manually [**19**], by the use of generic descriptive measures [**23**], or derived by transformation of the data learned by another machine learning model [**24**].

3. MATERIALS AND METHODS

In this section, the considered hydraulic system is described and the developed approach for generating fault classification models is presented. This includes the steps of data generation, feature extraction and selection, as well as the process of model setup, training and evaluation.

3.1. Hydraulic Press

Representative of an industrial hydraulic system, a hydraulic press is considered in this study. **Figure 2** shows the physical machine, along with the corresponding hydraulic circuit diagram. The system mainly comprises a position-controlled cylinder drive. The hydraulic power is supplied by a pump and conducted to the double-acting hydraulic cylinder (CY) through a proportional 4/3-way directional control valve (PV). Two switching valves (SV1 and SV2) allow to set the system to idle circulation and to lock the ram in position, respectively. The pump provides a constant volumetric flow of approximately 19 L/min at a maximal system pressure of 200 bar. The hydraulic cylinder has a nominal piston diameter of 80 mm and is coupled to a guided ram, which finally acts against a load force. The load force represents the reaction force of a workpiece and is provided by a secondary circuit in the physical test bench. During operation of the press, pressures are measured at the main cylinder and the power supply. Additionally, the position of the cylinder is recorded, as well as the volumetric flow entering the 4/3-way directional control valve and the control signals fed to the valves.

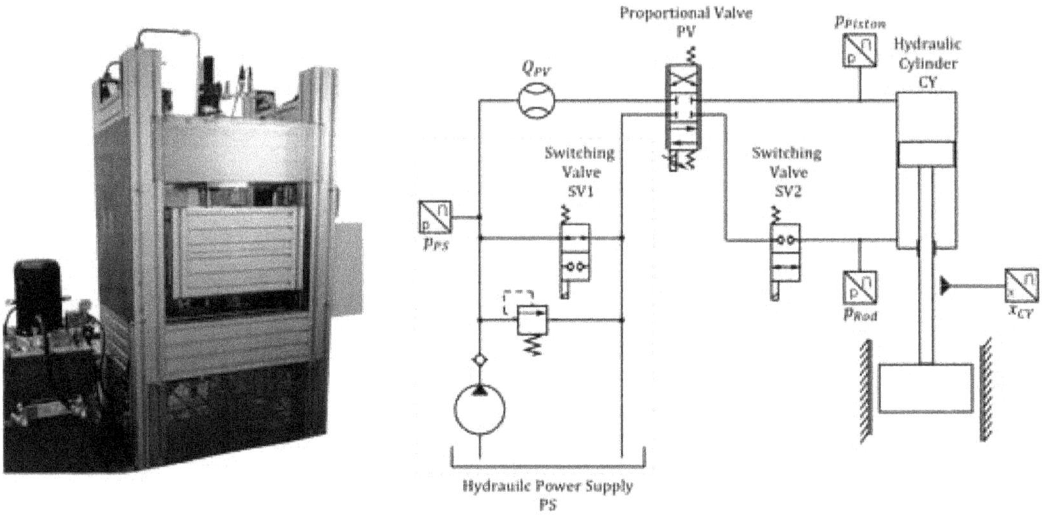

Figure 2. Hydraulic press test bench and its hydraulic circuit diagram.

The function of the system is to perform working cycles, in which the ram is moved to a defined position against the load force. At the beginning of the working cycle, the system is in idle mode, where the switching valve SV1 activates idle circulation of the pump and the switching valve SV2 locks the ram in the top position. Then, starting from the top position, SV1 and SV2 are switched and the main cylinder extends towards the workpiece in rapid traverse. During this phase, neither an external load nor a limitation of the velocity is active. As the ram approaches the workpiece and reaches a defined distance from it, the velocity is restricted through limiting the control signal of the valve PV and the volumetric flow to the cylinder, respectively. As the ram is in contact with the workpiece, an external load acts against the cylinder and increases linearly over the stroke of the cylinder. In this load phase, the system is given a fixed time to reach the set value, which depends on the desired target position of the ram. After this time, the controller initiates the backstroke in open-loop control and the cycle ends with another idle phase. **Figure 3** shows the progression of cylinder position, velocity and load force over one working cycle of the press.

For the generation of the large amounts of data required in this study, a corresponding simulation model of the system is set up as a lumped parameter model using the simulation software SimulationX (software version 4.2) **[25]**. Hydraulic components in the simulation model are parametrized according to nominal data sheet specifications. The pump is modeled with a pressure-dependent volumetric loss, while no information on hydro-mechanical parameters could be derived from the pump data sheet. Furthermore, the dynamic characteristics of the pump and the driving electric motor are neglected. A linear opening characteristic is defined for the proportional valve, while the relationship between the spool position and control signal is modeled by a second-order differential equation. Pressure losses in the piping are neglected. However, the capacitance of conductors is considered where hoses or long pipes are installed in the real system. This is especially the case for the hosing and piping from the hydraulic power supply to other parts of the system.

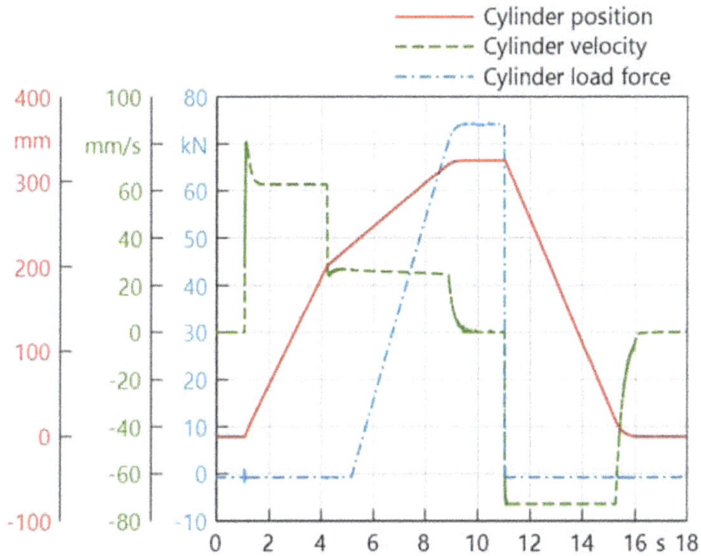

Figure 3. Simulated working cycle of the hydraulic press in regular operation.

To enable the effective training of a fault classification model from data, the training data ideally comprise all possible cases of operating scenarios and fault conditions. Hence, besides the regular operation of the press, operation under fault conditions must be emulated in the simulation model as well. Here, faults are emulated by introducing fault models into the lumped parameter simulation, which then are adjusted in intensity by varying the parameters of the fault models. The considered component faults are listed in **Table 1**. Leakages at the cylinder are modeled as pressure- and velocity-dependent flow through circular gaps, while the leakage at a connector of the hydraulic power supply is modeled by a pressure-dependent leakage factor. Increased friction at the main cylinder is modeled as an increased friction level, which is adjusted by setting the static friction force, whereas the dynamic friction force is lower than the static friction force by a constant factor. A worn valve spool is modeled by the valve spool overlap, while a sensor offset fault is introduced by adding an offset to the true sensor value. More detailed descriptions of the fault models used are given in [**26**].

Table 1. Faults of the hydraulic system considered in this study, along with the corresponding model parameters of the simulation model and their ranges of variation.

Fault	Symbol	Model Parameter	Parameter Range	Fault Threshold
Cylinder, Internal Leakage	QCY,Li	Gap height piston sealing	1–100 µm	>25 µm
Cylinder, External Leakage	QCY,Le	Gap height rod sealing	0.5–75 µm	>20 µm
Cylinder, Increased Friction	FCY,Fr	Static friction force	500–5000 N	>1500 N
Cylinder, Position Sensor Offset	xCY,off	Offset value	0–410 mm	>1 mm
Proportional Valve, Worn Control Edges	$y0, PV,$	Valve spool overlap	-(2–10)%	<-4%
Hydraulic Power Supply, External Leakage	QPS,Le	Leakage factor	0–0.1 L/(min·bar)	>0.0005 L/(min·bar)

3.2. Data Generation and Labeling

A dataset is generated from the simulation model by means of a variational study, where model parameters are varied. For this, a sampling scheme must be defined for the variable parameters of the simulation model. Here, the basis of the sampling scheme is a Latin Hypercube design. It is generated by creating ten random sampling schemes and choosing the design that best fulfills the maximin criterion. Accordingly, the scheme with the largest minimal distance between samples in the parameter space is selected. The eight-dimensional parameter space is formed by the six fault-indicating parameters and two process parameters, namely the target position of the press and the load force. Both process parameters are varied in three different levels of 290 mm, 310 mm and 340 mm and 40 kN/mm, 60 kN/mm and 80 kN/mm, respectively. Hence, in the obtained dataset, different fault intensities, fault combinations and operating conditions are represented. The temperature, however, is kept constant at 40 °C in all simulation runs.

While a Latin Hypercube design generally aims at sampling uniformly from the parameter space, different regions in the design space can be of different relevance for a fault classification. On the one hand, some fault cases and fault combinations can be of low practical relevance, such as the simultaneous occurrence of a high fault intensity for all considered faults. On the other hand, the severity of some of the considered faults is non-linearly related to the fault-indicating parameter. In this case, placing the samples uniformly over the range of the fault-indicating parameters, a non-uniform representation of fault severities would be obtained in the dataset. This can be illustrated by the example of the internal leakage at the main cylinder of the press, which is modeled as flow through a circular gap and adjusted by setting the height of the gap. As **Figure 4** shows, the leakage flow non-linearly increases with increasing gap height. As a result, the range of the gap height settings, which are considered to not represent a fault, is smaller than the range of settings that represents a fault. Consequently, in this case, uniform sampling leads to a higher representation of settings that represent a fault. Accordingly, an adjustment of the distribution of sampled points is required, to obtain a balanced representation of different fault states. For the given reasons, the density of samples is manually reduced in practically less relevant regions of the parameter space, while, in regions of low fault settings, the density of samples is increased by removing or adding random samples in these regions. Finally, by following the described sampling scheme, a variational study is performed with the simulation model and a dataset of 200,000 data instances is generated.

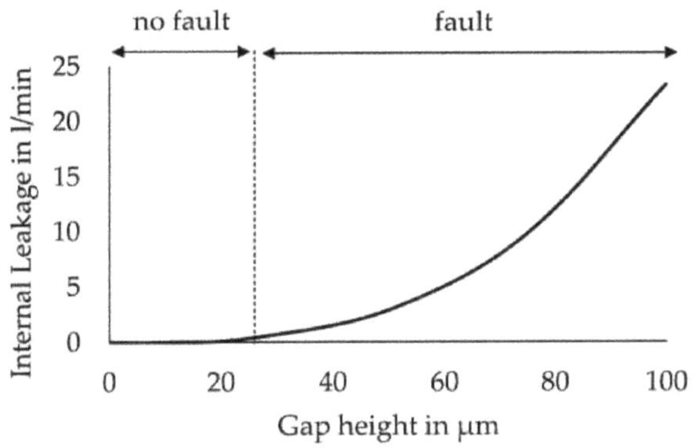

Figure 4. Example of non-uniform class distribution over the range of fault-indicating parameters.

For the application of methods of supervised learning, the generated data instances must be provided along with their class labels, which, in the framework of condition monitoring, indicate the fault state of the considered system. Here, the labels are defined for the single-target classification of binary classes. In the process of simulative data generation, continuous model parameters are used to indicate and adjust the fault intensities. Now, these fault-indicating parameters are transformed into categorical class labels by binning parameter ranges to classes of "fault" and "no fault". The threshold values, which define whether a certain degradation level is considered a fault or not, depend on the specific application and the tolerances allowed therein. In this study, the threshold values listed in **Table 1** are used. For the three different leakage faults, thresholds are set under consideration of the leakage flows resulting at the respective fault intensities. The defined thresholds of the gap heights, which mark the beginning of a fault for the internal and external leakage at the cylinder, produce volumetric leakage flows of 0.36 L/min and 0.1 L/min, respectively, at 40 °C and a differential pressure of 150 bar. The threshold value for the leakage factor, which marks the beginning of a fault of external leakage at the hydraulic power supply, is equivalent to a leakage flow of 0.1 L/min at 40 °C and the maximal pressure of 200 bar. For the other three faults, the thresholds are determined heuristically.

Another issue that arises during the variational study is the setting of the initial conditions for the simulation runs. Because model parameters are constantly varied between simulation runs, a single initial system condition, which is valid for all runs, cannot be determined. In order to circumvent this issue, the simulated press cycle is considered to represent one cycle within a series of multiple press cycles. As a result, the initial state of the system is determined by the state at the end of the previous cycle. To achieve this, every simulation run is started with the final second of the control sequence of the working cycle, instead of only simulating the actual working cycle. Afterwards, the first second of the simulated cycle is removed after the data are generated. Consequently, setting the initial conditions of the simulation is no longer critical while performing the variational study. The described approach and its effects are shown in **Figure 5** by the example of the pressure at the rod side of the hydraulic cylinder. When only simulating the actual working cycle, the progression of the rod-side pressure depends on the selection of the initial value, as **Figure 5**a shows. In contrast, **Figure 5**b shows that starting the simulation run with the ending sequence of a cycle allows to obtain results independent of the selected initial condition, when the first simulated second is removed afterwards.

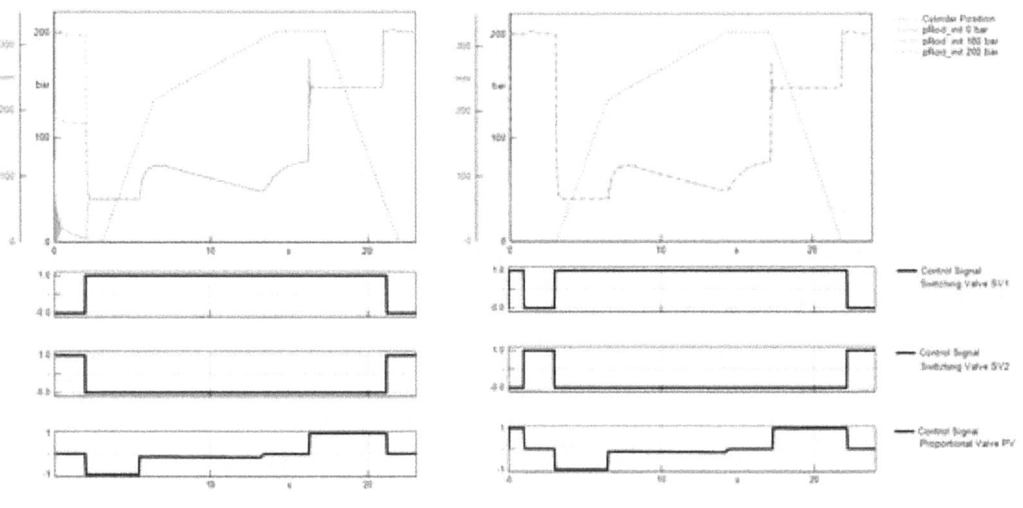

(a) (b)

Figure 5. Cylinder position, cylinder pressures and valve control signals simulated under fixed initial conditions while (a) starting the simulation with the actual control sequence of the press cycle and (b) starting the simulation with the ending sequence of a press cycle prior to the actual control sequence.

3.3. Feature Extraction and Selection

Before using the generated data for training a fault classification model, the raw time series are transformed into a fixed number of features. For this, the python-based open-source package tsfresh is used, which provides the calculation of a large variety of features for time series [**23**]. As the features must be calculated for each of the eight subsignals of the 200,000 multivariate time series, the following base set of 16 features with low computational costs is extracted:

- Minimum;
- Maximum;
- Mean;
- Median;
- Standard deviation;
- Variance;
- Skewness;
- Kurtosis;
- First location of minimum;
- First location of maximum;
- Last location of minimum;
- Last location of maximum;
- Sum of values;
- Absolute energy (sum over squared values);
- Absolute sum of changes (sum over absolute values of consecutive changes);
- Mean absolute change (mean over absolute values of consecutive changes).

Extracting this base set of features for a time series yields a feature vector of 128 features for one data instance. Generally, the resulting feature vectors only comprise subsignals recorded during data generation. However, derivations and transformations of the raw data often can reveal additional information about the characteristics of a signal. Moreover, the extracted features only represent characteristics over complete working cycles and thus only carry information on the aggregated effect of a fault on the observed data. In contrast, a conventional, visual inspection of sensor data by a human expert would typically involve a detailed inspection of anomalies in specific segments of the working cycle, such as checking for leakages in idle mode or the maximal cylinder velocity in rapid traverse. Such short-term effects can even be attenuated, when computing features over complete working cycles. Consequently, in this study, the state-of-the-art feature extraction is extended by incorporating further information into the feature extraction, which is typically available or can be acquired easily.

The first step of the extension consists in adding information about the progression of the recorded subsignals over time. This information is obtained by computing the first time derivatives of the subsignals and adding them to the time series as further subsignals. As a result, the number of subsignals is doubled and the number of extracted features per time series increases accordingly. The second step of the extension focusses on accentuating the characteristics of the subsignals in specific segments of the working cycle. For this, each time series is divided into segments of constant operation phases. These are the idle phases at the beginning and end of the cycle, rapid traverse, load phase and the backstroke. The transitions between operation phases are inferred from control signals given by the machine control. Subsequently, the same set of features that is extracted for the complete time series is again extracted for each subsignal in each of the defined segments, as **Figure 6** illustrates for the features "maximum" and "minimum". In the framework of this contribution, the originally generated feature set is referred to as global features, while the newly added features are referred to as segment features. The length l of the feature vector that represents a single data instance is again increased and now given by Equation (1), where $nsubsignals, nfeatures$ and $nsegments$ denote the number of subsignals, features and segments, respectively. Altogether, extracting features from the full time series and its five segments results in a set of 1536 features.

$$l = n_{subsignals} \cdot n_{features} \cdot (1 + n_{segments}) \quad (1)$$

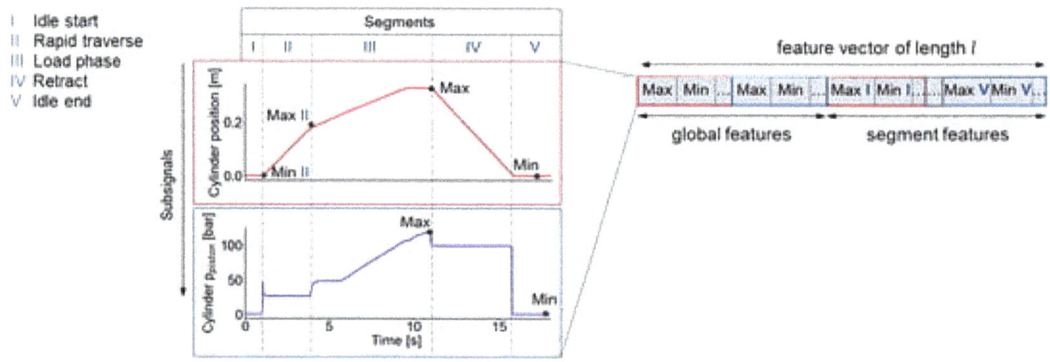

Figure 6. Concept of adding segment features to the base set of global features shown for two subsignals and the features "maximum" and "minimum".

As a result of the extended feature extraction, the number of features is increased, with the aim to also increase the amount of extracted information. However, not all the additionally generated characteristics can be assumed to be useful for fault classification. Concurrently, a concise set of features without irrelevant and redundant information is preferable for the efficient and robust training of most machine learning models and algorithms [19]. Moreover, a small set of only relevant features enables a human inspection and comprehension of the information fed to a classification model and consequently leads to higher transparency of the inference process. Therefore, feature selection follows, to reduce the set of features to those useful for the respective fault classification. This is achieved by using a wrapper method with decision tree classifiers. This means that a classification model is trained with the primary purpose of accessing the features implicitly selected within the model.

A decision tree classifier is a machine learning model that aims at hierarchically dividing a given dataset into groups of homogenous classes [27], which, in this case, are the classes "fault" and "no fault". The decision rules derived during training are based on defining threshold values for certain features, which then allow to classify a data instance. After training, the features and the corresponding threshold values of the different decision nodes can be inspected. Subsequently, it can be comprehended which features enable the effective separation of the data instances. **Figure 7** shows the first two levels of a decision tree for the classification of internal cylinder leakage. Moreover, decision trees, and their underlying algorithm, show low sensitivity of classification performance to the size of the feature sets that they are trained with [19], which makes them well suited for the task of processing the initially large feature set. To mitigate the tendency of decision trees to overfit, the maximum number of levels of the trees is limited. A maximum of five levels is heuristically found to provide a low variance of classification accuracy with no indication of overfitting.

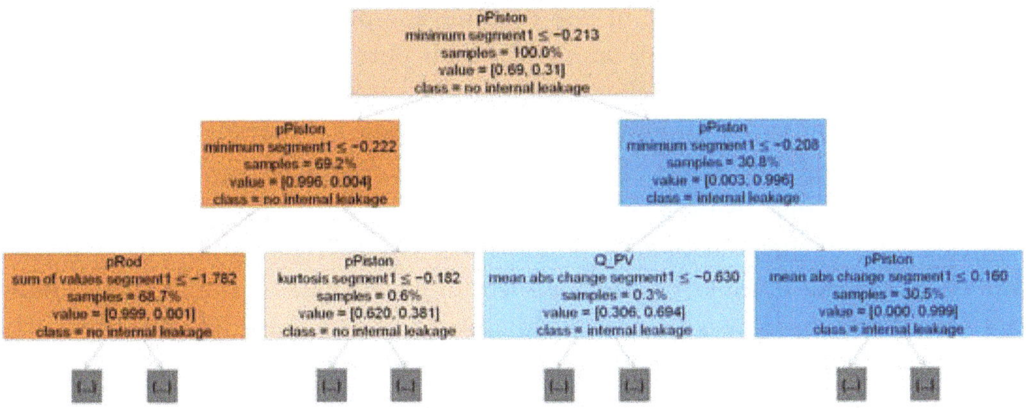

Figure 7. First two levels of a decision tree for the classification of internal leakage at the cylinder. Threshold values relate to features scaled by means of standardization.

As each of the considered component faults of the hydraulic press has a different impact on the machine behavior, the feature selection procedure is repeated for each of the considered faults. Furthermore, the resulting sets of selected features can depend on the specific set of training data used during selection. Random elements included in the CART algorithm, which is here used for training the decision trees, further add to the stochastic variation of the training results. Therefore, the feature selection is performed in the form of a 5-fold cross-validation. For this, the training and testing of each decision tree is repeated five times, using different excerpts of the data for training and testing in each repetition. Finally, the union set of features selected in all repetitions is formed and used for the training the final classification models. **Figure 8** shows the procedure of feature extraction and cross-validated feature selection. The data scaling is not illustrated therein, but standardization is performed within the cross-validation for each training and testing split individually.

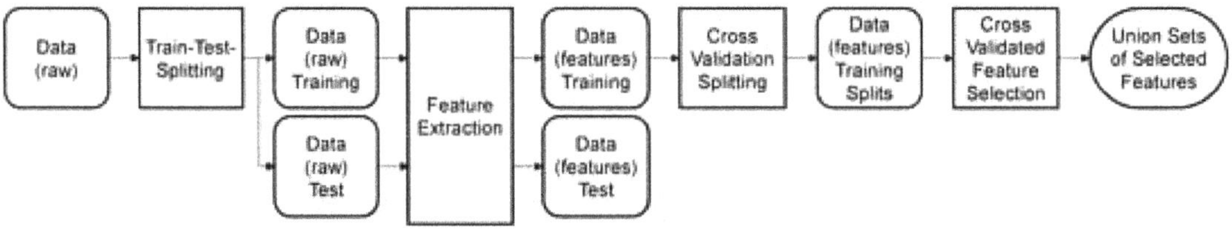

Figure 8. Process of feature extraction and cross-validated feature selection.

3.4. Model Setup, Training and Evaluation

Apart from the decision tree classifiers used during feature selection, the final fault classification is realized using feedforward neural networks. The problem of classifying the machine state is set up as a composition of single-target classifications with binary targets. This means that a model for binary classification is set up and trained for each of the considered component faults individually. This allows us to build smaller models, which potentially can be trained more efficiently and transparently. Moreover, modular fault classification models can be generated independently and on demand, while multi-target classification would require the a priori definition of all possible fault cases that can be expected in a system. Furthermore, using a separate classifier for each fault allows to obtain a diagnosis without the issue of an exponential increase in the number of class labels with an increasing number of faults, as would be the case in multi-target, multi-class classification. However, one disadvantage of the considered setup is that interdependencies between fault occurrences cannot be considered by the models.

The hyperparameters of the feedforward neural networks comprise the structural hyperparameters of the network, such as the number of hidden layers, number of nodes per layer and activation function of each layer, as well as fitting hyperparameters such as the loss function and the learning rate. As with the definition of the labels described previously in this section, the definition of the fitting hyperparameters depends on the definition of the problem at hand. Given that a binary classification is the targeted task, options for selecting the loss function are narrowed down to loss metrics, which are suitable for evaluating the outputs of the model against binary labels. Here, the binary cross-entropy is used as a loss metric. Furthermore, the selection of a learning rate can be simplified to only defining a starting learning rate when a state-of-the-art optimization algorithm with an adaptive learning rate is used. The Adam algorithm is used in this study, with a default setting of the starting leaning rate of 0.001.

Similarly, the numbers of nodes on the input layers and output layers of the neural networks are predetermined by the number of features fed to the models and the number of labels to predict, respectively. While the input layer is set to have a ReLU activation function, a sigmoid activation function is used in all hidden layers and the output layer. Further structural hyperparameters of the neural networks, which are the number of hidden layers and the number of nodes therein, are determined by means of a grid search. Different model structures are set up and trained so that the best model structure of the considered configurations can be identified for each fault case. The scheme of the grid search and the concept of varying the model structure are shown in **Figure 9**. The scheme shows that three different numbers of hidden layers and four different settings of the maximal number of nodes are varied. For a given number of hidden layers, the first hidden layer is constructed with the maximal number of nodes defined. Then, in each subsequent layer, the number of nodes is halved.

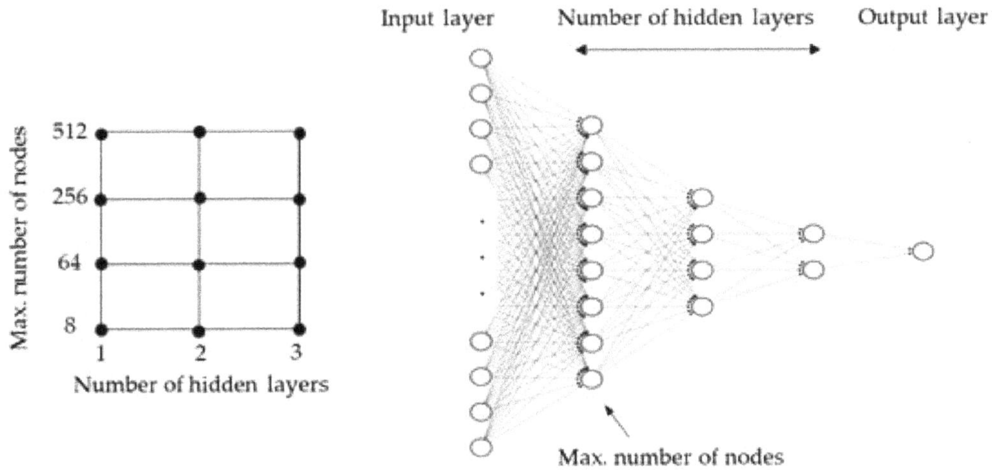

Figure 9. Applied scheme of grid search and corresponding model hyperparameters.

As with the feature selection step, each model configuration considered in the grid search is trained and tested with different excerpts of the data within a 5-fold cross-validation. A further test set, which is held out from the whole process of feature selection and model tuning, is finally used to evaluate the performance of the models selected in the grid search. The test set comprises around 40,000 working cycles, which is 20% of the available data. This form of validation procedure is referred to as the three-way holdout method and is required to obtain an unbiased, final estimate of the model performance, as the test data used in the cross-validation loop were already involved during feature selection and grid search [**28**]. The classification performance of the models is evaluated using the balanced accuracy. The balanced accuracy is defined as the mean value of the true-positive rate and the true-negative rate and thus gives a balanced estimate independent of the class distribution of the data [**29**]. The applied procedure of the cross-validated grid search is illustrated in **Figure 10.**

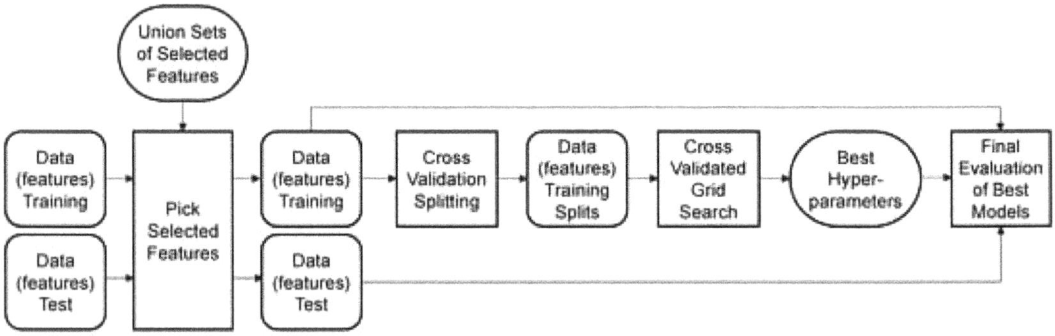

Figure 10. Procedure of cross-validated grid search and final evaluation of models following the three-way hold-out method.

4. RESULTS

4.1. Results of the Grid Search

The mean classification accuracies of the best models identified in the grid search are shown in **Table 2** for the different faults, for different feature sets, as well as with and without feature selection. **Table 3** lists the corresponding structures of these models. For the faults of internal cylinder leakage QCY,i, the position sensor offset xCY, off, and the external leakage at the hydraulic power supply QPS,Le, high classification accuracies around 99% are already obtained with the base set of global features. This range of accuracies is obtained for these faults with all tested structural configurations of the neural networks, with no clear dependence on model size and a maximal deviation between the best- and worst-performing models of less than 1% across all feature sets. Therefore, the smallest possible model structure (one hidden layer with eight nodes) is also considered the best-performing model for these faults. A marginal increase in the model performance can be observed for these three faults when extending the feature sets and applying feature selection. For the fault of worn control edges on the proportional valve $y0,V$, a similar trend is observed when increasing the feature set, while the feature selection does not always lead to improved maximal performance. Regarding the model size, however, the feature selection leads to the same effect as stated above for other faults, where the difference in performance between model sizes becomes marginal and the smallest models can be regarded as the best-performing models for a specific configuration. Using an increased set of features clearly improves the performance of the classifiers for the faults of increased friction at the cylinder $F_{CY,r}$, and an external leakage at the cylinder $Q_{CY,Le}$,. The prediction of increased friction appears to benefit from added derivative and segment features likewise. In contrast, for the external leakage, the information added by segment features appears to mainly induce the improvement, especially when no feature selection is applied. Moreover, for both latter faults, the best-performing models have larger network structures when feature selection is considered. While the achieved maximal accuracies can be considered high performance scores for the increased friction and the external cylinder leakage, the models cannot fully resolve the targeted input–output relationship with the given data and considered model structures.

Table 2. Mean classification accuracies obtained with the best model during the grid search with and without feature selection, for different feature sets and for different faults. Best results for each fault are highlighted in bold.

	Increased Cylinder Friction $F_{CY,Fr}$		Internal Cylinder Leakage $Q_{CY,Li}$		External Cylinder Leakage $Q_{CY,Le}$	
Feature Set	**Without Selection**	**With Selection**	**Without Selection**	**With Selection**	**Without Selection**	**With Selection**
Global	93.60%	93.57%	98.65%	99.29%	78.72%	78.81%
Global + derivative	95.51%	94.97%	98.44%	99.13%	78.68%	82.32%
Global + segment	96.87%	95.91%	99.26%	99.45%	88.69%	86.52%
Global + segment + deriv.	96.81%	**97.09%**	99.21%	**99.51%**	88.38%	**89.69%**
	Position Sensor Offset $x_{CY,off}$		**Worn Control Edges Prop. Valve $y_{0,PV}$**		**External Leakage at Hydr. Power Supply $Q_{PS,Le}$**	
Feature Set	**Without Selection**	**With Selection**	**Without Selection**	**With Selection**	**Without Selection**	**With Selection**
Global	98.51%	99.68%	96.62%	96.42%	99.69%	99.78%
Global + derivative	98.06%	99.64%	96.69%	96.44%	99.71%	99.82%
Global + segment	99.28%	**99.83%**	98.57%	98.93%	99.70%	99.83%
Global + segment + deriv.	99.10%	99.67%	98.38%	**98.94%**	99.70%	**99.85%**

Table 3. Structure of the hidden layers of the best models obtained from the grid search with and without feature selection, for different feature sets and for different faults.

	Increased Cylinder Friction FCY,Fr		Internal Cylinder Leakage QCY,Li		External Cylinder Leakage QCY,Le	
Feature Set	**Without Selection**	**With Selection**	**Without Selection**	**With Selection**	**Without Selection**	**With Selection**
Global	(256)	(512, 256)	(8)	(8)	(256)	(256, 128)
Global + derivative	(64)	(256)	(8)	(8)	(64, 32, 16)	(512, 256)
Global + segment	(64, 32)	(512, 256)	(8)	(8)	(64, 32)	(512, 256)
Global + segment + deriv.	(256, 128)	(256, 128)	(8)	(8)	(64, 32, 16)	(512, 256, 128)
	Position Sensor Offset xCY,off		**Worn Control Edges Prop. Valve $y0,PV$**		**External Leakage at Hydr. Power SupplyQPS,Le**	
Feature Set	**Without Selection**	**With Selection**	**Without Selection**	**With Selection**	**Without Selection**	**With Selection**
Global	(8)	(8)	(512, 256, 128)	(512, 256)	(8)	(8)
Global + derivative	(8)	(8)	(256)	(256)	(8)	(8)
Global + segment	(8)	(8)	(64, 32, 16)	(8)	(8)	(8)
Global + segment + deriv.	(8)	(8)	(8)	(8)	(8)	(8)

A final evaluation of the classification performance of the identified model structures is performed on a held-out test set, which was not involved during the cross-validated feature selection and grid search. This is carried out to validate that the trained models are not overfitted to the training data and that the features selected during feature selection are also important for classifying new, unseen data instances. For this, all the data, which were partitioned into five subsets during cross-validation, are jointly used to train the models that were identified as the best model structures during the grid search. The results are given in **Table 4** and show that, generally, the same scores are achieved as given in **Table 2** and estimated during the cross-validated grid search. Moreover, the scores obtained on the test set are mostly slightly higher, which can be attributed to the larger set of training data used in this final evaluation.

Table 4. Classification accuracies obtained with the best model structure on the held-out test set with and without feature selection, for different feature sets and for different faults. Best results for each fault are highlighted in bold.

	Increased Cylinder Friction $F_{CY,Fr}$		Internal Cylinder Leakage $Q_{CY,Li}$		External Cylinder Leakage $Q_{CY,Le}$	
Feature Set	**Without Selection**	**With Selection**	**Without Selection**	**With Selection**	**Without Selection**	**With Selection**
Global	94.51%	93.37%	99.08%	99.17%	81.30%	72.94%
Global + derivative	96.38%	94.73%	98.94%	99.07%	77.73%	74.04%
Global + segment	97.62%	96.59%	99.46%	99.47%	**91.89%**	87.11%
Global + segment + deriv.	**97.72%**	97.64%	99.43%	**99.47%**	90.63%	90.17%
	Position Sensor Offset $x_{CY,off}$		**Worn Control Edges Prop. Valve $y_{0,PV}$**		**External Leakage at Hydr. Power Supply $Q_{PS,Le}$**	
Feature Set	**Without Selection**	**With Selection**	**Without Selection**	**With Selection**	**Without Selection**	**With Selection**
Global	97.86%	98.98%	97.12%	97.13%	99.61%	99.66%
Global + derivative	98.53%	98.80%	98.00%	97.01%	99.68%	99.69%
Global + segment	**99.49%**	98.82%	**99.13%**	98.55%	99.75%	**99.86%**
Global + segment + deriv.	98.53%	98.74%	98.25%	98.79%	99.59%	99.76%

4.2. Results of the Feature Selection

Apart from the effects discussed in the previous subsection, further effects of the feature selection are described in the following. **Table 5** shows the number of features selected from the different feature sets for the different faults. Compared to the case of no feature selection, the number of features is significantly reduced after selection in all the listed cases. Moreover, considering the number of selected features across feature sets, for most faults, the number of selected features steadily increases when increasing the feature set before selection. The main increase in the number of selected features compared to the base set of global features is obtained by adding segment features, which is in alignment with the increase in performance induced by the segment features observed in the previous subsection. The exception is the fault of increased friction $F_{CY,r}$ where a higher number of features is selected that represent trend information of the time series.

Table 5. Number of features before and after feature selection for the different feature sets and different faults.

Feature Set	Number of Selected Features All Faults Before Selection	*FCY,Fr*	*QCY,Li*	*QCY,Le*	*xCY,off*	*y0,PV*	*QPS,Le*
Global	128	40	29	36	14	21	28
Global + derivative	256	38	36	39	14	26	39
Global + segment	738	25	57	46	20	46	42
Global + segment + deriv.	1536	39	59	52	22	51	42

A benefit of a reduced set of features is that it enables to inspect features manually. This allows us to gain an insight into the process of fault inference and into the general system behavior under faults. Moreover, using decision trees for feature selection, the selected features are provided with relative importance, which indicates which fraction of the data instances could be separated into classes of "fault" and "no fault" by using this feature. Accordingly, a ranking of the features is obtained, in addition to the reduction. However, as the feature selection is performed in a cross-validation loop and union sets of selected features are formed afterwards, the ranking information is not maintained after this step. Therefore, the ranking information is acquired from a single run of the cross-validation loop. In **Table 6**, the three most important features selected for each of the considered faults are listed. The subsignals can be found in the hydraulic circuit diagram in **Figure 2**.

Table 6. Three most important features identified through feature selection for each fault.

Fault	Subsignal	Features
FFr	*pPiston*	median—segment IV
	pRod	minimum of derivative—segment IV
	pRod	standard deviation of derivative—segment III
QCY,Li	*pPiston*	minimum—segment I
	QPV	maximum of derivative—segment I
	QPV	absolute sum of changes—segment V
QCY,Le	*pPiston*	minimum—segment III
	pRod	maximum—segment III
	pPiston	standard deviation—segment II

Fault	Subsignal	Features
$x_{CY,off}$	x_{CY}	minimum—segment I
	x_{CY}	median—segment I
	x_{CY}	minimum
$y_{0,PV}$	Q_{PV}	maximum—segment III
	Q_{PV}	maximum of derivative—segment III
	Q_{PV}	median
$Q_{PS,Le}$	Q_{PV}	minimum—segment I
	p_{PS}	absolute sum of changes—segment I
	Q_{PV}	mean absolute change—segment V

To check the plausibility of the selected features, an exemplary visual inspection of the respective subsignals is performed. **Figure 11** shows the progression of the pressures recorded at the hydraulic cylinder in regular operation, as well as under the influence of an internal cylinder leakage $Q_{CY,i}$. For the detection of this fault, the minimum of the pressure on the piston side of the cylinder in the idle phase at the beginning of the press cycle (segment I) is selected as the most important feature. This is in accordance with the observed subsignals shown in **Figure 11**. An internal leakage clearly alters the pressure progression in the idle phases at the beginning and end of the press cycle. Moreover, an internal leakage leads to an increased pressure level at the piston side and thus to an altered minimum value of this pressure. This increase is induced by the pressure equalization between the two cylinder chambers, which is enabled by the existence of a leakage gap. An example for the effect of an external cylinder leakage $Q_{CY,e}$ on both cylinder pressures is shown in **Figure 12**. For this fault, the minimum value of the pressure on the piston side during the load phase (segment III) is selected as the most important feature. Again, the selection of this feature appears plausible, as this characteristic of the rod-side pressure is uniquely impacted by the occurrence of the external leakage. Additionally, the drop in the rod-side pressure in segments I and V is strongly affected by the external leakage. However, the fact that this symptom is also observed with an internal cylinder leakage explains why this feature was not selected as important indicator of an external cylinder leakage.

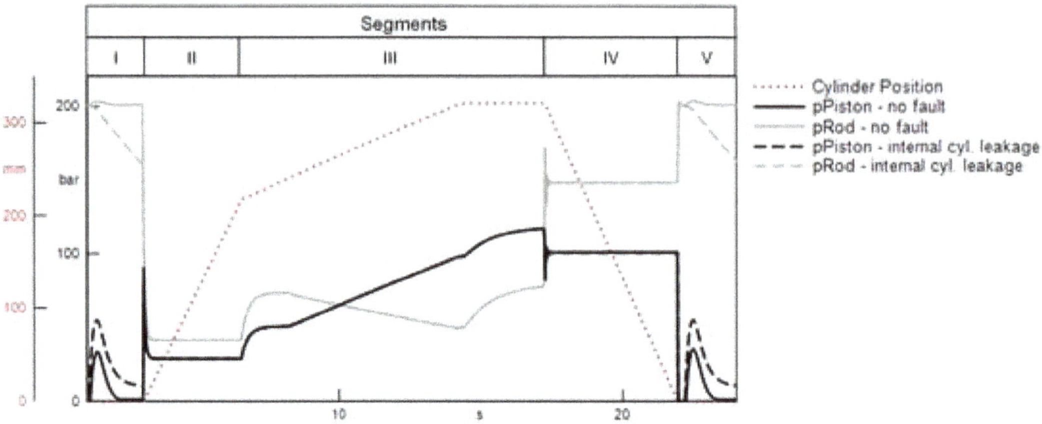

Figure 11. Simulated effect of an internal cylinder leakage on the cylinder pressures.

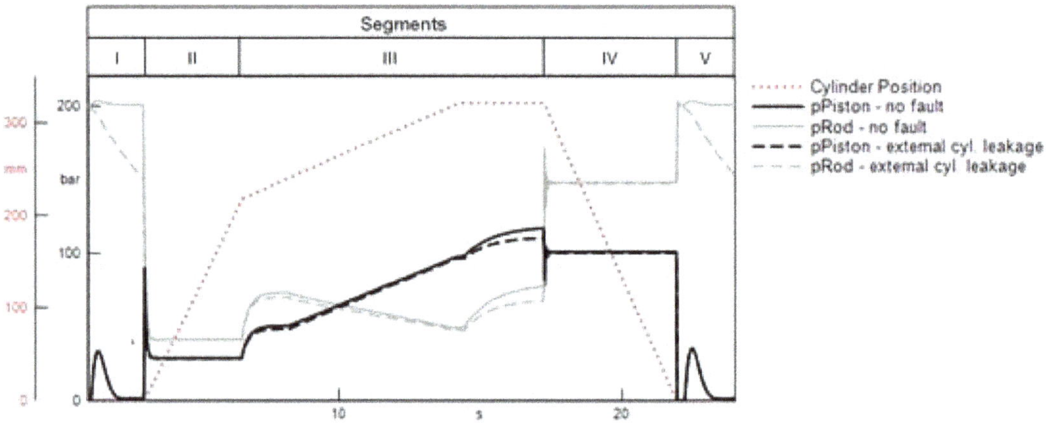

Figure 12. Simulated effect of an external cylinder leakage on the cylinder pressures.

5. DISCUSSION

The results demonstrate that the general development process used in this work effectively creates fault classification models for monitoring the state of a hydraulic press. The results indicate that various system failures necessitate distinct model architectures and information for accurate categorization. Internal leakage at the cylinder, offset at the position sensor, and external leakage at the hydraulic power supply could be detected using a basic set of common characteristics and simple neural network architectures. This suggests that these defects have a distinct effect on a specific group of features, regardless of other faults and operating conditions. The extended feature extraction strategy significantly influenced the performance of the classification models for the remaining faults. The disparity between the necessary features and the resulting model sizes reflects the effectiveness of the strategy used to conduct single-target classifications for each fault separately, rather than creating a multi-target job to predict all faults simultaneously.

Feature selection does not significantly impact the maximum model performance, regardless of the features and faults being examined, unlike feature set extension. Feature selection affects the sizes of the model structures of the top-performing models. The most effective models had bigger network topologies when feature selection was used, except in circumstances where the smallest model structure was adequate. The results show that feature selection leads to larger models when the connection between features and the presence of a problem is not completely understood. This is clearly demonstrated by the outcomes related to the issue of worn control edges of the proportional valve, as presented in Table 2 and Table 3. As the number of features increases and the classification accuracy nears 99%, the performance gap between various model architectures decreases, and smaller models can reach the same level of performance as larger ones. In the grid search, only the hidden layers' structure was changed, while the number of input layers was determined by the feature vector's size. Therefore, having more features at the input results in an increased number of model parameters initially.

The feature selection enhances the understanding of the development process and improves the clarity of data-driven development. This is particularly true when interpretable features are derived from the original time series data. The examples demonstrated that feature selection identified symptoms specific to particular problems in internal and exterior cylinder leaking.

6. AT A GLANCE

Automated assessment of machine conditions is crucial for effective maintenance scheduling. Data-driven techniques have been shown to facilitate the automatic identification of intricate patterns in sensor data that correspond to the condition of a system. Generalizable methods for creating these solutions inside industrial settings have not been established. This contribution outlines a method for creating data-based condition monitoring systems for industrial hydraulics by utilizing supervised learning and neural networks. The method includes extracting features and selecting them, then applying it to simulated data from a hydraulic press. The development process involves examining various stages to assess design choices and their effectiveness in fault classification jobs. The current approach can achieve high classification accuracies, however different faults may require specific setups of the classification models.

REFERENCES

1. ISO 17359:2018; Condition Monitoring and Diagnostics of Machines—General Guidelines (ISO 17359:2018). International Organization for Standardization: Geneva, Switzerland, 2018.
2. Watton, J. Modelling, Monitoring and Diagnostic Techniques for Fluid Power Systems; Springer: London, UK, 2007; ISBN 9781846283734. **[Google Scholar]**
3. Isermann, R. Fault-Diagnosis Systems; Springer: Berlin/Heidelberg, Germany, 2006; ISBN 978-3-540-24112-6. **[Google Scholar]**
4. Alt, R.; Murrenhoff, H.; Schmitz, K. A Survey of "Industrie 4.0" in the Field of Fluid Power—Challenges and Opportunities by the Example of Field Device Integration. In Proceedings of the 11th International Fluid Power Conference, Aachen, Germany, 19–21 March 2018; Murrenhoff, H., Ed.; RWTH Aachen University: Aachen, Germany, 2018. ISBN 3958862152. **[Google Scholar]**
5. Kuhn, M.; Johnson, K. Applied Predictive Modeling, Corrected at 5th Printing; Springer: New York, NY, USA, 2016; ISBN 978-1-4614-6848-6. **[Google Scholar]**
6. Döbel, I.; Leis, M.; Vogelsang, M.M.; Neustroev, D.; Petzka, H.; Rüping, S.; Voss, A.; Wegele, M.; Welz, J. Maschinelles Lernen—Kompetenzen, Anwendungen und Forschungsbedarf: Projektbericht. 2018. Available online: **https://www.bigdata-ai.fraunhofer.de/content/dam/bigdata/de/documents/Publikationen/BMBF_Fraunhofer_ML-Ergebnisbericht_Gesamt.pdf** (accessed on 28 July 2022).
7. Paulmann, G.; Mkadara, G. Condition monitoring of hydraulic pumps—Lessons learnt. In Proceedings of the 11th International Fluid Power Conference, Aachen, Germany, 19–21 March 2018; Murrenhoff, H., Ed.; RWTH Aachen University: Aachen, Germany, 2018. ISBN 3958862152. **[Google Scholar]**
8. Torikka, T.; Haack, S. Predictive Maintenance Service Powered by Machine Learning and Big Data. In Proceedings of the 11th International Fluid Power Conference, Aachen, Germany, 19–21 March 2018; Murrenhoff, H., Ed.; RWTH Aachen University: Aachen, Germany, 2018. ISBN 3958862152. **[Google Scholar]**
9. Schraft, J.P.; Becher, D.; Weber, J. Condition Monitoring Strategy for Pump-Driven Hydraulic Axes. In Proceedings of the ASME/BATH Symposium on Fluid Power and Motion Control, Online, 9–11 September 2020; The American Society of Mechanical Engineers: New York, NY, USA, 2020. ISBN 9780791883754. **[Google Scholar]**
10. Azeez, A.A.; Vuorinen, E.; Minav, T.; Casoli, P. AI-based condition monitoring of a variable displacement axial piston pump. In Proceedings of the 13th

International Fluid Power Conference, Aachen, Germany, 13–15 June 2022; Schmitz, K., Ed.; RWTH Aachen University: Aachen, Germany, 2022. [**Google Scholar**]

11. Wiens, T.; Fernandes, J. Application of Data Reduction Techniques to Dynamic Condition Monitoring of an Axial Piston Pump. In Proceedings of the ASME/BATH Symposium on Fluid Power and Motion Control, Longboat Key, FL, USA, 7–9 October 2019; The American Society of Mechanical Engineers: New York, NY, USA, 2020. ISBN 9780791859339. [**Google Scholar**]
12. Yao, J.; Li, X.; Wang, P.; Cheng, Y.; Yang, S. Data-driven Fault Diagnosis Based on Deep Learning for Multiple Failures of High Speed On/off Valve. In Proceedings of the 13th International Fluid Power Conference, Aachen, Germany, 13–15 June 2022; Schmitz, K., Ed.; RWTH Aachen University: Aachen, Germany, 2022. [**Google Scholar**]
13. Ali, E. Self-Learning Condition Monitoring for Smart Electrohydraulic Drives. Ph.D. Thesis, Technische Universität Dresden, Dresden, Germany, 2019. [**Google Scholar**]
14. Torikka, T. Bewertung von Analyseverfahren zur Zustandsüberwachung einer Axialkolbenpumpe. Ph.D. Thesis, Shaker, Aachen, Germany, 2011. [**Google Scholar**]
15. Duensing, Y.; Rodas Rivas, A.; Schmitz, K. Machine Learning for Failure Mode Detection in Mobile Machinery. In Proceedings of the 11th Kolloquium Mobilhydraulik, Karlsruhe, Germany, 10 September 2020; Geimer, M., Synek, P.-M., Eds.; KIT Scientific Publishing: Karlsruhe, Germany, 2020; pp. 1–25. [**Google Scholar**]
16. Helwig, N.J. Zustandsbewertung Industrieller Prozesse Mittels Multivariater Sensordatenanalyse am Beispiel Hydraulischer und Elektromechanischer Antriebssysteme. Ph.D. Thesis, Shaker, Aachen, Germany, 2019. [**Google Scholar**]
17. Adams, S.; Beling, P.; Farinholt, K.; Brown, N.; Polter, S.; Dong, Q. Condition Based Monitoring for a Hydraulic Actuator. In Proceedings of the Annual Conference of the PHM Society, Bilbao, Spain, 5–8 July 2016. [**Google Scholar**]
18. Makansi, F.; Schmitz, K. Feature Generation and Evaluation for Data-Based Condition Monitoring of a Hydraulic Press. In Proceedings of the 13th International Fluid Power Conference, Aachen, Germany, 13–15 June 2022; Schmitz, K., Ed.; RWTH Aachen University: Aachen, Germany, 2022. [**Google Scholar**]

19. Kuhn, M.; Johnson, K. Feature Engineering and Selection: A Practical Approach for Predictive Models; Chapman & Hall/CRC: Boca Raton, FL, USA, 2021; ISBN 978-1-03-209085-6. [**Google Scholar**]

20. Aggarwal, C.C. Neural Networks and Deep Learning: A Textbook; Springer International Publishing: Cham, Switzerland, 2018; ISBN 978 3 319 94462 3. [**Google Scholar**]

21. Hornik, K.; Stinchcombe, M.; White, H. Multilayer feedforward networks are universal approximators. Neural Netw. **1989**, 2, 359–366. [**Google Scholar**] [**CrossRef**]

22. Géron, A. Hands-on Machine Learning with Scikit-Learn, Keras, and TensorFlow; O'Reilly Media, Inc.: Sebastopol, CA, USA, 2019. [**Google Scholar**]

23. Christ, M.; Kempa-Liehr, A.W.; Feindt, M. Distributed and Parallel Time Series Feature Extraction for Industrial Big Data Applications. 2016. Available online: **http://arxiv.org/pdf/1610.07717v3** (accessed on 28 July 2022).

24. Radford, A.; Metz, L.; Chintala, S. Unsupervised Representation Learning with Deep Convolutional Generative Adversarial Networks. 2015. Available online: **http://arxiv.org/pdf/1511.06434v2** (accessed on 28 July 2022).

25. ESI Group. SimulationX. Available online: **https://www.esi-group.com/products/system-simulation** (accessed on 28 July 2022).

26. Makansi, F.; Schmitz, K. Simulation-Based Data Sampling for Condition Monitoring of Fluid Power Drives. IOP Conf. Ser. Mater. Sci. Eng. **2021**, 1097, 012018. [**Google Scholar**] [**CrossRef**]

27. Fürnkranz, J. Decision Tree. In Encyclopedia of Machine Learning and Data Mining; Sammut, C., Webb, G.I., Eds.; Springer: Boston, MA, USA, 2017; pp. 330–335. ISBN 978-1-4899-7685-7. [**Google Scholar**]

28. Raschka, S. Model Evaluation, Model Selection, and Algorithm Selection in Machine Learning. 2018. Available online: **http://arxiv.org/pdf/1811.12808v3** (accessed on 28 July 2022).

29. Brodersen, K.H.; Ong, C.S.; Stephan, K.E.; Buhmann, J.M. The Balanced Accuracy and Its Posterior Distribution. In Proceedings of the 2010 20th International Conference on Pattern Recognition (ICPR), Istanbul, Turkey, 23–26 August 2010; IEEE: Piscataway, NJ, USA, 2010; pp. 3121–3124, ISBN 978-1-4244-7542-1. [**Google Scholar**]

10 CONTROL OF AN OPEN CIRCUIT AXIAL PISTON PUMP USING LQG

1. INTRODUCTION: AN OVERVIEW

Axial piston displacement devices are commonly utilized in hydraulic drive systems. This is due to several advantages in comparison to conventional hydraulic displacement machines. The benefits include high power density per unit weight, compact size, low flow rate fluctuations, compatibility with high pressure, variable displacement control based on different laws, and the ability to operate in both open and closed-loop systems. The final advantage cited demonstrates its suitability for both mobile and industrial applications [2]. Their high cost is typically cited as a key negative due to the requirement for increased power of the hydraulic drive. The benefits mentioned have spurred scientific research in recent decades, leading to an increased scope of use for axial piston machines. Variable displacement axial piston machines are not only used but also of scientific interest.

The creation of software programs for simulating hydraulic elements, devices, and machines aids scientific study for constructive enhancements. Scientists from different research institutions and educational organizations have identified significant qualitative enhancements through detailed simulation studies. They have established a relationship between the structural characteristics and flow parameters in axial piston machines. There is limited study on enhancing the regulation of displacement volume, particularly in pumps of this kind [10]. The introduction of high response proportional electro-hydraulic spool valves, which replaced conventional hydraulic controllers, played a vital role in driving progress. Subsequently, proportional hydraulic technology advancements led to the development of secondary control systems. Axial piston machines are primarily utilized in these systems, and their efficiency is contingent on the control approach employed [15]. Non-traditional controllers have been increasingly used for controlling axial piston pumps in the past twenty years [16,17,18].

Research in the realm of technology mostly focuses on embedded systems and implementing intricate control principles using programmable platforms. The writers have expertise in implementing control systems for various types of plants, especially hydraulic drive systems. This work is motivated by the development of an embedded control system specifically built for a recognized type of open circuit axial piston pump. The solution is applied on a laboratory test equipment. An elaborate explanation of the hydraulic system regarding pump displacement control is provided, along with the designed system architecture for its regulation. The primary objective is to develop a linear-quadratic Gaussian (LQG) controller. The controller is created using a model derived from the "black box" system identification method, which relies on experimental data. The controller is validated by experimental research to analyze its performance.

2. SYSTEM DESCRIPTION

A laboratory test rig is created and utilized to research open circuit axial piston pumps with proportional valve control for system identification and embedded controller development [**4**]. The hydraulic circuit diagram of the test rig is shown in **Figure 1**, and its photo is presented in **Figure 2**. The system is based on a well-known model pump, A10VSO (Bosch Rexroth company). The regulation of the displacement volume of this type of pump is well known in the theory and practice of hydraulic drives [**2**]. It is based on two main approaches. The first consists of using a conventional hydraulic controller of type DR, DFR, DFLR, etc. These controllers are directly coupled to the pump casing and perform their regulating function (of pressure and flow rate) according to a certain law depending on their schematic and constructive solution. The other approach for their regulation consists of the use of a hydraulic valve that is specially developed for this type of pump with proportional electric control, which is mounted on the same plate of connection intended for the listed conventional controllers [**21**]. In this way, through integrated electronics or an external card type VT-5041, control of the spool of this valve is performed, and, in turn, the displacement volume of the pump is changed via the main energy parameters, pressure and flow rate [**21,22**]. In practice, this approach is designed to implement secondary control; however, it requires the use of a variant of the same pump that has a swash plate swivel angle sensor and the ability to change the angle of the swash plate in the opposite direction to work in the motor mode of operation. Inevitably, this modernization approach has a significantly higher cost but with an expanded range of applications. In this aspect, the developed control solution incorporates both regulation approaches. However, a pump is used without a swivel angle sensor and without the possibility of operation in motor mode. On the other hand, both types of controllers are used, connected in parallel by means of an additional developed mounting hydraulic block with a built-in check valve. It must be noted that this cannot be used for secondary control. Its advantage is that it can implement a hydraulic system with speed and load feedback based on a variable displacement axial piston pump of the most common type. A detailed description of the system was carried out in a previous publication [**4**]. Here, only a brief presentation is made, emphasizing the mentioned advantages and disadvantages compared to solutions imposed in practice.

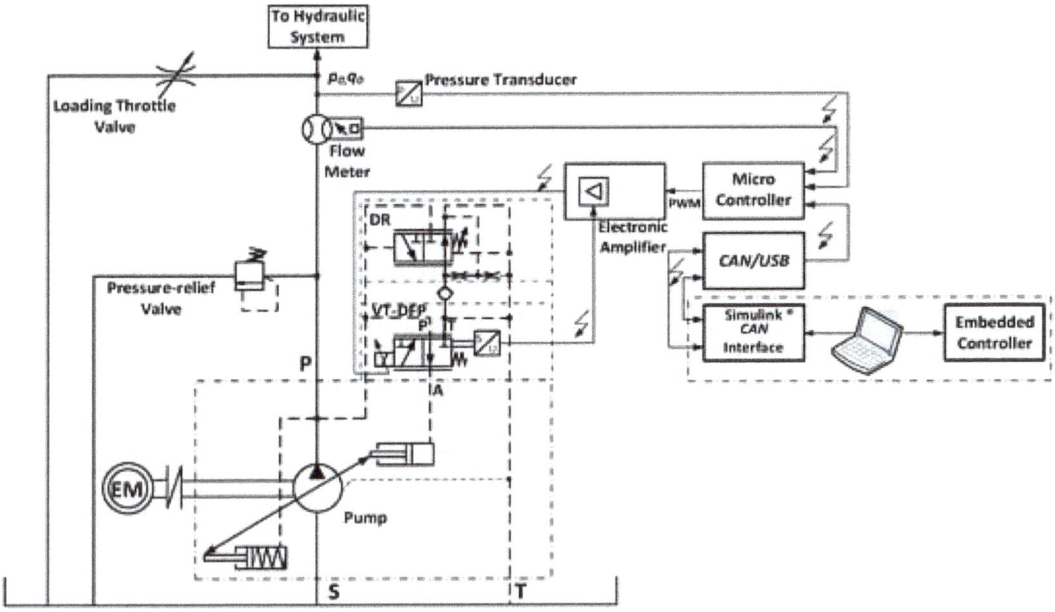

Figure 1. Hydraulic circuit diagram of a test rig for the axial piston pump investigation.

Figure 2. Photo of the test rig.

As mentioned above, the system consists of a variable displacement axial piston type A10VSO designed for open-circulation hydraulic drive systems. The pump has an 18 cubic centimeters displacement volume. It is equipped with a hydraulic pressure controller whose designation is better known as DR. Parallel to it is a proportional hydraulic valve VT-DFP (Bosch Rexroth Company) designed only for this type of pump and having a built-in spool position sensor. Control of this type of valve is realized by a selective external electronic amplifier type VT-5041, of which only the valve spool position feedback function and the power stage are used to power its proportional solenoid [**23**]. Classically, the system structure is presented in **Figure 3**. In the developed solution, an electronic amplifier is used to receive a reference signal from a PLC MC012 [**24**]. The system software is distributed between the industrial controller and the desktop workstation to allow rapid prototyping of the different types of identification experiments and embedded controllers. This is another advantage of the system as it makes it possible to realize advanced control laws, not just conventional PID controllers and their varieties established in hydraulic drives. The distributed system is based on a real-time communication channel between the PLC and the workstation running a Simulink® model (Mathworks). Communication is carried out via a USB/CAN network in blocking mode with a synchronization packet emitted from the industrial controller with a sample time of 10 milliseconds. This sampling period is fast enough to allow precise control of the pressure and flow rate.

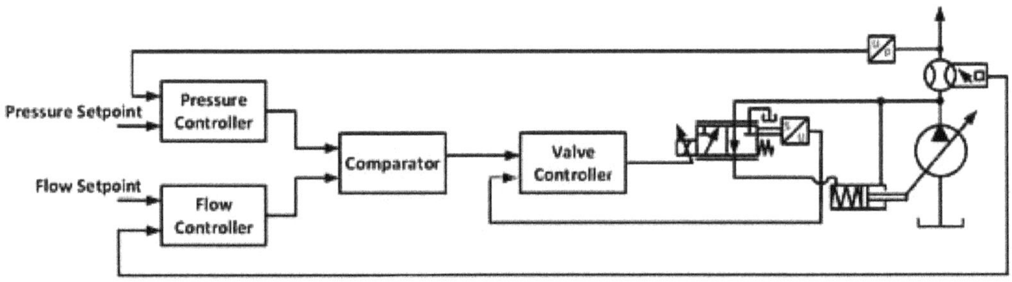

Figure 3. Hydraulic circuit diagram of a test rig for investigation of the axial piston pump.

3. SYSTEM IDENTIFICATION

To employ the linear quadratic control method, a linear time-invariant state space representation of the axial piston pump is required. Generally, two approaches are available to build a state space model, either through the application of first principle physical laws [**25**] and the consecutive estimation of model parameters ("gray box" model), or alternatively, the application of "black box" system identification techniques [**26,27**]. The "gray box" approach requires a detailed understanding of the internal construction of the axial piston pump, as well as the exact measurement of geometrical and physical variables. Even though that approach is beneficial for advancing construction aspects of the axial piston pump machine by reaching a high-fidelity simulation model, it does not guarantee successful control system design. As we know, the purpose of control-oriented models is to focus on the numerical signal transformations between manipulated and performance variables because the controller will aim to invert these transformations in order to achieve the desired behavior of the system. Certainly, input–output transformation can be obtained from a high-fidelity physical model of the system; however, a lot of non-control-related information will also be incorporated into the physical model. Hence, considering the amount of effort and experimental work necessary to achieve precise physical modelling, in most situations, control engineers prefer the estimation of black box functional models. Additionally, digital controllers require discrete-time representation of the system dynamics, which can be naturally achieved by selecting a discrete-time structure for the representation of the black box model. The key elements in black box system identification are the model structure, identification experiment and estimation method.

For the purpose of axial piston pump control, the selected model structure is a discrete-time state space model in controllability canonical form (Cauchy representation):

$$x(k+1) = Fx(k) + Gu(k) + Ke(k), \tag{1}$$

$$y(k) = Cx(k) + Du(k) + e(k), \tag{2}$$

where x=$(x1,x2,...,xn)T$ is a state vector, $u(k)$ is the PWM signal applied to a proportional spool valve, y=$(yq,yp)T$ is the output signal, yq is the pump flow rate, yp is the pump pressure and e=$(eq,ep)T$ is the residual error with respect to pump flow rate (eq) and ep pump pressure. F,,K,C,D are model parameters which are matrices with appropriate dimensions. To obtain model parameters, open loop identification experiments are performed for variable loading conditions according to the scheme presented in **Figure 4.** The discrete-time experiment is carried out with sample time Ts=0.01 s=0.01 s, which is sufficiently small for this type of dynamic system. The identification input signal is a white Gaussian noise that is applied as a PWM voltage signal to a proportional valve.

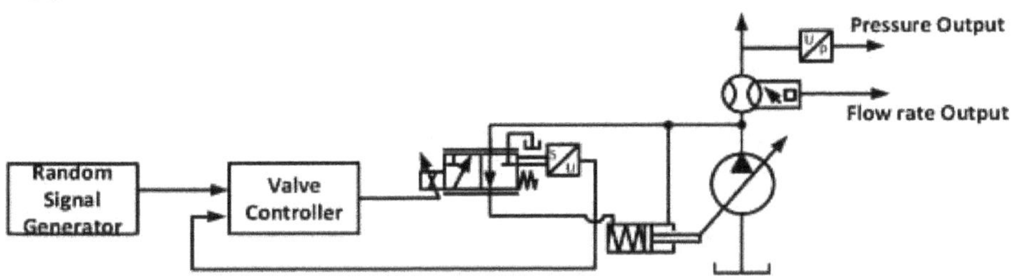

Figure 4. Structure scheme of open loop identification experiment.

The measured output signals are pump flow rate and pump pressure. The date set obtained after the experiment in the presence of full loading (the loading throttle valve depicted in **Figure 1** is approximately close) is depicted in **Figure 5**, **Figure 6** and **Figure 7**.

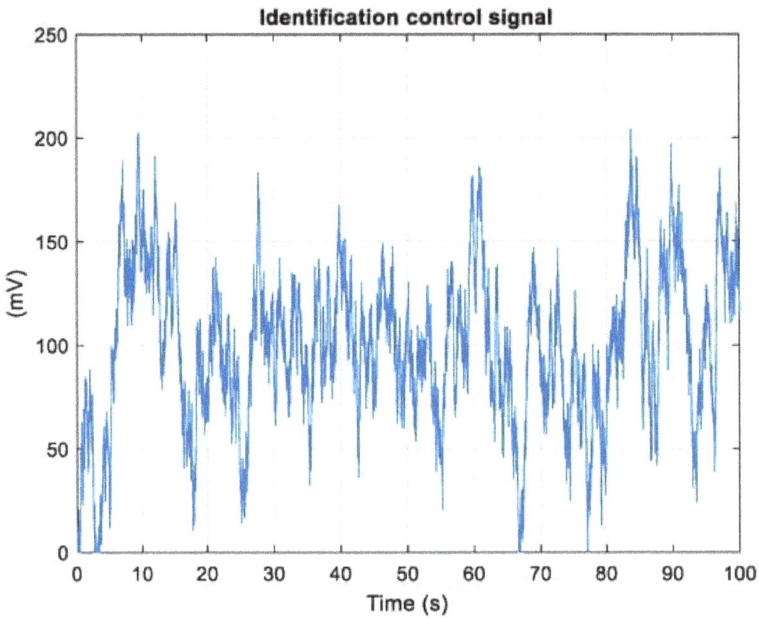

Figure 5. PWM voltage signal proportional to valve used for identification experiment.

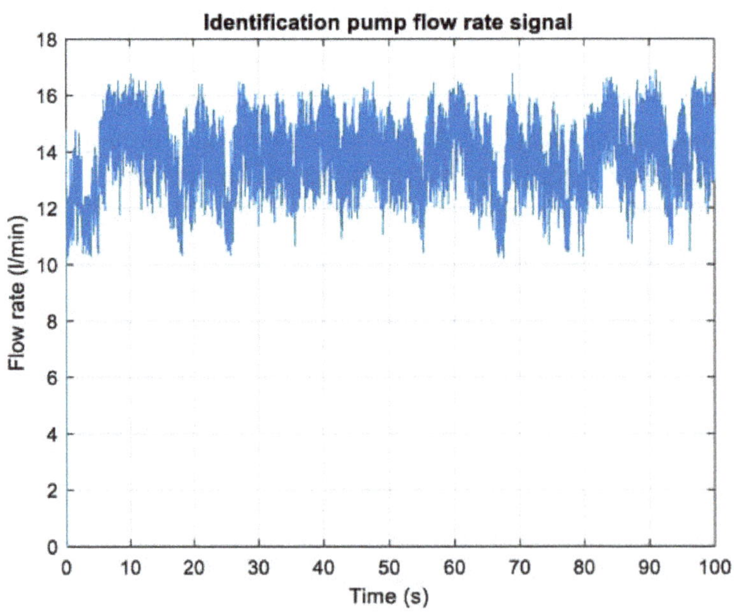

Figure 6. Pump flow rate signal used for identification experiment.

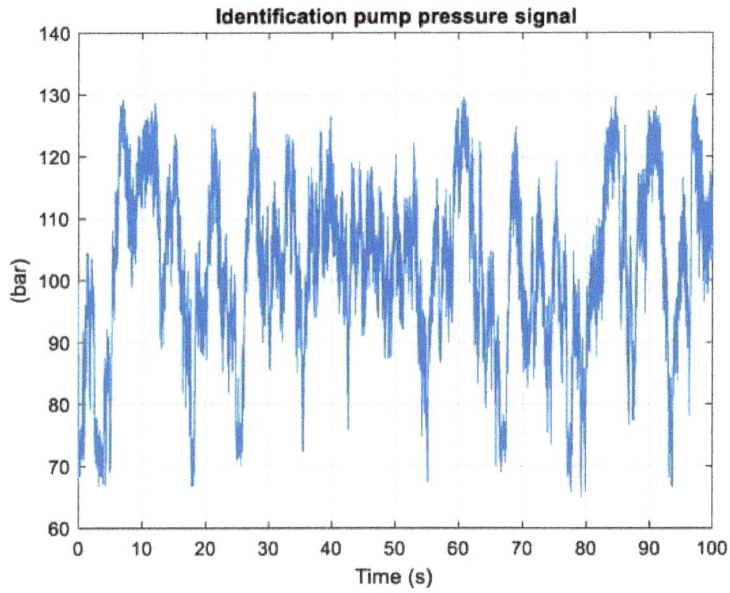

Figure 7. Pump pressure signal used for identification experiment.

The first 1000 points are removed from the data set in order to exclude transient responses. The remaining data are centered and divided into two data sets, the first is used for parameter estimation, and the second is used for model validation. Due to the absence of accurate information for the order of models (1) and (2), the models of the 1st to 10th order are estimated.

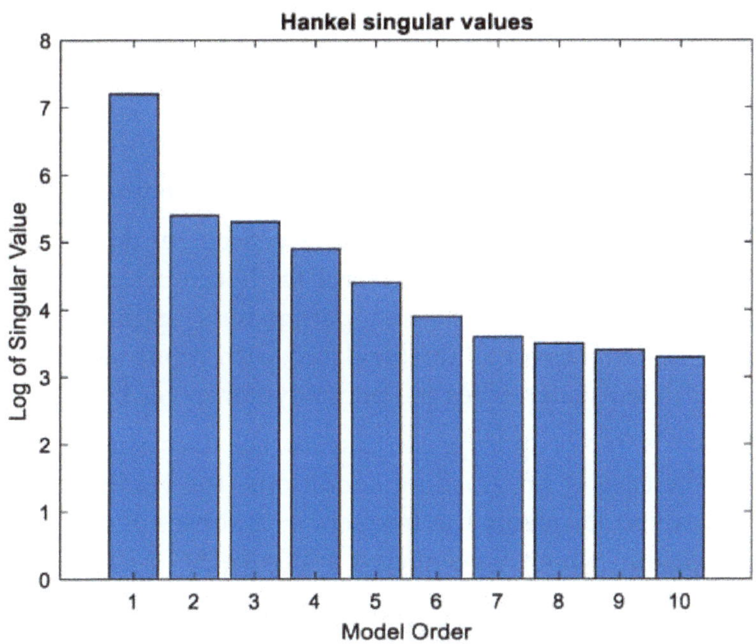

Figure 8. Hankel singular values of estimated models.

The Hankel singular values of models from the 1st to 10th order are depicted in **Figure 8**. As can be seen from the figure, the appropriate model order is the 5th, but the model of the 6th order has singular values close to the ones of the 5th order model. The Hankel singular values test provides only initial information about the supposed model order and cannot be used alone to determine the appropriate model. Thus, the models of the 4th, 5th, 6th and 8th orders are estimated. The comparison of measured outputs and those simulated by estimated model outputs are presented in **Figure 9**. As can be seen from the pump flow rate, the FIT between the measured and simulated signals is almost the same for all models. However, for the pump pressure, the FIT between the measured and simulated signal for the models of the 5th, 6th and 8th orders is significantly large compared to the one for the 4th order model.

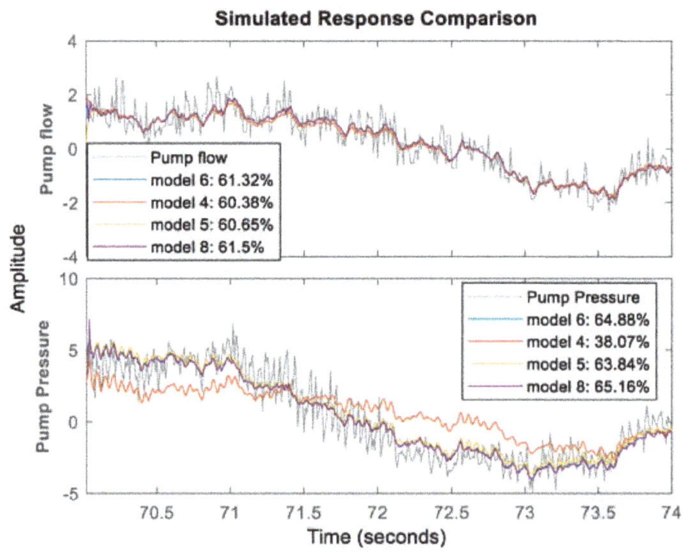

Figure 9. Comparison of model outputs and measured outputs.

The models of the 5th, 6th and 8th orders passed residual correlation tests (**Figure 10**). As can be seen, the residual error for both outputs is white Gaussian noise. This indicates that the obtained parameter's estimates are unbiased. Additionally, the residual error is uncorrelated with input signals which shows that the model sufficiently describes the pump dynamics for both channels.

The 6th order is most appropriate due to the fact that it also has good performance in the case of non-loading conditions (the loading throttle valve is fully open). This can be seen from the frequency responses of estimated models in non-loading conditions, which are depicted in **Figure 11**. It is seen that in a high-frequency range, the responses of the 6th and 8th model orders are close, whereas the response of the 5th order model is approximately the same as the 4th order model.

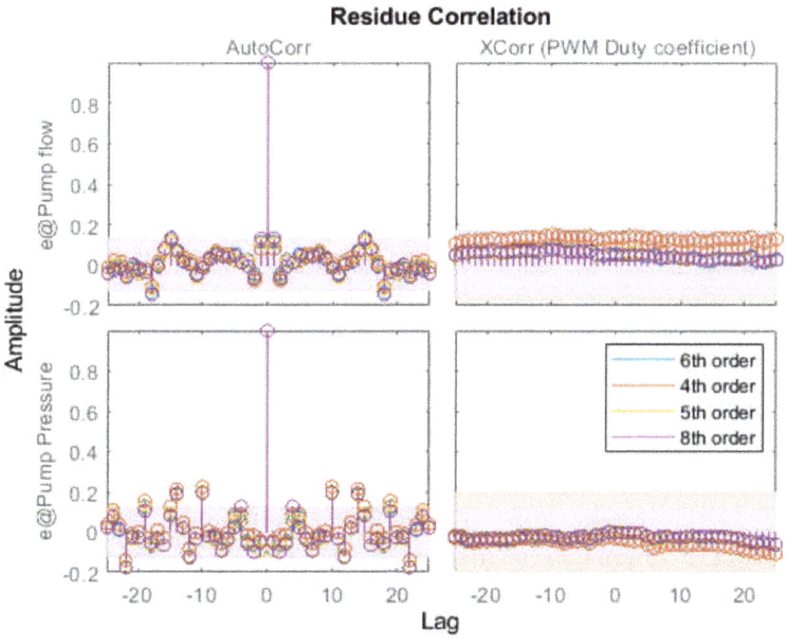

Figure 10. Residual correlation test of 6th order model.

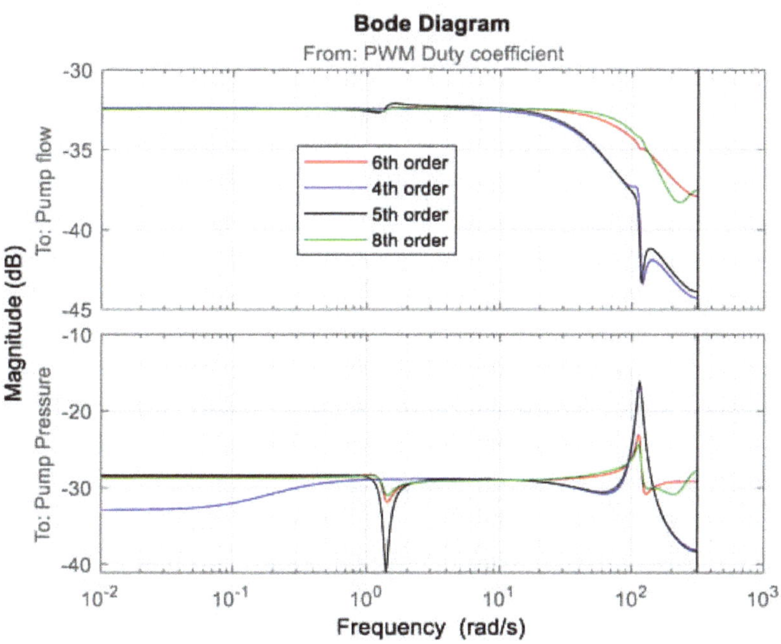

Figure 11. Comparison of frequency responses of 4th, 5th, 6th and 8th order model in case of non-loading.

The step responses along their confidence intervals with respect to pump flow rate and pump pressure for the 4th and 6th model orders are depicted in **Figure 12.** It is seen that the confidence intervals are small.

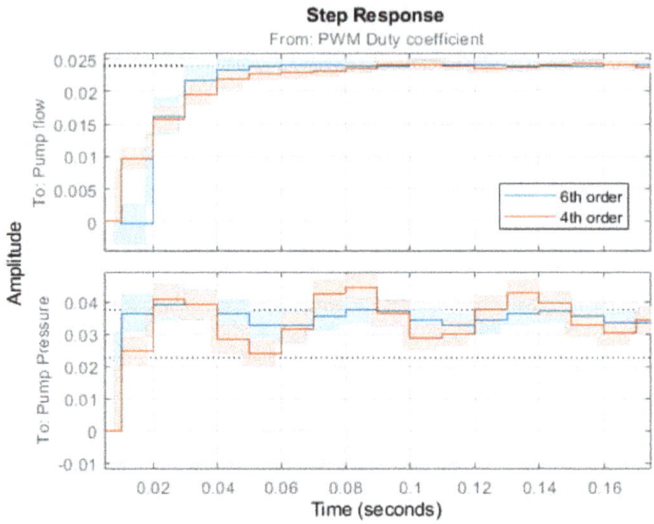

Figure 12. Comparison of magnitude frequency of 6th order model.

The matrices of the estimated 6th order model take the values:

$$A = \begin{pmatrix} 0 & 1 & 0 & 0 & 0 & 0 \\ 0 & 0 & 1 & 0 & 0 & 0 \\ -0.3 & 0.43 & 0.84 & -0.01 & 0.04 & -0.03 \\ 0 & 0 & 0 & 0 & 1 & 0 \\ 0 & 0 & 0 & 0 & 0 & 1 \\ 0.57 & -11.17 & 10.83 & 0.72 & -1.62 & 1.9 \end{pmatrix}, B = \begin{pmatrix} 0.005 \\ 0.161 \\ 0 \\ 0.209 \\ 0.046 \\ 0.058 \end{pmatrix},$$
$$K = \begin{pmatrix} 0 & 0.075 \\ 0.082 & -0.015 \\ 0.031 & 0.031 \\ 0.423 & 0.099 \\ -0.033 & 0.259 \\ 0.141 & -0.029 \end{pmatrix}, C = \begin{pmatrix} 1 & 0 & 0 & 0 & 0 & 0 \\ 0 & 0 & 0 & 1 & 0 & 0 \end{pmatrix} \text{ and } D = \begin{pmatrix} 0 \\ 0 \end{pmatrix}. \quad (3)$$

4. LQG CONTROL DESIGN

The structure of the designed control system is depicted in **Figure 13.** The purpose of the controller design in this research is to regulate the flow rate *yq* of the axial piston pump to the desired reference value *rq* with maximal steady-state accuracy and minimal settling time. Several problems arise in the control of axial piston pumps. First, the pump is based on a single input multiple output (SIMO) model. It can be established through the controllability test that we can regulate both pressure *yp* and the flow rate *yq* to desired values, if and only if they lie in a common equilibrium manifold $yp=f(yq)$. However, due to the uncertainty in the system dynamics, the exact analytical form of the manifold is not established; hence, we can either regulate the pressure or the flow rate alternately. The second problem arising from the control design is the inherent uncertainty in the identified pump model. This uncertainty can be characterized by the limited size of the identification data, the presence of uncontrollable disturbances acting on the plant and the non-stationarity of the pump dynamics with the selected operating point. The operating point is selected in our experiment by the level of closing of the precise resistance throttle valve bypassing the pump. Hence, the third problem with the present system is that due to uncertainty, it requires a feedback controller. However, due to the static nature of the plant model, we need to extend the feedback loop with an additional integrator element to guarantee reference tracking in the presence of output loading disturbances.

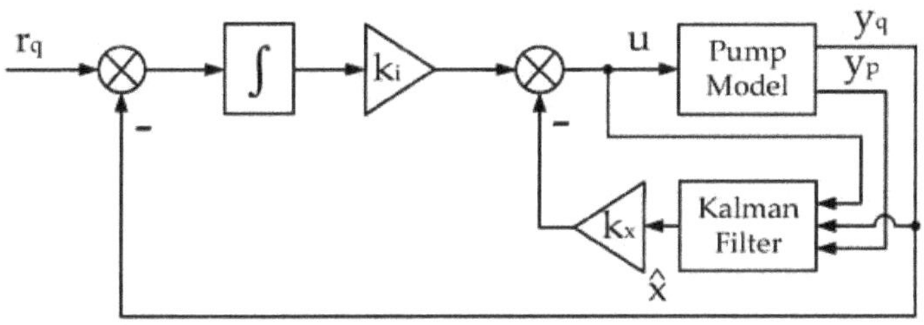

Figure 13. Structure of closed-loop system with LQG controller and Kalman filter.

The linear quadratic control approach is derived from the key results of optimal control theory, where a quadratic cost functional is minimized with respect to the realization of the control signal. The benefit of the LQR optimal solution is that it can be represented as a function of the Lagrange multiplier costate vector which is coupled to the system state vector through the solution of the differential Ricatti equation **[28]**. The quadratic cost function for the closed-loop pump model is selected as

$$J(u) = \sum_{k=0}^{\infty} x^T[k]Q_x x[k] + ru[k]^2 + w_{I_y} I_y[k], \tag{4}$$

where $\boldsymbol{x}[k]=(\boldsymbol{x}q[k]\boldsymbol{x}2[k]\boldsymbol{x}3[k]\boldsymbol{x}p[k]\boldsymbol{x}5[k]\boldsymbol{x}6[k])T$ is the state vector of the identified pump model, $Qx\in\mathbb{R}6\times6$ is the positive definite real matrix weight of the state vector components and their correlations and $r\in\mathbb{R}$ is the weight gain for the amplitude of the control signal. In the cost function, we also include the signal $I[k]$ weighted with $Qy\in\mathbb{R}$, which represents the integrated output error with respect to the flow rate:

$$I_y[k] = \sum_{i=0}^{k} T_S(r_q[k] - y_q[k]) = \sum_{i=0}^{k} T_S e_q[k]. \tag{5}$$

For calculation purposes, the integrated signal is represented as a finite difference equation for a first-order Euler approximation of the integration operation in continuous time as

$$I_y[k] = I_y[k-1] + T_S e_q[k]. \tag{6}$$

The weighting matrix Qx, together with r and wIy, are tunable parameters reflecting performance requirements for the closed-loop system. However, due to the implicit representation of the physical state variables in the state vector, it is not obvious how to select the components of the matrix Qx. Hence, we employ mapping with matrix C between the output and the state variables to translate the tuning of Qx into the output space as

$$y^T[k]Q_y y[k] = x^T[k]\begin{pmatrix} C_q{}^T & 0 \\ 0 & C_p{}^T \end{pmatrix} Q_y \begin{pmatrix} C_q & 0 \\ 0 & C_p \end{pmatrix} x[k] = x^T[k]Q_x x[k], \quad (7)$$

where the matrix $Qy\in\mathbb{R}2\times2\in\mathbb{R}2\times2$ is a corresponding weighting matrix in the output space of the model. Since the physical meaning of the output signals is established as $y=(yqyp)T$, we can specify the matrix $Qy=diag(wq,wp)$, where wq represents the cost weight for the flow rate channel and wp represents the cost weight for the pump pressure channel. In this mapping, the matrix

$$Q_x = \begin{pmatrix} C_q{}^T w_q C_q & 0 \\ 0 & C_p{}^T w_p C_p \end{pmatrix}. \quad (8)$$

Considering that in the identified model, matrices Cq and Cp contain just one non-zero element, i.e., a unit corresponding to the state for the output signal, we see that with such a weighting matrix, we put weight on only two of the states and the rest are left outside of the cost function, or we do not care for their amplitude, since they do not carry an explicit physical meaning.

Equivalently, we can extend the state vector of the plant model, including the integrator state $I[k]$, as

$$x_{ext} = \begin{pmatrix} I_y \\ x \end{pmatrix}, \quad (9)$$

and then the plant model matrices F and G will become

$$F = \begin{pmatrix} 1 & -T_S C_q & 0 \\ 0 & F_q & F_{qp} \\ 0 & F_{pq} & F_p \end{pmatrix}, G = \begin{pmatrix} 0 \\ G_q \\ G_p \end{pmatrix}. \quad (10)$$

Additionally, the state weighing matrix for the extended state vector will become

$$Q_{ext} = \begin{pmatrix} w_{I_y} & 0 \\ 0 & Q_x \end{pmatrix}. \quad (11)$$

Following the methodology for LQR design, we look for a linear state feedback controller gain such that the control signal is

$$u[k] = -K_p x[k] + k_i e_q[k] = -K_{ext} x_{ext}[k], \quad (12)$$

where $Kp\in\mathbb{R}1\times6$ is a row vector with proportional gains multiplying the state vector, and $ki\in\mathbb{R}$ is the integral gain multiplying the reference tracking error $eq[k]$ for the flow rate, then $Kext=(kiKp)$. According to ref. **[28]**, these gains can be obtained as

$$K_{ext} = \frac{G^T PF}{G^T PG + r}, \quad (13)$$

where the denominator is scalar because the system has a single input. The matrix P is a positive definite solution of the infinite time algebraic Ricatti equation

$$F^T PF - P - F^T PG\left(G^T PB + r\right)^{-1} G^T PF + Q_{ext} = 0. \quad (14)$$

The numerical values we select for the integrated error wIy=500, the flow rate channel wq=0.001, the pump pressure wp=0.001.001 and the control signal amplitude r=0.001 lead to the following solution for the state feedback

$$k_i = 637.7478, \quad (15)$$

$$K_p = (-1.27, 19.9, 28.91, -0.88, 1.77, -0.92). \quad (16)$$

As can be seen, the integral gain of 637 dominates the controller response, followed by the gain of 28.9 for the third state variable, representing the delayed flow rate output $yq[k\text{-}2]$, followed by the gain of 19.9 for the $yq[k\text{-}1]$.

The key problem with the practical implementation of the LQR controller is that the internal state vector of the system $[k]$ is generally inaccessible for direct measurement. Hence, it must be estimated by a calculated value $\hat{\boldsymbol{x}}[k]$ from the previous values of the measurable system signals (the input and output plant channels). To produce the state vector estimate, we employ the stationary Kalman filter algorithm, which minimizes the variance $E((\boldsymbol{x}[k]\text{-}\hat{\boldsymbol{x}}[k])2)$ of the estimation error over all initial conditions and realizations of the model residuals $eq[k]$ and $ep[k]$, which are assumed to behave as Gaussian white-noise processes with unit variance. In addition, an unbiased estimate is usually required, meaning

$$E\{\hat{x}[k]\} = E\{x[k]\}. \quad (17)$$

The Kalman filter algorithm has many modifications **[29]** depending on the model of the process and the assumptions about the statistical distribution of the noise and disturbances. The model of the axial piston pump is a linear time-invariant system with stationary noise action on the state and output. Hence, we design the state observer algorithm as

$$\hat{x}[k+1] = F\hat{x}[k] + Gu[k] + L(y[k+1] - CF\hat{x}[k] - CG\hat{x}[k]), \quad (18)$$

where L∈ℝ6×2 is the filter gain matrix calculated to minimize the state error variance. The filter gain is calculated as

$$L = DC^T\left(CDC^T + 0.01I_2\right)^{-1}, \quad (19)$$

where Ve=E(eeT)∈ℝ2×2 is the variance of models' (1) and (2) residuals, and D is the error covariance matrix calculated as a solution of the algebraic Ricatti equation,

$$D = FD_eF^T + KV_eK^T. \quad (20)$$

The resultant filter gain matrix obtained from the identified model becomes

$$L = \begin{pmatrix} 0 & 0.07 \\ 0.08 & -0.02 \\ 0.03 & 0.03 \\ 0.42 & 0.1 \\ -0.03 & 0.26 \\ 0.14 & -0.03 \end{pmatrix}. \tag{21}$$

The first column from the filter gain matrix corresponds to the flow rate channel and the second column corresponds to the pressure output signal. The filter gain depends on how much information is carried from each state to the output signal and how much we can trust the output signal given the present level of external noise variance. Therefore, the higher gains in matrix L indicate that the respective state variable will be more strongly compensated by the output estimation error. The optimal error covariance matrix obtained from the solution of the Ricatti equation is

$$D = 10^{-4} \times \begin{pmatrix} 2 & 0 & 1 & -2 & 2 & -1 \\ 0 & 2 & 1 & 3 & -1 & 0 \\ 1 & 1 & 1 & 0 & 0 & 0 \\ -2 & 3 & 0 & 40 & 11 & 13 \\ 2 & -1 & 0 & 11 & 24 & 12 \\ -1 & 0 & 0 & 13 & 21 & 21 \end{pmatrix}. \tag{22}$$

As can be seen from the diagonal elements, the variance of the first three states determining the flow rate output signal of the model is 10 times lower than the variances of the pressure determining states. The same is true for the cross-correlations between the errors of the estimated states. The fundamental mechanical constraints can explain that in the pump, when provided with a fixed input signal to the proportional valve, the swash plate swivel angle will determine a fixed displacement volume leading to a fixed flow rate in dependence on the shaft rotation speed. On the other hand, the pressure in the system will be more strongly dependent on the externally attached loading system to the pump and not so much on the input signal (t). Hence, the variance in the estimated pressure states will become higher. Another observation from the state covariance matrix is that the 3rd and 6th states have lower variances than the 1st and 2nd or 4th and 5th states. This can be related to the casual relationships between those states if one observes the structure of matrix F.

Figure 14 presents the closed loop system frequency responses with the designed linear quadratic regulator and Kalman filter. The output sensitivity function characterizes the ability of the closed loop to attenuate the loading disturbances. We see that the designed regulator dampens the amplitudes of the output disturbances up to 100 times in the low-frequency domain and around 10 times in the middle frequency up to 2 rad/s. From the complementary sensitivity function, we can say that the bandwidth of the closed loop is around 10 rad/s which is acceptable for a regulated pump, considering the capabilities of the employed actuator and flow meter. **Figure 14** also shows the control signal sensitivity. As expected, it is higher in the low-frequency range and lowers down in the high frequencies to prevent amplification of measurement noise in the actuator command.

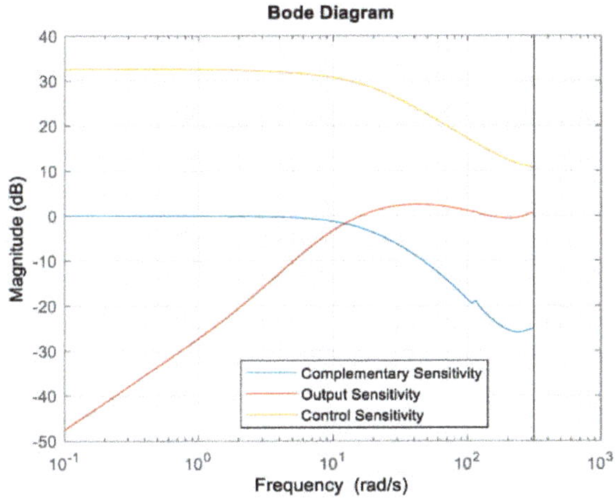

Figure 14. Sensitivities of closed-loop system with LQG controller and Kalman filter.

5. EXPERIMENTAL VALIDATION OF DESIGNED CONTROLLER

This section presents the experimental results measured with the designed controller in a hardware in the loop (HIL) experiment with the MC012 microcontroller. The loading system for the pump is represented by a variable throttle valve bypassing the output of the pump to the tank, which is presented in the hydraulic circuit diagram in **Figure 1**.

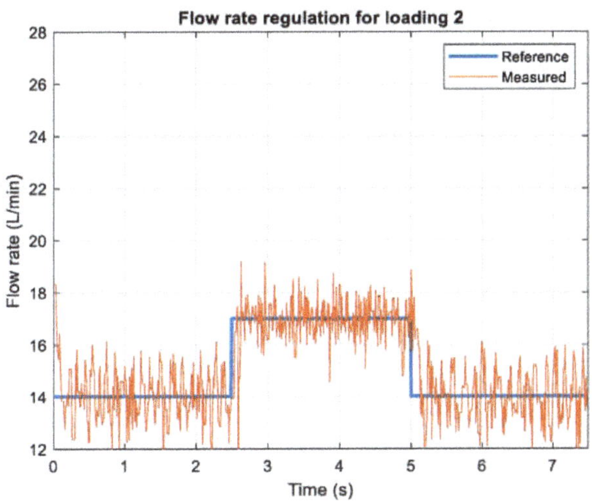

Figure 15. Experimental results for loading condition 2.

The opening of the loading throttle valve can be manually selected by a precision micrometer scale from 0 to 5 mm. We test the performance of the LQG controller for loading throttle valve openings of 2 mm, 3 mm, 4 mm and 5 mm. The recorded signals are the output pump pressure, output flow rate, LVDT feedback from the pump proportional valve and control signal sent to the electronic amplifier. The experimental results are summarized in **Figure 15**, **Figure 16**, **Figure 17** and **Figure 18**.

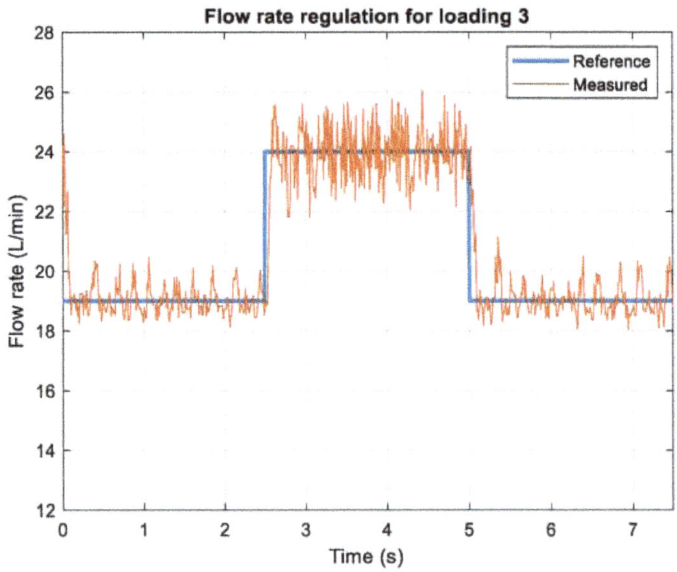

Figure 16. Experimental results for loading condition 3.

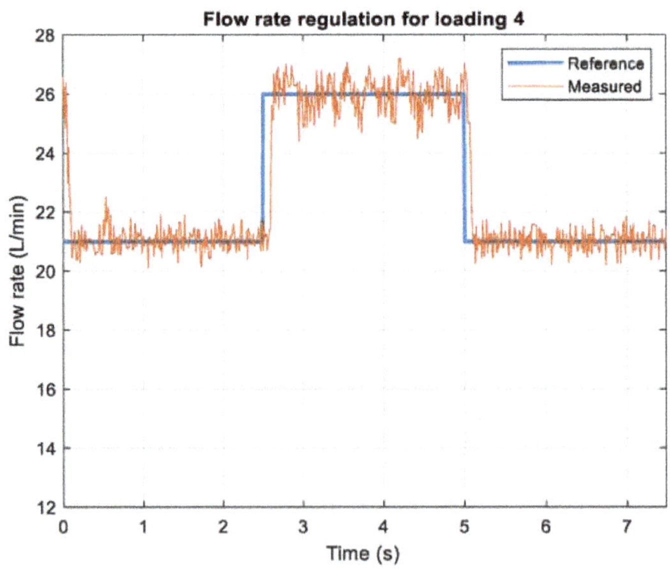

Figure 17. Experimental results for loading condition 4.

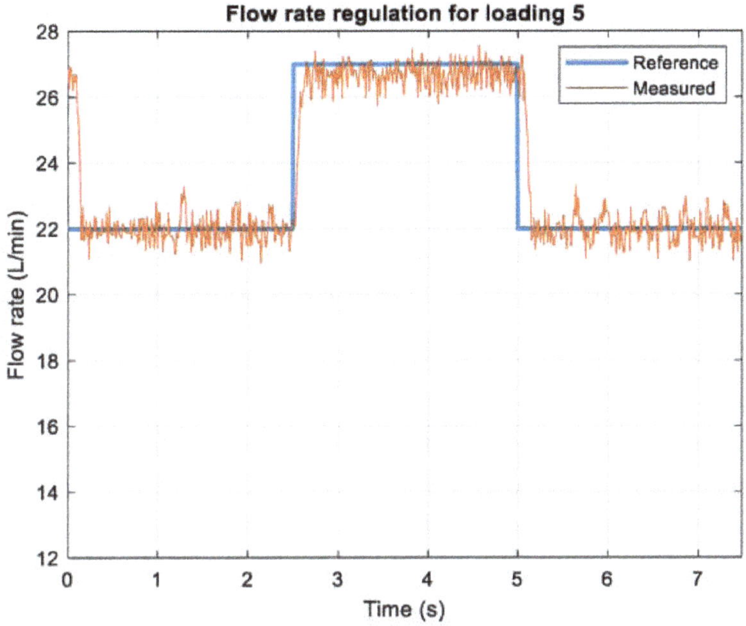

Figure 18. Experimental results for loading condition 5.

For the minimal opening of the loading valve of 2 mm, the closed loop pump system will experience maximal loading as pressure. In such conditions, the reaction force on the swash plate from the output pressure will be maximal, and the control effort from the actual will be required to compensate for the variations in the flow rate. To examine the system during such conditions, we apply a reference flow rate of 14 L per minute and a step reference jump to 17 L per minute for a period of 2.5 s. As can be seen in the figure, we note a very fast system response below 100 ms. The flow rate is held around the reference value with random deviations of +/- 0.5 L per min. Part of these observed variations can be attributed to the sensor noise and imprecision. As noted in the introduction, we employ a low-cost flow rate meter based on an orbital hydraulic motor and a hall sensor.

Measurement is based on the capturing of the impulses period from the Hall sensor with a precision of 100 ns. Hence, this measurement technique will introduce some limitations to the maximal accuracy we can obtain for the flow rate. Another component of the observed flow rate variations can be attributed to the actual irregularities in the flow through the gear flow meter. During these high loading experiments, we observe very demanding conditions for the system with pressure in the range of 80 to 120 bar, providing a power output of around 2 kW. Since the electrical motor driving the pump shaft is an unregulated three-phase induction machine, such a high load will lead to variation in the rotational speed of the motor, reflecting directly on the generated flow rate by the pump. Even in such high loading conditions, the designed control system is able to keep the mean value of the flow rate signal around the reference value without much amplified noise in the control signal.

Figure 16 examines the transient behavior for lower loading corresponding to the opening of the loading throttle valve of 3 mm. The pressure during this experiment is around 50 bars, which corresponds to the output power of the pump of around 1.5 kW. The reference signal of the controller is a step from 19 L per minute up to 24 L per minute. The transient response and accuracy of the regulations are comparable to the case of a 1.5 kW operating point with a settling time below 100 ms. It is interesting to note that with an increase in flow rate through the pump, increasing the pressure and the output power, the amplitude of flow rate variations increases. This can be taken as a confirmation of the hypothesis that flow rate variations are due to increased load on the induction motor causing the instability of its rotational speed.

Figure 17 examines the reference tracking performance for the case of a 4 mm opening of the loading throttle valve. In this case, we set the reference step levels between 21 L/min and 26 L/min. The operating pressure in these conditions is between 20 to 30 bars, providing a power output for the pump of around 1 kW. In these conditions, the settling time is increased to around 100 ms. Due to decreased pressure in the system, the reaction of the swash plate to change in the actuating piston becomes slower; hence, the transient responses are delayed. However, such a scale of the response is within sufficient limits for the practical application of the controller. The flow rate variations, in this case, are comparable to the case with the opening of the loading valve up to 3 mm. Similar to the previous case, the amplitude of the variations increases with the increasing flow rate, hence the increased power output of the pump.

The last examined loading condition (**Figure 18**) corresponds to a 5 mm opening of the loading valve, which can be characterized as minimal loading. During the experiment, the reference step is from 22 L per minute to 27 L per minute. The settling time is around 200 ms, which again is increased because of the decrease in the pressure in the system, reducing the forces acting on the action valve to move the swash plate. It is worth noting that, in these cases of loading, minimized variations of the stationary flow rate around the reference are observed. The pressure in the system is around 20 bar which corresponds to power output below 1 kW. In this case, the loading of the induction motor driving the pump is reduced, and its rotation speed is more stable, which minimizes the variations in the flow rate.

Experimental results for the flow rate regulation prove the experimental stability and performance of the designed LQG controller. The controller is examined for four operating points in the full range of loading conditions for the pump between 14 to 26 L per minute. Even though the identified model is for a single operating point, the closed loop system is able to maintain the reference tracking performance, which is an indication of its robustness.

Figure 19 compares the resultant pressures in the system for the four operating points selected by opening the loading throttle valve. As can be expected with the increase in flow rate, the pressure in the system is increased too. It is notable that for the highest loading conditions, the oscillations in pressure are much more pronounced. Hence, it is not recommended to operate the closed loop system at such high pressure for prolonged intervals. The working envelope of the designed closed loop system can be characterized by a flow rate range between 14 and 26 L/min and a pressure range between 20 to 80 bar. Another thing which can be noted in the figure is that there is a nonlinear relationship between the opening of the loading valve and the mean value of the pressure in the system.

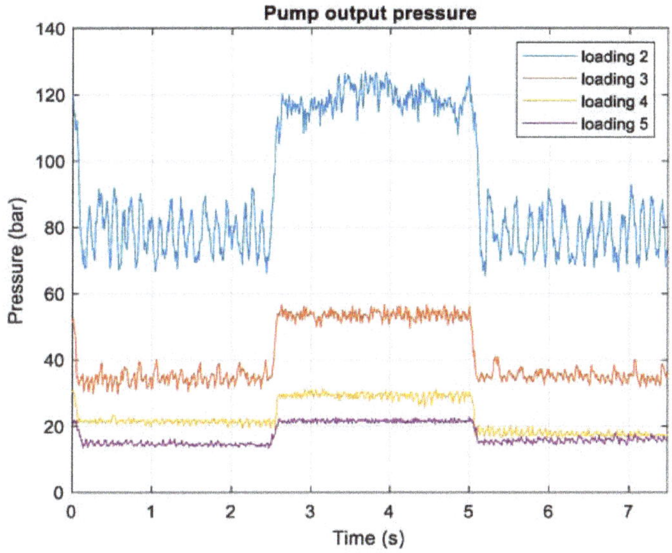

Figure 19. Comparison of pump output pressure for different loading conditions.

Figure 20 shows the control system from the LQG controller during the experiments. The control signal sent to the valve is from 0 to 10V. Therefore, the negative values in the figures are saturated at 0. The most important feature of the examined control signals is that they do not amplify the sensor noise present in the output. This is because of the application of the noise model from the system identification and because of the design of the Kalman filter. Such output noises can be easily amplified by some classical control algorithms, such as PID, leading to the overloading of the actuator valve. It is interesting to note that the average value of the control signals for the selected operating points is not much different, which is an indication that the identified model is able to carry information not only for its operating point but for others too.

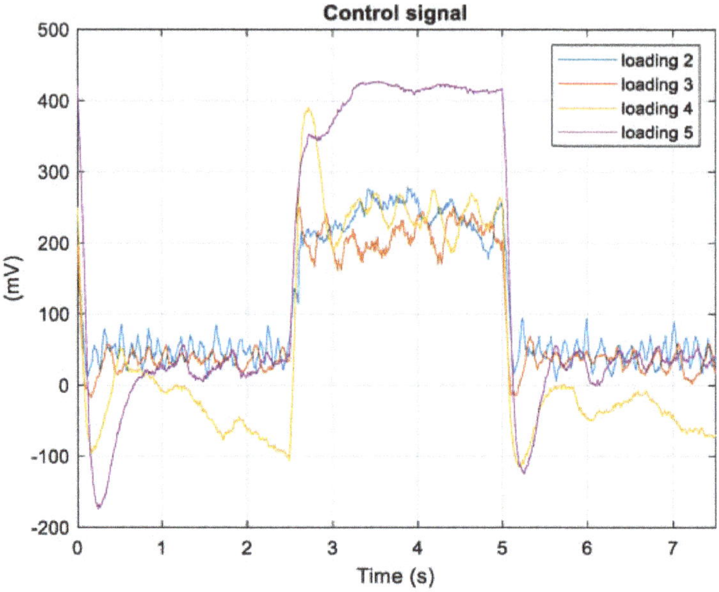

Figure 20. Comparison of control signals for different loading conditions.

Additionally, in **Figure 21**, we present the LVDT signal for the proportional valve position during the examined loading conditions. The valve we use is bidirectional; hence, we limit its range from 0 to 50% because in the pump we test, the swash plate can work only on positive swivel angles. Alternatively, if the pump can be inverted to work as a motor, the proportional valve operating range will be from 0 to 100%. In the figure, it is important to note that the proportional valve is fast enough to follow the dynamics of the generated control signal from the controller.

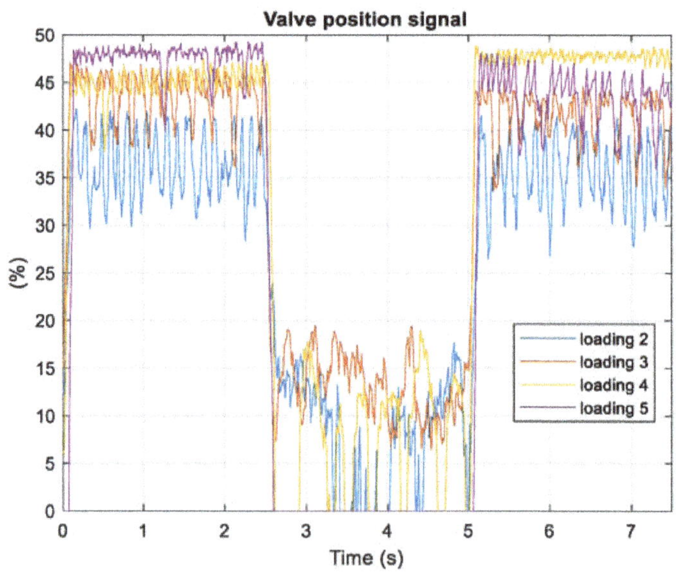

Figure 21. Comparison of LVDT signals for different loading conditions.

As an additional experiment, we examine the performance of the pump for a fixed reference flow of 22 L per minute and with variable (dynamic) loading conditions simulating a real application operation. The loading is modeled by manually decreasing the opening of the loading valve from 5 mm to 1 mm. **Figure 22** and **Figure 23** present the results of this experiment. As can be seen in **Figure 22**, the flow rate is held steady by the control system at 22 L/min for all loading pressures in **Figure 23** from 20 to 100 bar. With the increase in the loading of the pump and the required output power from the induction motor, the oscillations in the flow rate begin to increase too.

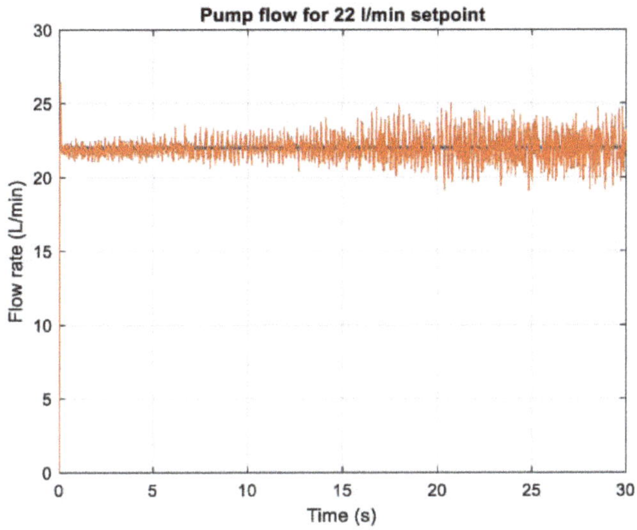

Figure 22. Pump flow rate for 22 L/min setpoints.

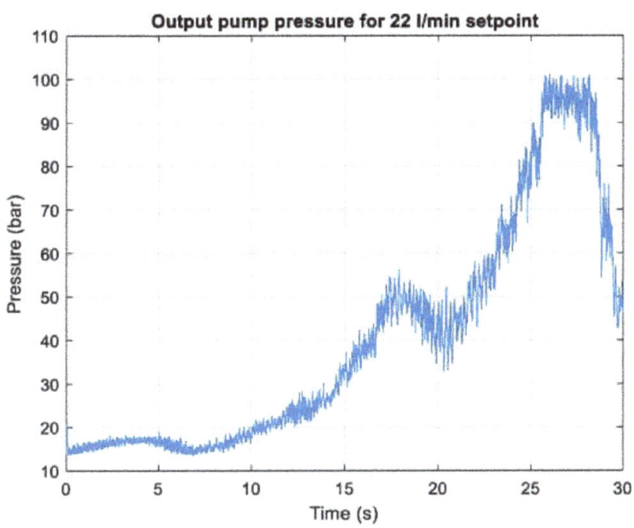

Figure 23. Output pump pressure for 22 L/min setpoints.

6. CONCISE SUMMARY

The article's primary focus is on creating an embedded system designed to control an axial piston pump equipped with a proportional valve.This work introduces an embedded system that utilizes a pump without a swash plate swivel angle sensor and rotating speed sensor, unlike current systems.This approach is suitable for flow and pressure control in opencircuit hydraulic drive systems but not for secondary contol.The proposed control solutions allow for the implementation of both traditional analogue PID controllers and more sophisticated control algorithms. An estimation of the "black box" model of the axial piston pump with proportional valve control is derived from experimental data.The validation tests demonstrate the model's correctness and suitability for controller design . he control system is constructed using a linear-quadratic Gaussian (LQG) controller with a time-invariant Kalman filter and is applied to a laboratory test rig.

An indepth explanation of the hydraulic system regarding pump displacement control is provided, along with the system architecture designed for its control. The control system's design is confirmed by experimental investigations under different loading circumstances to assess its performance.

7. AT A GLANCE

Recently, advancements in hydraulic variable displacement axial piston machines have concentrated on enhancing their design and refining their displacement control techniques. Both orientations aim to enhance the efficiency of the machinery. Enhancing their performance is crucial for enhancing the overall efficiency of the hydraulic system, whether they function as pumps or hydraulic motors. This work is motivated by the development of an embedded control system specifically built for a recognized type of open circuit axial piston pump. The solution is applied on a laboratory test equipment. An in-depth explanation of the hydraulic system regarding pump displacement control is provided, along with the designed system architecture for its regulation. The control system utilizes a linear-quadratic Gaussian (LQG) controller. The controller is created using a model derived from identification using experimental data. The controller's design is confirmed through experimental tests, allowing for the evaluation of its performance.

REFERENCES

1. Manring, N. Fluid Power Pumps and Motors: Analysis, Design, and Control; McGraw-Hill Education: New York, NY, USA, 2013. [**Google Scholar**]
2. Frankenfield, T. Using Industrial Hydraulics. In Rexroth Worldwide Hydraulics; Penton Publishing Inc.: Belfast, UK, 1984. [**Google Scholar**]

3. Fresia, P.; Rundo, M. Lumped parameter model and experimental tests on a pressure limiter for variable displacement pumps. E3S Web. Conf. **2020**, 197, 07005. **[Google Scholar] [CrossRef]**

4. Mitov, A.; Kralev, J.; Slavov, T.; Angelov, I. Design of Embedded Control System for Open Circuit Axial Piston Pump. In Proceedings of the 22st International Symposium on Electrical Apparatus and Technologies (SIELA 2022), Bourgas, Bulgaria, 1–4 June 2022. in press. **[Google Scholar]**

5. Casoli, P.; Pastori, M.; Scolari, F. Swash plate design for pressure ripple reduction—A theoretical analysis. AIP Conf. Proc. **2019**, 2191, 020038. **[Google Scholar]**

6. Karpenko, M.; Marijonas, B. Investigation of hydrodynamic processes in the system—Axial piston pumps–pipeline–fittings. In Transport Problems 2018, X International Scientific Conference, VII International Symposium of Young Researchers: Proceedings, Katowice, Wisła, 27–29 June 2018; Silesian University of Technology: Gliwice, Poland, 2018; pp. 832–843. **[Google Scholar]**

7. Ivantysyn, J.; Ivantysynova, M. Hydrostatic Pumps and Drives; Vogel: Wurzburg, Germany, 1993. **[Google Scholar]**

8. Zeiger, G.; Akers, A. Dynamic Analysis of an Axial Piston Pump Swashplate Control. Proc. Inst. Mech. Eng. Part C J. Mech. Eng. Sci. **1986**, 200, 49–58. **[Google Scholar] [CrossRef]**

9. Manring, N.; Mehta, V. Physical limitations for the bandwidth frequency of a pressure controlled, axial-piston pump. J. Dyn. Syst. Meas. Control. Trans. ASME **2011**, 133, 1–12. **[Google Scholar] [CrossRef]**

10. Murrenhoff, H. Regelung von Verstellbaren Verdrängereinheiten am Konstantdrucknetz. Ph.D. Thesis, RWTH Aachen, Aachen, Germany, 1983. **[Google Scholar]**

11. Lin, S.S.; Akers, A. Optimal control theory applied to pressure-controlled axial piston pump design. J. Dyn. Syst. Meas. Control. Trans. ASME **1990**, 112, 475–481. **[Google Scholar] [CrossRef]**

12. Akers, A.; Lin, S. Optimal control theory applied to a pump with single-stage electrohydraulic servovalve. J. Dyn. Syst. Meas. Control. Trans. ASME **1988**, 110, 120–125. **[Google Scholar] [CrossRef]**

13. Kordak, R. Hydrostatic Drives with Control of the Secondary Unit. In The Hydraulic Trainer; Mannesmann Rexroth GmbH: Lohr a. Main, Germany, 1996; Volume 6. **[Google Scholar]**

14. Belan, H.C.; Locateli, C.C.; Lantto, B.; Krus, P.; de Negri, V.J. Digital secondary control architecture for aircraft application. In Proceedings of the Seventh Workshop on Digital Fluid Power, Linz, Austria, 26–27 February 2015. **[Google Scholar]**
15. Berg, H.; Ivantysynova, M. Design and testing of a robust linear controller for secondary controlled hydraulic drive. Proc. Inst. Mech. Eng. Part I J. Syst. Control. Eng. **1999**, 213, 375–385. **[Google Scholar] [CrossRef]**
16. Kemmetmüller, W.; Fuchshumer, F.; Kugi, A. Nonlinear pressure control of self-supplied variable displacement axial piston pumps. Control. Eng. Pract. **2010**, 18, 84–93. **[Google Scholar] [CrossRef]**
17. Park, S.; Lee, J.; Kim, J. Robust control of the pressure in a control-cylinder with direct drive valve for the variable displacement axial piston pump. Proc. Inst. Mech. Eng. Part I J. Syst. Control. Eng. **2009**, 223, 455–465. **[Google Scholar] [CrossRef]**
18. Wei, J.; Guo, K.; Fang, J.; Tian, Q. Nonlinear supply pressure control for a variable displacement axial piston pump. Proc. Inst. Mech. Eng. Part I J. Syst. Control. Eng. **2015**, 229, 614–624. **[Google Scholar] [CrossRef]**
19. Kralev, J.K.; Petkov, P.H.; Slavov, T.N. Design of Embedded Robust Control Systems Using MATLAB®/Simulink®; IET Control: London, UK, 2018. **[Google Scholar]**
20. Forrai, A. Embedded Control System Design: A Model Based Approach; Springer: Warsaw, Poland, 2013. **[Google Scholar]**
21. Rexroth Bosch Group. Pressure and Flow Control System; Technical Data Sheet, RE 30630; Rexroth Bosch Group: Lohr a. Main, Germany, 2015. **[Google Scholar]**
22. Rexroth Bosch Group. Proportional Directional Valves, Direct Operated, with Electrical Position Feedback as Pilot Control Valve for Control Systems SY(H)DFE; Technical Data Sheet, RE 29016; Rexroth Bosch Group: Lohr a. Main, Germany, 2019. **[Google Scholar]**
23. Tonyan, M. Electronically Controlled Proportional Valves; Marcel Dekker, Inc.: New York, NY, USA, 1985. **[Google Scholar]**
24. Danfoss. Plus+1 Controllers MC012-020 and 022; Data Sheet, 11077167; Danfoss: Nordborg, Denmark, 2013. **[Google Scholar]**
25. Merrit, H. Hydraulic Control Systems; John Wiley & Sons Inc.: Hoboken, NJ, USA, 1967. **[Google Scholar]**
26. Ljung, L. System Identification: Theory for the User, 2nd ed.; Prentice Hall: Hoboken, NJ, USA, 1999. **[Google Scholar]**

27. Slavov, T.; Mitov, A.; Kralev, J. Advanced Embedded Control of Electrohydraulic Power Steering System. Cybern. Inf. Technol. **2020**, 20, 105–121. **[Google Scholar]** **[CrossRef]**

28. Zhou, K.; Doyle, J. Robust and Optimal Control; Prentice Hall International: Upper Saddle River, NJ, USA, 1996. **[Google Scholar]**

29. Grewal, M.; Andrews, A. Kalman Filtering: Theory and Practice with MATLAB®: Fourth Edition; Wiley-IEEE Press: Hoboken, NJ, USA, 2014. **[Google Scholar]**

11 UTILIZING TAGUCHI TECHNIQUES AS A SIMULATION-BASED STRATEGY TO DESIGN NUMERICAL SIMULATIONS FOR EVALUATING CONTACT STRESS IN GEROTOR PUMPS

1. INTRODUCTION: AN OVERVIEW

Fluid technology has emerged as a significant topic in relation to environmental concerns within human activities. Hydraulic machines are crucial in fluid technology and are expected to advance their transition to more environmentally friendly practices through enhanced models and simulations in the near future[1]. The gerotor pump is designed to contribute to cutting-edge research technologies in industrial and aeronautical fields, mechanical engineering, pharmacy, medicine, and more recently, in the development of a virtual prototype for use in a Digital Twin tool [2].

In modern times, there is a significant emphasis on minimizing energy waste and enhancing efficiency, as even a little percentage of these factors has become essential in all technologies. Enhancing electricity usage, considered a crucial measure in decreasing transportation's influence on climate change, can be illustrated using Figure 1 from Kawamata et al.'s research. Friction in the motor-based drive system of electric vehicles accounts for about 10% of the world's total energy usage [3]. Aerodynamic losses are more significant in electric vehicles (EVs) than in internal combustion engine (ICE) vehicles. However, energy consumption can still be improved by addressing friction losses.

Contact stresses and friction are crucial performance indicators of a trochoidal gear set functioning as a gerotor pump. From Colbourne's 1976 study on reducing contact stress in internal gear pumps to recent research on gap estimation in ORBIT motors by Roy et al. and the detailed design concept of hydraulic displacement units by Sliwinski, a significant amount of literature has been published on contact, deformation, clearance, and stress in gerotor pump teeth. “Contact” and “stress” were among the top ten most often used keywords in publications from 2009 to 2019. Ivanović and Biernacki have undertaken significant contact stress investigations on gerotor pumps during the past twelve years. Ivanović and colleagues have investigated analytical methodologies, wear rate, friction, and stress index evaluation in this complex field [9,10,11,12,13,14,15,16].

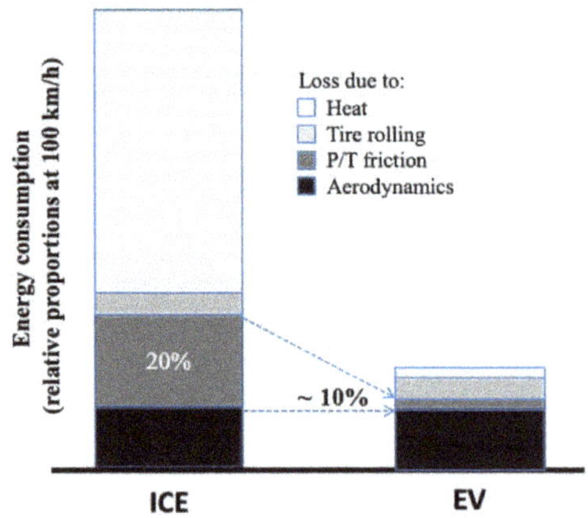

Figure 1. Comparison of energy consumption loss between ICE vehicle and EV (adapted from Kawamata et al. [3]).

Biernacki and colleagues conducted research focusing on novel materials and innovative design approaches. Both research groups have excelled in the field. When considering how to select the gear set to enhance performance indices, optimization can provide assistance. Robinson and Vacca [24,25] and De Martin et al. [26] have recently published comprehensive studies on multi-objective optimization for classic circular-toothed profiles within the last five years. The technique identified that the industry may discover optimal solutions nearly equivalent to the state-of-the-art pump designs. Tessari et al. [27] and Pulliti et al. [28] conducted a comprehensive analysis of a company’s commercial gear sets, enabling them to derive an empirical formula for eccentricity. This formula serves as the beginning parameter for the optimization technique aimed at maximizing volumetric efficiency.

When determining the geometrical parameters for designing a new trochoidal gear pump, the process is not simple. An intermediate step between using commercial gear sets and optimization procedures would be beneficial for both academia and industry. The designer aims to streamline the analysis setup process and eliminate the costs involved with running trials. Numerical simulation serves as a design assistance. When conducting a novel numerical simulation, one may encounter challenges related to parameter evaluation, determining the number of simulations to do, estimating parameter combinations, and connecting mechanical and fluid-dynamic features. This work aims to answer the researcher's issues from a practitioner's viewpoint by utilizing the Taguchi method to integrate information from commercial gear sets into numerical simulations. The Taguchi method is easily accessible to any end-user, making it an open tool that is not limited by software.

Design of Experiments (DoE) is a widely recognized and potent tool. The Department of Energy (DoE) procedure is segmented into three primary phases: The experiment involves three main phases: planning, conducting, and analysis. In the planning phase, factors and levels are chosen. During the conducting phase, test results are gathered. Finally, in the analysis phase, positive or negative information about the selected factors and levels is produced based on the data acquired in the first two phases. A full-factorial experiment with an equal number of data points for each factor level is suitable for investigating a few factors but not recommended for studying numerous factors.

The Taguchi methodology is a comprehensive method that utilizes orthogonal arrays (OA) to carry out tiny, highly fractional factorial experiments that may be scaled up to bigger, full-factorial studies [31]. Another option is to utilize Monte Carlo simulation, which necessitates a significant number of testing scenarios to achieve a precise calculation of the mean and variance. However, this process can be costly and time-intensive [29,32]. Each optical amplifier has a specified maximum achievable resolution [33]. The choice of an orthogonal array (OA) to utilize mostly relies on: (i) the quantity of factors and interactions being studied, (ii) the number of levels for the factors, and (iii) the preferred experimental resolution or budget constraints. This method will reduce the total number of tests required while still providing valuable insights. The limited connection between academics and industrial specialists has hindered the widespread implementation of the Taguchi method [34]. Dr. Taguchi invented the signal-to-noise (S/N) ratio as a performance measure to assess data and identify the optimal values of selected components that may effectively manage noise impacts. Various uncontrolled and unknown causes that fluctuate in intensity contribute to variances between data points, resulting in noise effects. Taguchi parameter design uses repetitions to find which control elements and their levels can reduce variation, instead of assuming that error variation is a combination of all noise effects equally distributed among treatment conditions.

The Taguchi approach has been utilized to analyze volumetric properties in a gerotor pump to different extents. Jung et al. [35] applied the concept to create lobes with different profiles. The design parameters achieved are promising in the experimental findings of volumetric performance. However, the description of the implementation is lacking, and noise effects are not addressed. Ivanović et al. investigated the determination of the best parameters for the volumetric properties of a gerotor pump, namely flow rate and volumetric efficiency, utilizing factorial design, response surface approach, and the Taguchi method. Three geometries were examined at varying rotation speeds and pressures, identifying them as significant parameters. These experimental efforts do not provide a report on a chosen set of noise factors. The submitted work's significant addition is utilizing many input parameters as noise components to evaluate the contact stress characteristics in a gerotor pump.

The Taguchi approach has been utilized in software and computer design. An example is the demonstrated support in managing errors found in software development by quantifying performance (38). The systematic production of a representative subset of parameters in Computer Aided Design (CAD) and Computer Aided Manufacturing (CAM) benefits the number of options available, as demonstrated in bevel gear machining simulations [39]. The mechanical behavior can be theoretically represented using the finite element approach, making contact stress an ideal variable for experimental design within software packages. Choudhury et al. [40] showed that it is possible to optimize the mechanical design of an epicyclic gear train by identifying a set of input variables that minimize stresses and decrease the risk of tooth failure. The simulated experiments can replicate noise effects by utilizing operational conditions as prescribed factors in the software within an outer array, together with an inner array dedicated to control factors, representing adjustable parameters in the simulation. Therefore, employing the Taguchi approach in numerical simulation to evaluate the contact stress characteristics of a gerotor pump is another significant addition in the current study.

This work's technique consists of three sequential phases: establishing the trochoidal gear set backdrop, validating the contact stress numerical simulation, and applying the Taguchi strategy to a gerotor pump. Creating the appropriate blueprint is a distinct task for each stage. The initial phase defines a trochoidal profile, explains the function of a gerotor pump, selects gears from relevant literature, generates geometry, and incorporates nomenclature and notation.

The second phase focuses on establishing a strong foundation for implementing the Finite Element Model (FEM) to assess contact stresses, with the support of verification. A collection of simulations and their geometries were chosen and collected from existing literature. The primary goal was to accurately replicate the results due to the lack of technical data in certain published studies. The goal of this strategy is to recreate the simulations and confirm the accuracy of the results within acceptable limits. The last step consists of a sequential simulation-based design method with four Taguchi experiments resulting in 128 numerical simulations. The experiments range from a screening scenario with two levels and seven parameters to multiple levels with four parameters and three noise factors. This novel application of the Taguchi technique involves analyzing contact stress through a software package capable of simulating noise effects. The study focuses on a series of gerotor geometries with consistent volumetric capacity and dimensional limitations. Therefore, both external noise (environmental factors like temperature) and internal noise (functional factors like torque and friction coefficient) are important in relation to the noise impacts of the Taguchi method.

This research aims to demonstrate the dependability of the suggested methodology for predicting optimal settings in the geometry of gerotor pumps and identifying gear sets in industrial designs. This study aims to develop a unique methodology for determining the optimal gear set configuration and forecasting the measurable performance of a functional gerotor pump. It incorporates design best practices to assist designers in making informed decisions during the numerical simulation process.

2. TROCHOIDAL GEAR SET BACKGROUND

A trochoidal gear set consists of two gears, an inner/internal gear and an outer/external gear, with trochoidal tooth profiles. Refer to Figure 2 and the Nomenclature section for further details. The two gears are designed to ensure that each tooth or lobe of the internal gear is constantly in sliding contact with a tooth of the exterior gear, allowing for inter-tooth contact. These sites are referred to as contact points. The pumping motion occurs due to the changing volume between two adjacent contact points, which expands and contracts as the gears rotate. The internal gear pump is commonly referred to as the gerotor pump. Theoretical analysis and formulas are omitted in this work for simplicity and can be found extensively in the literature. For further details, go to the mentioned work [41].

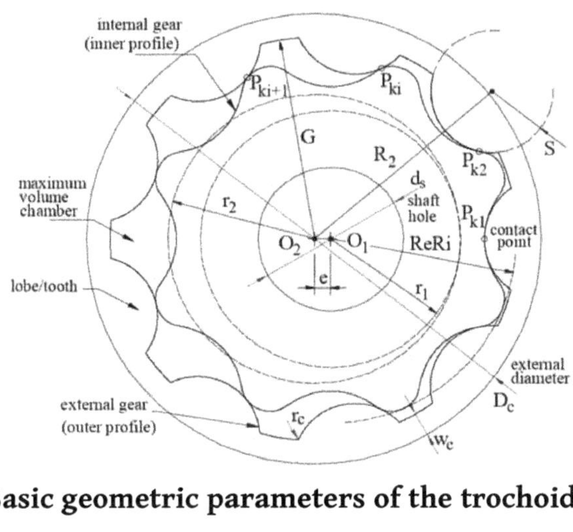

Figure 2. Basic geometric parameters of the trochoidal gear set.

2.1. Gear Sets under Study

Works from the literature were selected to validate the implemented numerical model. The main criterion of election was that selected works should contain numerical mechanical analyses, including an evaluation of the contact stress level. In addition, the researchers of these selected works ought to have a long and solid career in the mechanical analysis of trochoidal gear sets and gerotor pumps. Hence, the researchers Ivanović and Biernacki fulfil more than enough of the above-mentioned selection criteria, and their studies selected have been [**13,14,17,23**], respectively. In addition, a reference geometry with theoretical, numerical, and experimental work is also added to the study [**42**].

The codification used in the present study to unify all gear sets follows the same sequence as Per___: PZXeYYY. X is a figure coding the number of teeth of the outer gear and YYY are three figures coding the non-decimal eccentricity value of the gear set (the three digits without the decimal dot). This codification is illustrated in **Table 1**, along with supplementary information.

Table 1. Gear sets under study (PZXeYYY, where X is the number of teeth of the outer gear and YYY are the non-decimal eccentricity value).

Gear Set	Researchers' Work	Analysis-Environment	Observation
PZ7e377	Biernacki [**23**]	FEM-ABAQUS	The keyway is removed
PZ6e356/1375 PZ6e356/1575	Ivanović et al. [**14**]	FEM-CATIA/FEMAP	$\lambda' = 1/\lambda = 1.375$ $\lambda' = 1/\lambda = 1.575$
PZ9e285	Gamez-Montero et al. [**42**]	FEM-GiD/COMET	The shaft hole is included
MZ9e855		Photoelasticity model	Scaled 3:1 of PZ9e285

2.2. Nomenclature and Notation

Nomenclature and notation of the basic geometric parameters are not standardized, and each researcher uses their own. The present study works with the GeroLAB notation [**43**], then the researchers' geometries under analysis are contextualized in this specific nomenclature and presented in **Table 2**.

Table 2. Geometrical and technical parameters of gear sets (please refer to Nomenclature section and Figure 2).

Input Gear Set Variable/ Parameter	PZ7e377	PZ6e356/1375 PZ6e356/1575	PZ9e285	MZ9e885
GeroLAB				3:1 (PZ9e285)
Z [-]	7	6	9	

e [mm]	3.77	3.56	2.85	
S [mm]	11.11	9.79 (/1375) 14.06 (/1575)	10.85	
DeRi [mm]	53.24	46.28	65.45	
G [mm]	30.50	26.94	35.80	
r_c [mm]	0	0	0	
H [mm]	10.40	16.46	9.25	
Geometrical				3:1 (PZ9e285)
w_c [mm]	7.0	4.1	4.2	
D_c (O_2) [mm]	75	62	80	
d_s (O_1) [mm]	25	16	44	
Material properties	POM *	Steel	Sintered metallic powder	Epoxy
Young's module [GPa]	3	200	115	3
Density [kg/m^3]	1410	-	6800	1160
Poisson's coefficient [-]	0.43	0.30	0.25	0.35
Friction coefficient, *m* [-]	0.40	0.40	0.40	0.30
Torque, *T* [N·m]	7.16	0.621 (/1375) 0.632 (/1575)	18.75	37.5
* Polyoxymethylene.				

2.3. Geometry Generation

Once basic geometrical parameters are decoded, as presented in **Table 2**, the full set of trochoidal gears has to be generated. The well-known formulae for calculating the trochoidal profiles of a gear set can be used by introducing these geometrical parameters into specialized free software, such as GeroLAB (Terrassa, Spain) [**44**], which will calculate and provide the technical drawings in CAD format. By importing the drawing into CAD software (Vélizy-Villacoublay, France), a planar body of the geometry is created and modified with the appropriate operations, such as splitting, adding, and assembling, to be able to be imported into a mechanical simulation software.

3. VERIFICATION OF CONTACT STRESS NUMERICAL SIMULATION

3.1. Numerical Model Definition

Two main series of numerical simulations are considered. The first one is aimed at validating the implemented Finite Element Model (FEM) and consists of simulating real cases of gerotor systems published in the literature, which include mechanical data. The second series is aimed at analyzing the influence of different parameters according to the Taguchi method.

All simulations are run in ANSYS® 19.2R (Canonsburg, PA, USA). This general-purpose software environment was selected because of its wide availability among practitioners and also because most of the definitions in terms of mechanical description are analogous and comparable to those of other FEM simulation packages. Moreover, the multiphysics capability of ANSYS® would allow future enhancement of the model applicability including fluid or thermal transitive simulations, although these steps are far beyond the aim of the current research. In addition, many other researchers have used this software package to study gear contact stresses before, such as Ram Kumar et al. **[45]**, Karthick et. Al. **[46]**, Benaïcha et al. **[47]**, Rao **[48]**, Lahtivirta and Lehtovaara **[49]**, or Lisle et al. **[50]**. The most relevant information about the implemented numerical models is described in the following subsections.

3.1.1. Geometry and Material Properties

The geometry in all cases consists of two parts: the internal and the external gear. For every reference case, two angular positions of the gear sets are simulated: 0^0 and 25^0. One additional case at 40^0 is considered for the PZ9e285 case. The keyway is not allocated in any geometry. Hence, since two specific positions in the gear set are only required, the Static Structural module is the most suitable selection.

Despite the fact that all the simulations chosen from the literature share common characteristics, two of them are computed with a 3D geometry. On the one hand, the specific reasons why these researchers decided to implement 3D are unknown. On the other hand, although these simulations were run in 3D, the results are shown in 2D, in the front plane of the gear sets. Then, this fact endorsed the decision to assume a 2D plain stress hypothesis with gear thickness as input data in this study. Two main advantages are added: the remarkable diminution of mesh elements and the contact definition as edge–edge and not surface–surface, which helps the program to be more precise.

For all simulations, isotropic linear elastic material is considered. Thus, the only properties required for the numerical simulation are Young's modulus and Poisson's and friction coefficients (see **Table 2**).

3.1.2. Loads, Contacts, and Boundary Conditions

The boundary conditions consist of fixing the edge of the external diameter of the external gear and imposing a restrained and fixed displacement on the edge of the shaft hole of the internal gear, while allowing its rotation. The load is performed by a counter-clockwise torque applied on the edge of the shaft hole of the internal gear. For the validation cases, the value of the torque is set according to the corresponding reference in the researchers' works.

Finally, contacts are described as frictional through the friction coefficient definition. All contacts are set between the specific edges of the teeth (internal and external) in interaction. Thus, it is required to divide the continuous internal and external edges in the trochoidal tooth profiles into as many parts as the corresponding number of teeth. This definition is essential to correctly set the contact areas to allow an independent analysis of the several contact points at the same time between gears: the quasi-static condition.

Using frictional contact rather than frictionless is preferred because it more accurately denotes real cases and allows one to study the possible life-long variation of the friction coefficient as an external factor on the contact stress, which is addressed later. Frictional contacts [**5**,**51**] and frictionless contacts [**49**,**50**] are both previously used by other researchers in gear contact studies.

3.1.3. Elements and Mesh

PLANE183 elements are used to mesh both internal and external gears. CONTA172 and TARGE169 elements are used to define the contacts. The average size of the mesh is 5 mm. This dimension is reduced to an edge sizing of 1 mm near the contact areas by imposing this seeding distance on the edges that define the contacts. Since the goal is to validate the model by obtaining results akin to the researchers' work, an equivalent configuration must be used. The quality of the mesh is checked with the orthogonal quality indicator, with an average value of over 0.9, and also with the skewness indicator, with an average value of below 0.3, for all cases. Those indicators fulfilled the recommended quality mesh requirements in [**52**].

3.1.4. Analysis

The full Newton–Raphson solution procedure is employed to perform the implicit calculation of the implemented non-linear model. Average solution time is approximately ten minutes in an Intel® Core ™ i7-9700 K CPU @ 3.60 GHz with 16 GB RAM memory.

The use of equivalent von Mises stress is widely accepted and used in several research studies [**45**,**46**,**47**,**48**], although some of the research [**51**] points out that von Mises stress values could be approximately one half of the theoretical Hertz contact stress. In any case, the proposed model is validated through comparison with other accepted models that used von Mises's stress. Finally, the influence of the analyzed parameters is also studied in terms of maximum von Mises's stress at contacts, being mostly a comparative than a quantitative analysis.

3.2. Simulation Results: Validation and Discussion

The FEM validation aims to assess the recreated authors' results from the researchers' works aforementioned. If the validation is positive, meaning that the values obtained with the simulations match those published within reasonable bounds, the methodology will prove to be valid for future simulations using the same model applied up to here.

The maximum equivalent von Mises stress among all contact zones is the variable for comparison with reference stresses. **Figure 3** graphically gathers authors' FEM results of maximum contact stress in each gear set from researchers' works at two angular positions, 0° and 25° (the angular position of each gear set in the researcher's reference). A worthwhile comparison of results could have come from the deformation performance of the gear sets. However, limited information is available in the available literature and, consequently, it was decided to concentrate on contact stress results, which are the ultimate goal.

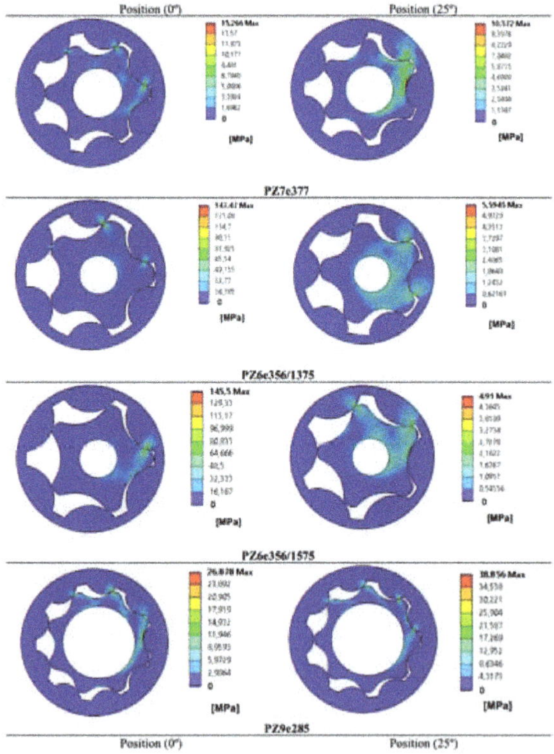

Figure 3. Authors' FEM results of maximum contact stress [MPa] in each gear set from researchers' works at 0° and 25° angular positions. (The commas in this specific figure represents decimal dots. For interpretation of the references to color in this figure legend, the reader is referred to the Web version of this article.).

For the PZ9e285(25°) case, the mesh convergence analysis was performed obtaining the results plotted in **Figure 4**. The figure shows a clear contact stress value trend when reducing the contact edge sizing from 5 mm (11.5% variation) to 2 mm and from 2 mm (0.4% variation) to 1 mm. However, for smaller mesh sizes (0.50 mm and 0.25 mm), local stress concentration effects were observed to increase the maximum contact stress in a fictitious way. Thus, 2 mm edge size was precise enough to carry out the research, but 1 mm was chosen because it was more precise and was not a significantly greater calculation effort than the 2 mm case. For a 1 mm mesh size case, the PZ9e285 model included 7164 nodes, 2218 2D elements, and 420 contact elements.

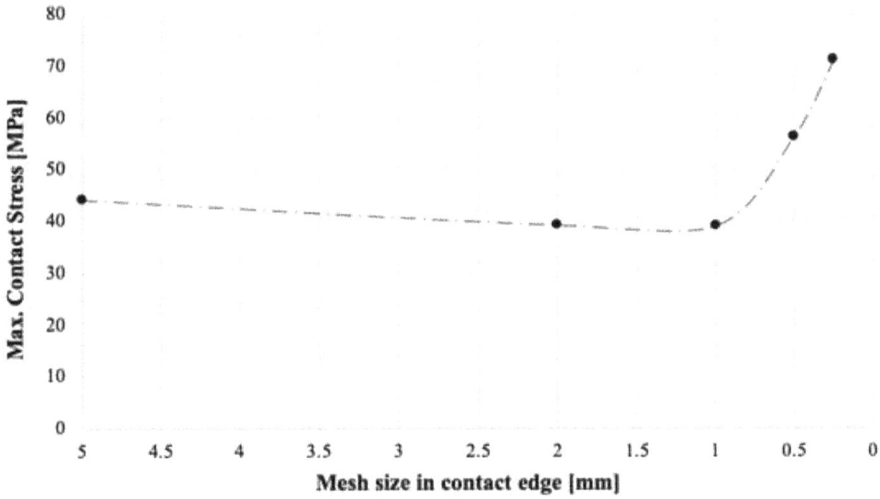

Figure 4. Mesh convergence analysis of PZ9e285(25°) case.

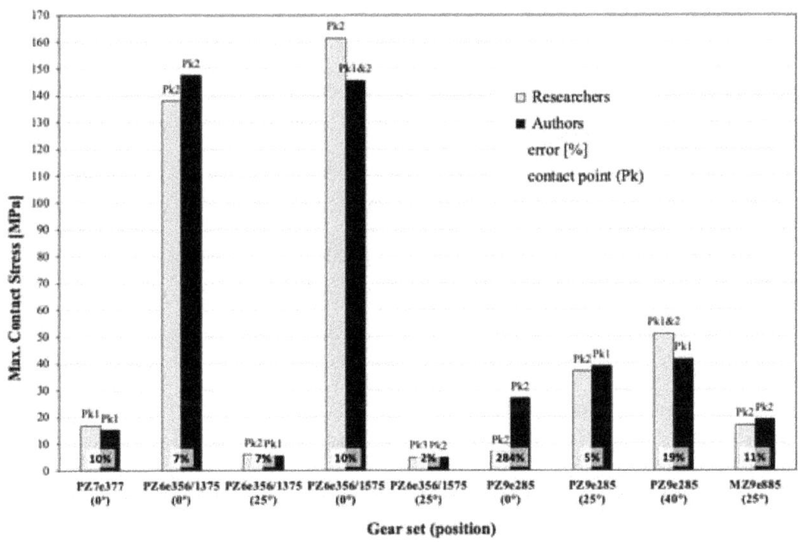

Figure 5. Results comparison Researchers vs. Authors, relative error [%] in maximum contact stress [MPa] and contact point (Pk) of maximum contact stress in each gear set for corresponding angular position in brackets (degree °).

The main results of the nine simulations are summarized in **Figure 5**. When researchers' and authors' results are contrasted at the maximum contact stress, a 9% average error is obtained when the PZ9e285(0°) is not included (this specific angular position of this gear set shows an unexpected value, and it is considered an outlier because clear stress concentration effects were detected in this case. In fact, a 40° position was also simulated to check the stability of the PZ9e285 case and to overcome this limitation). Since there is a restricted amount of available information about mechanical stresses in gerotors, this accuracy is judged to be adequate to validate the implemented model for describing the mechanical response of gerotor sets.

With regard to the location of the contact point at maximum stress, two clear tendencies can be pointed out:

- Maximum volume chamber position corresponding to PZ7e377(25°), PZ6e356/1375(0°), PZ6e356/1575(0°), and PZ9e285(0°). The location of the contact point at maximum stress is Pk2 for all gear sets, according to researchers' and authors' results. (Note: PZ7e377(25°) gear set in reference [**17**] does not provide an exact value but it can be gathered from the graphical figures);
- Minimum volume chamber position corresponding to PZ7e377(0°), PZ6e356/1375(25°), PZ6e356/1575(25°), PZ9e285(25°), and MZ9e885(25°). The location of the contact point at maximum stress are Pk and Pki + 1 for all results, researchers' and author's, respectively. The exceptions are PZ7e377(0°) and MZ9e885(25°).
- Hence, it reveals the importance of having either an odd or even number of external gear teeth in the set-up of the simulation. In addition, the impossibility of simulating exactly the same geometries as the researchers must be remembered, as does how the unnecessary parts of the original geometry were neglected, such as the keyway in the shaft hole.
- From these results, the plausibility of the proposed FEM model of the authors is sustained, as well as providing an opening for making progress toward the Taguchi approach. This decision is also justified by the fact that the final purpose of the implemented model is to study the significance and influence of the different design and environmental variables on the contact stress level more than providing a completely accurate result of the stress value.

4. THE TAGUCHI APPROACH TO A GEROTOR PUMP

The aim of parameter design outlined by Dr. Taguchi is to obtain a gear set that maximises the safety factor (SF) defined as the ratio between the reference yield strength of the material and the maximum contact stress. This gear set will be pursuing a twofold goal of (i) detaching the linkage of contact stress with volumetric capacity and dimensional constraints and (ii) enhancing the quality of contact stress assessment without controlling or removing the cause of variation contributed to by the numerical simulation's input settings to generate robust FEM results against noise effects. This approach is an iterative process that will be carried out sequentially in four experiments. The means to achieve a certain average contact stress will not be addressed. The Minitab® 19 software package (State College, PA, USA) complements the data processing.

4.1. The Volumetric Capacity Target and the Dimensional Constraints of the Gerotor Pump

Volumetric characteristics and contact stress performance of a gerotor pump are linked by the geometric parameters [**53**]. Then, in order to soften this link, the contact stress, the volumetric capacity, and several dimensional constraints of the gerotor pump are fixed constants in each gear set for all experiments as follows:

- The PZ9e285 gear set is the chosen prototypical gerotor because it is well-known by the authors regarding, among others, its fluid dynamic performance [**54**];
- The benchmark gerotor is named D80d40cv1 and the parameters to keep constant are summarized in **Table 3** and depicted in **Figure 6** based on [**55**];
- The material properties remain unchangeable in all the experiments: Young's module, density, and Poisson's coefficient of the PZ9e285 gear set (refer to **Table 2**);
- The shaft keyway is designed with dimensions Ls = 30 mm (parallel $0.75{\cdot}d_s$) shown in **Figure 6** as in [**55**];

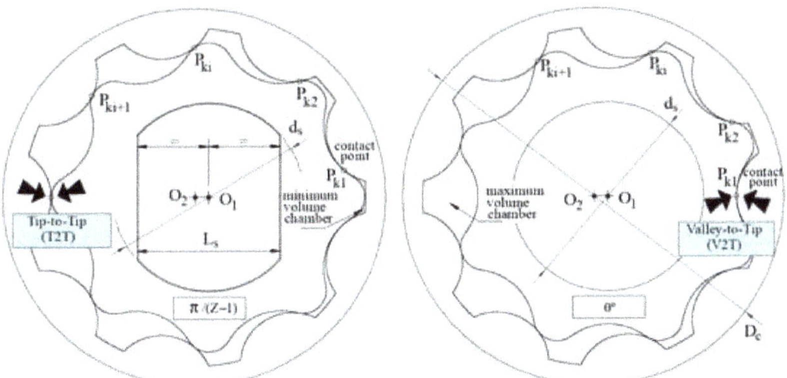

Figure 6. The benchmark gerotor in working function: the Tip-to-Tip (T2T) position (left) and the Valley-to-Tip (V2T) position (right), the shaft keyway ('yes' shaft keyway, left) and the shaft hole without keyway ('no' shaft keyway, right).

- Two specific angular positions as working functions will be under the study: Tip-to-Tip (T2T) and Valley-to-Tip (V2T), both depicted in **Figure 6**. The V2T corresponds to the maximum volume chamber and the T2T corresponds to the minimum volume chamber. This labeling presumes to enhance the comprehension of the contact points' action;
- The geometries and the analytical values of volumetric capacity and flow irregularity of each experimental gear set are provided by using GeroLAB [**43,44**].

Table 3. D80d40cv1 benchmark gerotor. (Please refer to Nomenclature section and Figure 2).

Parameter	Value	Significance/Target
$c_{v,H} = c_v/H$ [cc/(rev·mm)]	1	Volumetric capacity/Flow rate
D_c [mm]	80	Housing/Dimensional constraint
d_s [mm]	40	Internal diameter located in the inner gear to accommodate the shaft/Through-shaft application
H [mm]	9.25	Casing/Dimensional constraint

4.2. The Taguchi Method: The Designed Experiments

The designed experiments by the Taguchi approach will follow these specific features:

- The number of control factors. The number of parameters is chosen based on the trochoidal gear profile (Z, e, S), external gear (r_c, w_c), and working function (reference position RP, shaft keyway SK);
- The number of levels. The levels of each parameter are chosen to be significant in the study, from the benchmark gerotor, the literature, the previously presented FEM validation, and the authors' know-how. In addition, all level combinations have to accomplish the feasible geometry of a trochoidal gear set practicable to be generated with GeroLAB, which has become a challenging task. The number of levels is selected as the Taguchi approach is going forward, from two-levels to a mixed-level design;
- The noise factors. The mechanical operating conditions of the gear pump are described by three factors: material temperature (θ), torque (T), and material friction coefficient (μ), which can be tuned by surface modification methods [**56,57**]. The designed experiments can use these operating conditions as the input of FEM conditions as prescribed by an outer array only for noise factors (T, μ and θ);

- Improve the quality of contact stress assessment. The goal of the experiments is to maximize the safety factor (SF) as the response of the signal-to-noise ratio (S/N) set to Higher-is-Better (HB) where:

$$\text{HB S/N} = -10\log\left(\frac{1}{n}\sum_{i=1}^{n}\frac{1}{y_i^2}\right)$$

where n = the number of tests in a trial (number of repetitions regardless of noise levels) and yi = each observed value (data points). In all cases, the target is to maximize the HB S/N ratio. The development of a family of matrices based on orthogonal arrays (OA) and the signal-to-noise (S/N) definition as an indicator of the ratio of the mean to the standard deviation are great contributions of the Taguchi method. As a result, they will be used to evaluate and discuss the results of each Taguchi experiment:

- Statistical treatment. In all experiments, the significance (alpha) level was set as 0.05, and means and standard deviation were calculated for all volumetric capacities. The p-value inferior to alpha concludes that there is a statistically significant association between the response characteristic and the term. The p-value between the significance levels of 0.05 and 0.1 can be used for evaluating terms, and it can be considered with practical significance. A higher p-value will conclude that there were no statistically significant differences observed.

4.3. First Taguchi Experiment: The Screening Case

The first round, referred to as the screening case, is used to find the few important, influential factors out of the many possible factors involved with a process design. The recommended strategy is to start with the smallest orthogonal array that will accommodate the typically large number of actors under simultaneous evaluation at two levels.

Hence, the $L_8(2^7)$ OA inner array with a resolution number of 1-low (A and BxC are in the same column) without noise factors is selected and presented in **Table 4**. The resolution number is a measure of the amount of confounding in a column [**33**]. As the assignment of factors is to all columns, unavoidably, many interactions are confounded (mixed) with the main effects. This is the major compromise of using fractional factorial experiments: by reducing the number of tests, some information must be surrendered.

Table 4. The parameter design table of the screening case of the first Taguchi experiment. (Please refer to Nomenclature section and Figure 2).

8 gear sets in common							
Vol. capacity target:	$\underline{c}v,H$ = 1 cc·rev^{-1}·mm^{-1} (0.093), mean (standard deviation)						
Dimensional constraints:	D_c = 80 mm; d_s = 40 mm; H = 9.25 mm						
FEM conditions:	T = 18.75 N·m; μ = 0.4; θ = 20°						
L8 OA inner array							
(control factors)	A	B	C	D	E	F	G
Factor parameter (column) Trial gear set::Taguchi experiment (row)	Z	e [mm]	Reference position (*RP*) **(Figure 6)**	S [mm]	r_c [mm]	Shaft keyway (*SK*) **(Figure 6)**	w_c [mm]
1::1	8	2.50	T2T	5.86	0.0	No	3.0
2::1	8	2.50	T2T	13.96	3.0	Yes	6.0
3::1	8	2.97	V2T	5.86	0.0	Yes	6.0
4::1	8	2.97	V2T	13.96	3.0	No	3.0
5::1	9	2.50	V2T	5.86	3.0	No	6.0
6::1	9	2.50	V2T	13.96	0.0	Yes	3.0
7::1	9	2.97	T2T	5.86	3.0	Yes	3.0
8::1	9	2.97	T2T	13.96	0.0	No	6.0

The meaning of each number in the OA nomenclature $L_8(2^7)$ is: 8 FEM simulations in this screening case (inner array with 8 gear sets trials, without noise factors, corresponding to each row in **Table 4**), 7 factor parameters (Z, e, RP, S, r_c, SK, and w_c corresponding to each column in **Table 4**) and 2 levels (low and high values corresponding to the columns of the chosen Taguchi OA family in **Table 4**). In addition, the table head of **Table 4** indicates the volumetric capacity target and the dimensional constraints to remain constant in the gerotor pump, together with the FEM conditions used in the simulation.

The meaning of OA nomenclature and the arrangement of **Table 4** will be retained for all incoming Taguchi experiments.

By the inspection of **Figure 7**, an even number of teeth produces a higher maximum contact stress in all gear sets, with the exception of the 5:1 gear set owing to the selected reference position. The location of the contact point in the second tooth prevails in five out of eight cases. The 8::1 gear set shows the minimum value of contact stress, which is almost half of the 4::1 gear set, indicating the significance of the geometry and the reference position.

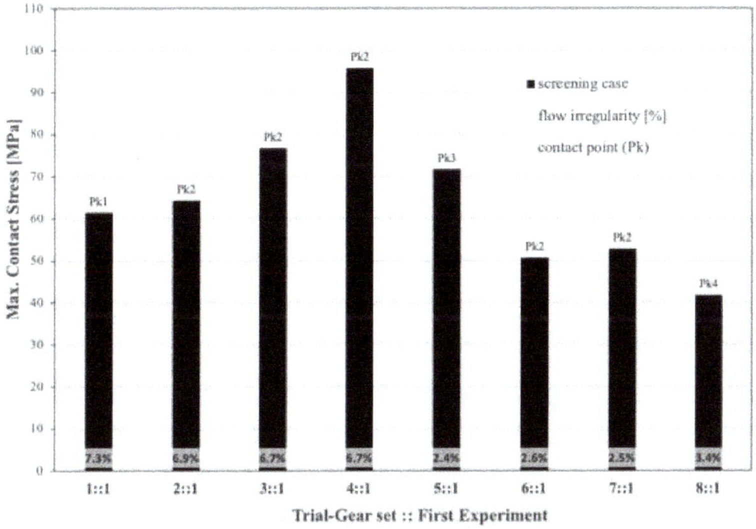

Figure 7. Results of FEM maximum contact stress [MPa], GeroLAB flow irregularity [%], and contact point (Pk) of maximum contact stress in each gear set for the first Taguchi experiment (see Table 4).

Figure 8 illustrates the main effects plot for the screening case. The true critical characteristics (control factors) must be identified and minimized in number. This graphic aids in a better picture of the importance of the chosen control parameters. The larger the vertical increment between the two levels and the higher the slope in each factor, the more emphasis is placed on this parameter: the larger the HB S/N ratio, the better. The figure indicates that the parameters Z, RP, and r_c are the strength factors. Conversely, the parameter w_c, with a behavior almost horizontal between the two levels, is the first factor to be discarded, together with the parameter SK.

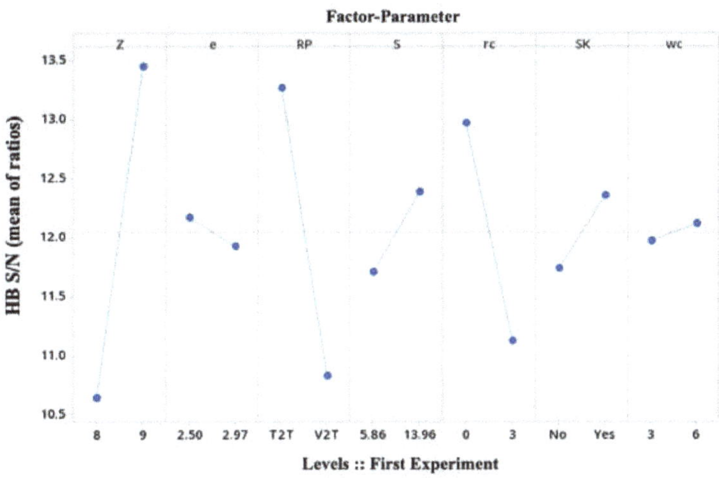

Figure 8. Main effects plot for the first Taguchi experiment (see Table 4).

Obviously, the resolution is low in the first round of experiments (a small fractional factorial) and it will progress to higher-resolution experiments (a large fractional or full factorial) as few factors are going to be identified as influential.

In addition, this screening case has mixed geometrical basic parameters to define the trochoidal gear set, such as Z, e, S, and external gear r_c, with working functions, such as SK and RP, where the former shows less dominance than the latter, and it will be discarded. Owing to the higher effect of the level 'Yes' in the shaft keyway, which is closer to real working function, all the gear sets in the incoming Taguchi experiments have a shaft keyway.

4.4. Second Taguchi Experiment: The Two-Levels-Four-Geometric-Basic-Parameters Case and One Noise Factor

The Taguchi approach to parameter design includes noise factors as related elements of the process that cannot be controlled or removed as a measure of the robust levels of the control factors. Thus, from the screening case, working function related to reference position is going to be used as a non-controlled process noise effect in this second Taguchi experiment.

$L_8(2^4)$ OA inner array with a resolution number of 2 (A and BxCxD, or AxB and CxD are in the same column), with one noise factor at the outer array L1 OA. Here, 16 FEM simulations were carried out in this second Taguchi (8 gear sets trials, 2 working functions noise factors). This second experiment encompasses less factor parameters (Z, e, S, and r_c) corresponding to each column in **Table 5**. The outer array accommodates the noise factor at two levels, T2T and V2T, corresponding to columns N1 and N2 in **Table 5**.

Table 5. The parameter design table of the second Taguchi experiment. (Please refer to Nomenclature section and Figure 2).

8 gear sets in common							
Vol. capacity target: Dimensional constraints: FEM conditions:	cv,H = 0.997 $cc \cdot rev^{-1} \cdot mm^{-1}$ (0.023), mean (standard deviation) D_c = 80 mm; d_s = 40 mm; H = 9.25 mm; (SK = yes) T = 18.75 N·m; μ = 0.4; θ = 20°						
L8 OA inner array (control factors)	A	B	D	G	L1 OA outer array (noise factor)	N1	N2
Factor parameter (column) Trial gear set::Taguchi experiment (row)	Z	r_c [mm]	S [mm]	e [mm]	Reference position **(Figure 6)** (column)	T2T	V2T

8 gear sets in common							
1::2	8	0.0	5.86	2.50			
2::2	8	0.0	13.96	2.97			
3::2	8	3.0	5.86	2.97			
4::2	8	3.0	13.96	2.50			
5::2	9	0.0	5.86	2.97			
6::2	9	0.0	13.96	2.50			
7::2	9	3.0	5.86	2.50			
8::2	9	3.0	13.96	2.97			

As in the case of FEM validation, the corresponding concordances reveal the important effect of the reference position in the numerical simulation. Comparing the gear set 8:1 in **Figure 7** and the gear set 8:2 in **Figure 9**, both of them present the lower value of maximum contact stress with an acceptable value of 3.4% of flow irregularity. Since the 8:2 gear set uses the reference position as a noise factor to expose the robust levels of the control factors, these parameters will produce successively better approximations to converge to the sought gear, which is more suitable regarding contact stresses. Considering the location of contact points, it stays the same or within a unit difference.

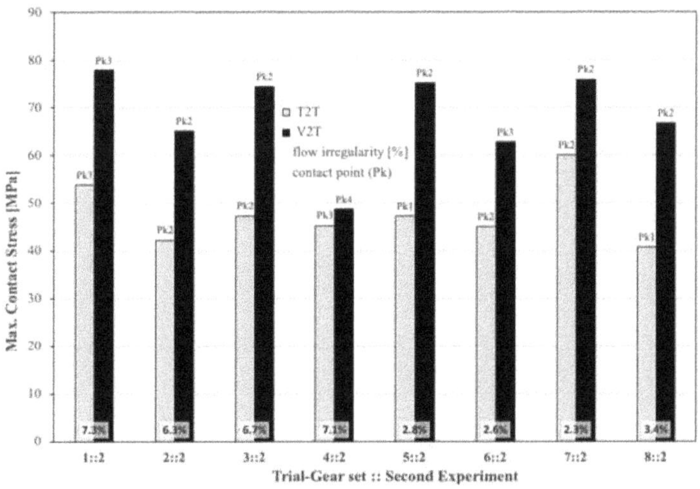

Figure 9. Results of FEM maximum contact stress [MPa], GeroLAB flow irregularity [%], and contact point (Pk) of maximum contact stress in each gear set with the working function noise factor for the second Taguchi experiment (see Table 5).

Since the slope in the graphic is an indicator of the importance of a main effect, **Figure 10** shows that the most dominant factor is the arc radius of the external gear tooth, S. In addition, this geometrical parameter S is considered with practical significance in the analysis of variance for HB S/N ratios (p = 0.083).

The mutual dependency between a pair of factors is an interaction of those factors. The interaction plot of the main effect factors Z and S is constructed by plotting the average response HB S/N values of each factor level combination and is presented in **Figure 11**. The effect of parameter S is estimated to be low since it would depend upon which level of parameter Z is being used. Being almost parallel lines is an indication of the absence of interaction between both factors. In addition, for this experiment, none of the studied interaction effects are statistically significant.

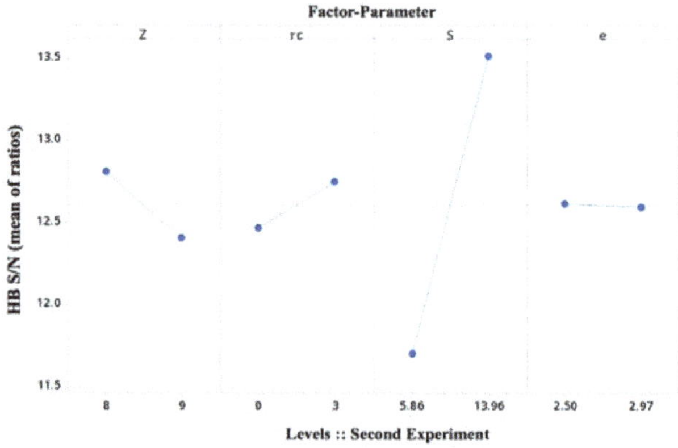

Figure 10. Main effects plot for the second Taguchi experiment (see Table 5).

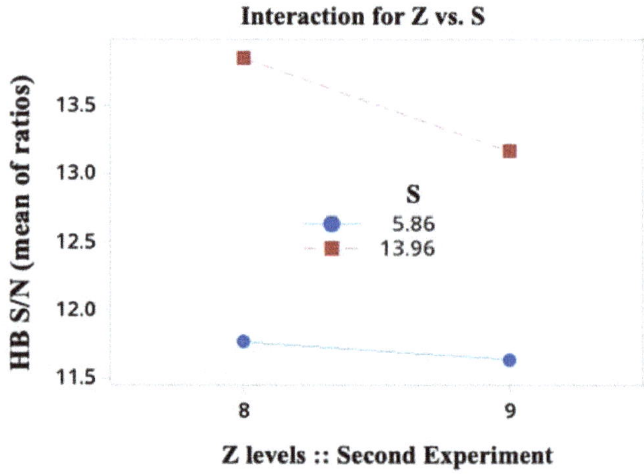

Figure 11. Interaction plot between Z and S for the second Taguchi experiment (see Table 5).

On the other hand, these results are quite confusing, since it is difficult to accept Z as having a similar effect as r_c in the gear set contact stresses as well as no significant interaction effect. One might add the well-known opposite effect on the volumetric characteristics of an even Z value, as it can be seen in the numerals of the flow irregularity in **Figure 9**. Upon debating this controversial effect, it is decided to go with a full factorial experiment by using the main geometric basic parameters: the number of external teeth, Z, the eccentricity, e, and the arc radius of the external gear tooth, S.

4.5. Third Taguchi Experiment: The Two-Levels-Three-Geometric-Basic-Parameters Case and Three Noise Factors

As previously mentioned, Taguchi parameter design is used to dampen the effect of noise, which includes parameters that are either uncountable or are too expensive to control, by choosing the proper level of control parameters. Then, once more, the question arises: why bother to assign these noise factors when they cannot be controlled or eliminated? The main reason, in this work, is related to the numerical simulations: several settings in the software environment could act as noise since they are associated with input operating conditions in the FEM simulation.

Then, the strategy is to force noise effects into the numerical simulation with an experimental layout based on an outer array only for operating conditions: a combination of outer noise (environmental-related, such as temperature, θ) and inner noise (function-related, such as torque, T, and friction coefficient, μ). If all of these noise factors were mixed together in an inner array for control factors, it would be a traditional cause of detection experiment: this is one of the major contributions of the Taguchi approach to noise emulation in FEM methodology.

Table 6 illustrates the $L_8(2^3)$ OA inner array with a resolution number of 4-high (full factorial, all items are in separate columns), with 3 noise factors. Here, 32 FEM simulations were carried out (8 gear sets trials, 4 FEM conditions noise factors). This third experiment encompasses three factor parameters (Z, e, and S) corresponding to each column. Owing to the low effect of r_c, it was removed from the inner array, avoiding interactions confounded with the main effects and gaining resolution. The outer array accommodates the 3 noise factors at two levels, corresponding to columns N1, N2, N3, and N4 (refer to **Table 6**).

Table 6. The parameter design table of the third Taguchi experiment. (Please refer to Nomenclature section and Figure 2).

8 gear sets in common								
Vol. capacity target: Dimensional constraints: FEM conditions:	$\underline{c}v,H$ = 0.997 cc·rev^{-1}·mm^{-1} (0.023), mean (standard deviation) D_c = 80 mm; d_s = 40 mm; H = 9.25 mm; (*SK* = yes) *Noise factors*							
L8 OA inner array (control factors)	A	B	D	L4 OA outer array (noise factors)	N1	N2	N3	N4
Factor parameter (column) Trial gear et::Taguchi experiment (row)	*Z*	*e* [mm]	*S* [mm]	*T* [N·m] *μ* [-] θ [°C] (column)	15.0 0.01 20	22.5 0.50 20	22.5 0.01 40	15.0 0.50 40

1::3	8	2.50	5.86					
2::3	8	2.50	13.96					
3::3	8	2.97	5.86					
4::3	8	2.97	13.96					
5::3	9	2.50	5.86					
6::3	9	2.50	13.96					
7::3	9	2.97	5.86					
8::3	9	2.97	13.96					

In **Figure 11** and **Figure 12**, results referring to the third experiment are reported. **Figure 12** illustrates an important effect not influenced by the noise FEM conditions: the location of the maximum contact stress at the same contact point for all operating conditions once the three geometrical basic parameters are taken into account. It is also worth noting that the combination of a higher torque and a lower frictional factor results in higher contact stress, regardless of temperature.

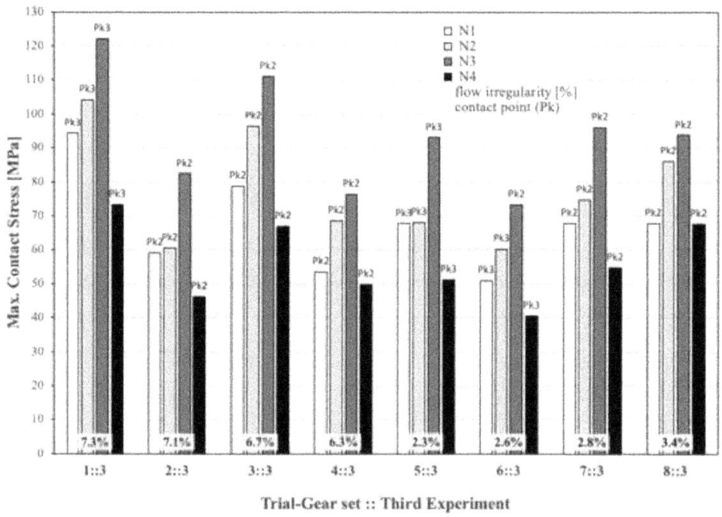

Figure 12. Results of FEM maximum contact stress [MPa], GeroLAB flow irregularity [%], and contact point (Pk) of maximum contact stress in each gear set with 4 FEM conditions noise factors for the third Taguchi experiment (see Table 6).

Figure 13, as previously in **Figure 10**, shows that the most dominant factor is the arc radius of the external gear tooth, S, with practical significance in the analysis of variance for HB S/N ratios ($p = 0.080$). This is an expected result, since removing the parameter cutting radius shows its minimal effect. However, the absence of interactions confounded with the main effects in this full factorial experiment and the inclusion of the operating conditions reinforces this outcome. Now, the odd number of external gear teeth shows a dominant effect, and the eccentricity normalises its intrinsic interest and value.

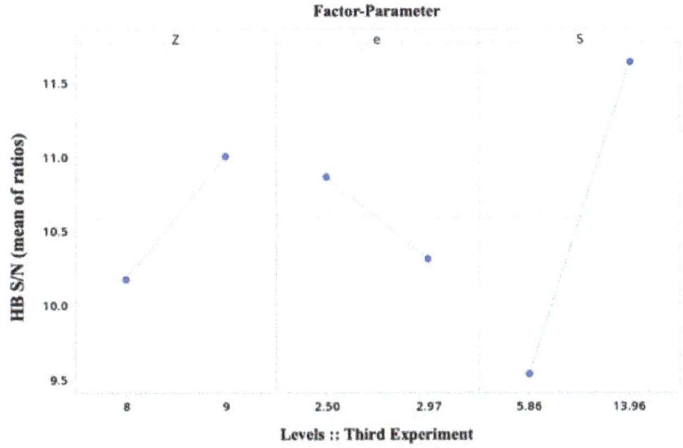

Figure 13. Main effects plot for the third Taguchi experiment (see Table 6).

The interaction plot of the main effect factors Z and S is again computed and presented in **Figure 14.** The lines' loss of parallelism and the effect of an odd number value for Z are indicated, although it is not statistically significant yet. The OA size and the interaction potential of interest to the experimenter are the key factors; as a general recommendation, it is preferred to study more factors than to study interactions [**33**].

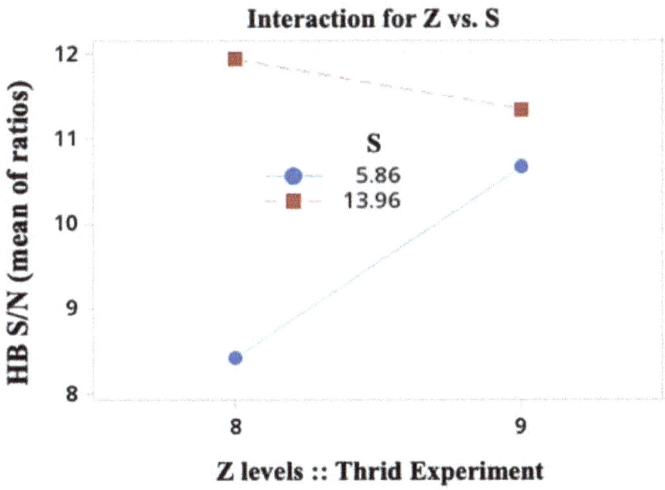

Figure 14. Interaction plot between Z and S for the second Taguchi experiment (see Table 5).

4.6. Fourth Taguchi Experiment: The Mixed-Level-Design-Case with Three Noise Factors

The last experiment in this iterative process is designed to elucidate the ability of the Taguchi method to pinpoint the most suitable gear set configuration in front of FEM contact stress. With the learning from the previous three Taguchi experiments, it is decided to complete the study with an OA inner array with four control factors:

the trochoidal gear set parameters (Z, e, and S) and the working function (RP). Note that it is decided to include the RP in the inner array, since its effects have been shown to be of noticeable importance. The FEM conditions are kept to the three operating conditions (T, μ, and θ) in the outer array to emulate noise effects. To complete the experiment, one more level is added to the trochoidal gear set parameters in a mixed-level design case. The Taguchi approach prefers using three or more levels of design parameters to estimate non-linear effects rather than two levels of classical DoE [**30**].

Table 7 illustrates the $L_{18}(2^1\text{–}3^3)$ mixed-level OA inner array with a resolution number of 1-low (no specific interaction columns available, except between AxB by using a specific layout). Here, 72 FEM simulations were carried out (18 gear sets trials, 4 FEM conditions noise factors). The outer array accommodates the three operating conditions (environmental-related, q, and function-related, T and μ) at two levels, corresponding to each column N1, N2, N3, and N4. In this mixed-level design, the parameters of the cutting radius and the wall width of the external gear are kept constant for the 18 gear sets, and factor interactions are not studied based on the results of the previous experiments.

Table 7. The parameter design table of the fourth Taguchi experiment. (Please refer to Nomenclature section and Figure 2).

18 gear sets in common									
Vol. capacity target:	$\overline{C}_{v,H}$ = 0.996 cc·rev^{-1}·mm^{-1} (0.041), mean (standard deviation)								
Dimensional constraints:	D_c = 80 mm; d_s = 40 mm; H = 9.25 mm; (*SK* = yes)								
FEM conditions:	*Noise factors*								
L18 OA inner array (control factors)	A	B	C	D	L2 OA outer array (noise facts)	N1	N2	N3	N4
Factor parameter (column) Trial gear set:Taguchi experiment (row)	Reference position (*RP*) (**Figure 6**)	*Z*	*S* [mm]	*e* [mm]	*T* [N·m] *μ* [-] θ [°C] (column)	15.0 0.0 ** 20	22.5 0.4 20	22.5 0.0 ** 40	15.0 0.4 40
1::4	V2T	7	5.8	2.61					
2::4	V2T	7	8.0	2.77					
3::4	V2T	7	10.4	2.93					
4::4	V2T	9	5.8	2.61					
5::4	V2T	9	8.0	2.77					
6::4	V2T	9	10.4	2.93					
7::4	V2T	11	5.8	2.77					
8::4	V2T	11	8.0	2.93					

18 gear sets in common									
9::4	V2T	11	10.4	2.61					
10::4	T2T	7	5.8	2.93					
11::4	T2T	7	8.0	2.61					
12::4	T2T	7	10.4	2.77					
13::4	T2T	9	5.8	2.77					
14::4	T2T	9	8.0	2.93					
15::4	T2T	9	10.4	2.61					
16::4	T2T	11	5.8	2.93					
17::4	T2T	11	8.0	2.61					
18::4	T2T	11	10.4	2.77					
** Frictionless.									

In **Figure 15**, the diversity of operating conditions is studied by crossing the inner array of control parameters by the outer array of noise factors. The slope of the HB S/N ratios once more illustrates the dominance of the three trochoidal gear set parameters (Z, e, and S) in front of the working function (RP). In addition, the main effects plots for means and standard deviations confirm the same tendency.

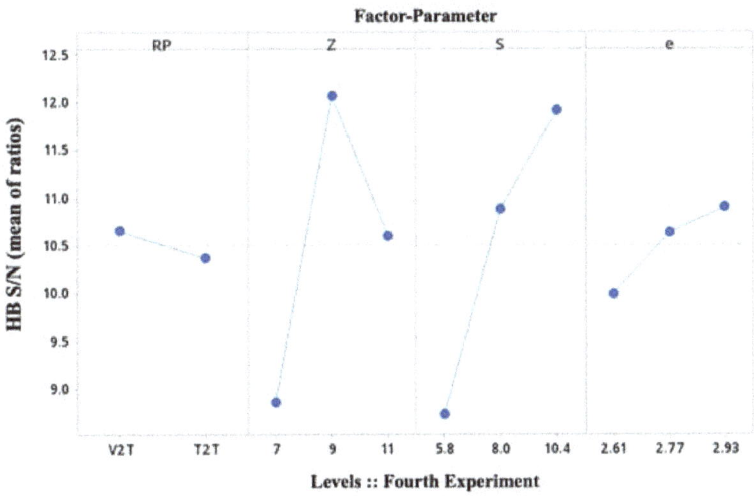

Figure 15. Main effects plot for the fourth Taguchi experiment (see Table 7).

The statistical significance for Z and S supports this result. This fourth experiment shows statistical significance at an alpha level of 0.05 for Z and S, in particular, $p = 0.053$ and $p = 0.050$, respectively. Conversely, none of the interaction effects are significant.

For the contact stress experiment, the optimal control factor settings based on the highest HB S/N have been determined by the inspection of **Figure 15.** In order to decide which level is best to maximize the safety factor, the HB S/N ratios at each level (low, medium, and high) of each factor are compared. Hence, the levels are:

Z = 9 (level 2-medium); S = 10.4 mm (level 3-high); e = 2.93 mm (level 3-high) >>> Gear set 6::4.

The Taguchi technique for parametric design indicates that the optimal settings are found in the row corresponding to the gear set 6::4 in Table 7. Furthermore, its flow irregularity is 3%, which is considered satisfactory. If the anticipated optimal parameters did not match any of the rows, a confirmation trial would be conducted as a last step. The reference location has a negligible impact as a control variable. These parameters optimize safety for all noise configurations. Comparing the three trochoidal gear set parameters of the PZ9e285 gear set in Table 2 with the projected optimum settings demonstrates the methodology's reliability in assessing the real-world geometry of a gerotor pump.

5. CONCISE SUMMARY

This chapter discussed efforts to demonstrate the reliability of Taguchi techniques as a viable simulation-based strategy in designing numerical simulations to evaluate contact stresses in trochoidal gears used in gerotor pumps. The Finite Element Model suggested in this research was initially validated using published results from the literature. Ivanović and Biernacki, two experienced specialists in the field of mechanical analysis of gerotor pumps, were chosen for the study. The research in this paper combined and unified nomenclature and notation. The average of the findings from the nine simulations revealed a 9% variance in the maximum contact stress at contact locations. This inaccuracy is considered acceptable given the available technical data and the constraints of replicating the identical geometries used by the researchers. Furthermore, the position of the contact points was connected in every instance. The numerical model was validated within acceptable limits and utilized in simulations based on the Taguchi method.

The Taguchi approach utilized a sequential simulation-based design strategy involving four tests to allow for a manageable amount of gear sets. The 128 numerical simulations achieved the difficult task of maintaining consistent volumetric capacity and dimensional constraints, such as gear thickness, external diameter of the external gear, and the internal diameter of the shaft hole in the inner gear, in order to isolate contact stress performance from flow rate and casing. This unique approach helped avoid the numerous geometric options in FEM numerical simulations and the 'black box' effect in the software input operating conditions, such as temperature and torque related to function and friction coefficient. The progressive simulation-based planning reduced the number of parameters and decreased the solution space under settings that mimic noise effects.

The results of contact stress indicated that the arc radius of the external gear tooth was the most influential parameter. The study revealed that the uneven number of exterior gear teeth was the second most influential factor, while the eccentricity added to its inherent significance and worth. Both dominant parameters exhibited statistical significance, confirming this finding. Nevertheless, none of the interaction effects reached statistical significance.

The study showed that the suggested methodology can accurately identify the best gear set configuration for minimizing FEM contact stress and predicting optimal settings in the geometry of real-working gerotor pumps in industrial designs. The statement was verified using the particular values of a trochoidal gear set: 9 external gear teeth, an arc radius of 10.4 mm, and an eccentricity of 2.93 mm. The optimized contact stress gear set achieves a desired volumetric capacity of 1 cc·rev-1 per gear thickness and a flow irregularity of 3%. This approach shows promising results in improving the quality of contact stress assessment without adjusting or eliminating the variations caused by the input settings of the numerical simulation. Taguchi resilient design is a simpler method compared to conventional optimization. The accessibility of this open tool to end-users and its approach are closely connected to real production, goods in industry, quality, and costs. This study has developed a unique method in the area to enhance the numerical simulation process in designing and configuring the gear set of a functional gerotor pump. The next phase of this research will involve identifying the control factors that influence the mean to reach a predetermined contact stress target and forecasting the average response for different control factor settings using the model for the signal-to-noise ratio response.

6. AT A GLANCE

The issue of contact in a trochoidal gear is a recognized drawback and a key performance indicator of a gerotor pump. Numerical simulations help assess contact stress, but the designer is responsible for establishing geometrical parameters, operating circumstances, and the number of simulations to execute. This work introduces Taguchi techniques as a simulation-based strategy to limit down geometrical parameter combinations, minimizing the solution space and optimizing the number of runs. The work initially aims to validate the proposed numerical model using contact stress measurements from established researchers in the field and to standardize nomenclature and notation. The Taguchi approach involves a series of four trials, starting from a screening case with two levels and seven parameters, and progressing to multiple levels with four parameters and three software input operating conditions (temperature, torque, and friction coefficient) to simulate noise effects. Analyzed were the contact stresses of 128 gear sets with similar volumetric capacity and dimensions limitations to isolate mechanical performance from flow rate and casing. The suggested methodology's practicality is confirmed by the identification of the most appropriate gear set design and the prediction of the quantitative performances of a functioning gerotor pump.

REFERENCES

1. Rundo, M. Models for flow rate simulation in gear pumps: A review. Energies **2017**, 10, 1261. **[Google Scholar] [CrossRef]**
2. Pareja-Corcho, J.; Moreno, A.; Simoes, B.; Pedrera-Busselo, A.; San-Jose, E.; Ruiz-Salguero, O.; Posada, J. A Virtual Prototype for Fast Design and Visualization of Gerotor Pumps. Appl. Sci. **2021**, 11, 1190. **[Google Scholar] [CrossRef]**
3. Kawamata, H.; Kuroda, S.; Tanaka, S.; Oshima, M. Improvement of Practical Electric Consumption by Drag Reducing under Cross Wind; SAE Technical Paper 2016-01-1626; SAE International: Warrendale, PA, USA, 2016; pp. 1–10. **[Google Scholar] [CrossRef]**
4. Colbourne, J.R. Reduction of contact stress in internal gear pumps. J. Eng. Ind. **1976**, 98, 1296–1300. **[Google Scholar] [CrossRef]**
5. Roy, D. Mechanics and FEM estimation of deformation, gaps and stresses generated in star-ring active contacts of GEROTOR units during operation. J. Br. Soc. Mech. Sci. Eng. **2021**, 43, 429. **[Google Scholar] [CrossRef]**
6. Roy, D.; Maiti, R.; Das, P.K. Mechanics and FEM estimation of gaps generated in star-ring active contacts of ORBIT motor during operation. Int. J. Mech. Mater. Des. **2020**, 16, 69–89. **[Google Scholar] [CrossRef]**
7. Sliwinski, P. The methodology of design of satellite working mechanism of positive displacement machine. Sci. Rep. **2022**, 12, 1–22. **[Google Scholar] [CrossRef]**
8. Gamez-Montero, P.J.; Codina, E.; Castilla, R. A Review of Gerotor Technology in Hydraulic Machines. Energies **2019**, 12, 2423. **[Google Scholar] [CrossRef]**
9. Ivanović, L.; Devedžić, G.; Ćuković, S.; Mirić, N. Modeling of the Meshing of Trochoidal Profiles with Clearances. J. Mech. Des. **2012**, 134, 041003. **[Google Scholar] [CrossRef]**
10. Ivanović, L.; Josifović, D.; Ilić, A. Modelling of trochoidal gearing at the gerotor pump. Mech. Mach. Sci. **2013**, 13, 553–562. **[Google Scholar] [CrossRef]**
11. Ivanović, L.; Ilić, A.; Miloradović, D.; Josifović, D. Modelling and simulation of the load in the epicyclic rotary pump with trochoidal gear profiles. IOP Conf. Ser. Mater. Sci. Eng. **2018**, 393, 1–9. **[Google Scholar] [CrossRef]**

12. Ivanović, L.; Devedžić, G.; Mirić, N.; Ćuković, S. Analysis of forces and moments in gerotor pumps. Proc. Inst. Mech. Eng. C **2010**, 224, 2257–2269. **[Google Scholar] [CrossRef]**

13. Ivanović, L. Reduction of the maximum contact stresses by changing geometric parameters of the trochoidal gearing teeth profile. Meccanica **2016**, 51, 2243–2257. **[Google Scholar] [CrossRef]**

14. Ivanović, L.; Rakić, B.; Stojanović, B.; Matejić, M. Comparative analysis of analytical and numerical calculations of contact stresses at rotational elements of gerotor pumps. Appl. Eng. Lett. **2016**, 1, 1–7. **[Google Scholar]**

15. Ivanović, L.; Mackić, T.; Stojanović, B. Analysis of the instantaneous friction coefficient of the trochoidal gear pair. J. Balk. Tribol. Assoc. **2016**, 22, 281–293. **[Google Scholar]**

16. Ivanović, L.; Matejić, M. Improving gerotor pump performance through design, modelling and simulation. Int. J. Fluid Power **2021**, 21, 327–346. **[Google Scholar] [CrossRef]**

17. Biernacki, K.; Stryczek, J. Analysis of stress and deformation in plastic gears used in gerotor pumps. J. Strain Anal. Eng. Des. **2010**, 45, 465–479. **[Google Scholar] [CrossRef]**

18. Biernacki, K. Selection of the optimum tooth profile for plastic cycloidal gears. Proc. Inst. Mech. Eng. C **2014**, 228, 3395–3404. **[Google Scholar] [CrossRef]**

19. Stryczek, J.; Bednarczyk, S.; Biernacki, K. Strength analysis of the polyoxymethylene cycloidal gears of the gerotor pump. Arch. Civ. Mech. Eng. **2014**, 14, 647–660. **[Google Scholar] [CrossRef]**

20. Stryczek, J.; Bednarczyk, S.; Biernacki, K. Gerotor pump with POM gears: Design, production technology, research. Arch. Civ. Mech. Eng. **2014**, 14, 391–397. **[Google Scholar] [CrossRef]**

21. Biernacki, K. Analysis of the material and design modifications influence on strength of the cycloidal gear system. Int. J. Adv. Manuf. Technol. **2015**, 16, 537–546. **[Google Scholar] [CrossRef]**

22. Biernacki, K. New concept of power transmission for gerotor hydraulic machines. Proc. Inst. Mech. Eng. Part C J. Mech. Eng. Sci. **2021**, 235, 4873–4883. **[Google Scholar] [CrossRef]**

23. Biernacki, K. New construction of cycloidal gear unit made of plastics. Proc. Inst. Mech. Eng. Part C J. Mech. Eng. Sci. **2021**, 235, 800–811. **[Google Scholar] [CrossRef]**

24. Robison, A.; Vacca, A. Multi-objective optimization of circular-toothed gerotors for kinematics and wear by genetic algorithm. Mech. Mach. Theory **2018**, 128, 150–168. **[Google Scholar] [CrossRef]**

25. Robison, A.; Vacca, A. Performance comparison of epitrochoidal, hypotrochoidal, and cycloidal gerotor gear profiles. Mech. Mach. Theory **2021**, 158, 104228. **[Google Scholar] [CrossRef]**
26. De Martin, A.; Jacazio, G.; Sorli, M. Optimization of Gerotor Pumps with Asymmetric Profiles through an Evolutionary Strategy Algorithm. Machines **2019**, 7, 17. **[Google Scholar] [CrossRef]**
27. Tessari, F.; Galluzzi, R.; Amati, N. Efficiency-driven design methodology of gerotor hydraulic units. J. Mech. Des. **2020**, 142, 063501. **[Google Scholar] [CrossRef]**
28. Puliti, M.; Tessari, F.; Galluzzi, R.; Tonoli, A.; Amati, N. Design methodology of gerotor hydraulic machines for mechatronic applications. In Proceedings of the ASME International Mechanical Engineering Congress and Exposition, IMECE2021, Virtual, Online, 5 November 2021. **[Google Scholar]**
29. Phadke, M.S. Quality Engineering Using Robust Design; Prentice Hall International: Englewood Cliffs, NJ, USA, 1989. **[Google Scholar]**
30. Antony, J. Taguchi or classical design of experiments: A perspective from a practitioner. Sens. Rev. **2006**, 26, 227–230. **[Google Scholar] [CrossRef]**
31. Taguchi, G. Introduction to Quality Engineering: Designing Quality into Products and Processes; Asian Productivity Organization: Tokyo, Japan, 1986. **[Google Scholar]**
32. Unal, R.; Dean, E.B. Taguchi approach to design optimization for quality and cost: An overview. In Proceedings of the 1991 Annual Conference of the International Society of Parametric Analysts, Hampton, VA, USA, 1 January 1991. **[Google Scholar]**
33. Ross, R. Taguchi Techniques for Quality Engineering, 2nd ed.; McGraw-Hill: New York, NY, USA, 1996. **[Google Scholar]**
34. Antony, J.; Jiju Antony, F. Teaching the Taguchi method to industrial engineers. Work Study **2001**, 50, 141–149. **[Google Scholar] [CrossRef]**
35. Jung, S.-Y.; Bae, J.-H.; Kim, M.-S.; Kim, C. Development of a new gerotor for oil pumps with multiple profiles. Int. J. Precis. Eng. Manuf. **2011**, 12, 835–841. **[Google Scholar] [CrossRef]**
36. Ivanović, L.T.; Veličković, S.N.; Stojanović, B.Ž.; Kandeva, M.; Jakimovska, K. The selection of optimal parameters of gerotor pump by application of factorial experimental design. FME Trans. **2017**, 45, 159–164. **[Google Scholar] [CrossRef]**
37. Ivanović, L.; Stojanović, B.; Blagojević, J.; Bogdanović, G.; Marinković, A. Analysis of the flow rate and the volumetric efficiency of the trochoidal pump

by application of Taguchi method. Teh. Vjesn. **2017**, 24, 265–270. [**Google Scholar**] [**CrossRef**] [**Green Version**]

38. Lazić, L.; Milinković, S. Reducing software defects removal cost via design of experiments using Taguchi approach. Software Qual. J. **2015**, 23, 267–295. [**Google Scholar**] [**CrossRef**]

39. Vosniakos, G.C.; Kalattas, A.; Siasos, A. Optimal process planning for helical bevel gears using Taguchi design of simulated machining experiments. Proc. IMechE Part B J Eng. Manuf. **2018**, 232, 2627–2640. [**Google Scholar**] [**CrossRef**]

40. Choudhury, K.; Mandol, S.; Dan, P.K. Optimization of Operations in Epicyclic Gear Train Modelled Through Computer Aided Design. Asia-Pac. J. Manag. Res. Innov. **2014**, 10, 323–335. [**Google Scholar**] [**CrossRef**]

41. Stryczek, J.; Stryczek, P. Synthetic approach to the design, manufacturing and examination of gerotor and orbital hydraulic machines. Energies **2021**, 14, 624. [**Google Scholar**] [**CrossRef**]

42. Gamez-Montero, P.J.; Castilla, R.; Khamashta, M.; Codina, E. Contact problems of a trochoidal-gear pump. Int. J. Mech. Sci. **2006**, 48, 14711480. [**Google Scholar**] [**CrossRef**]

43. Gamez-Montero, P.J.; Castilla, R.; Mujal, R.; Khamashta, M.; Codina, E. GEROLAB package system: Innovative tool to design a trochoidal-gear pump. J. Mech. Des. **2009**, 131, 074502. [**Google Scholar**] [**CrossRef**]

44. GeroLAB Package System. Available online: **http://www.gerolab.es** (accessed on 1 February 2022).

45. Ram Kumar, A.C.; Mohammed Raffic, N.; Ganesh Babu, K.; Selvakumar, S. Static structural analysis of spur gear using ANSYS 15.0 and material selection by COPRAS, MOORA techniques. Mater. Today Proc. **2021**, 47, 25–36. [**Google Scholar**] [**CrossRef**]

46. Karthick, L.; Shanmugasundaram, A.; Jagadish, C.A.; Prabhu, T.; Vidya Prakash, S. Numerical analysis of Modified tooth in Spur Gear for increasing the performance using fool proofing technology. Mater. Today Proc. **2021**, 47, 5862–5868. [**Google Scholar**] [**CrossRef**]

47. Benaïcha, Y.; Perret-Liaudet, J.; Beley, J.D.; Rigaud, E.; Thouverez, F. On a flexible multibody modelling approach using FE-based contact formulation for describing gear transmission error. Mech. Mach. Theory **2022**, 167, 104505. [**Google Scholar**] [**CrossRef**]

48. Rao, P.S. Contact Stress and Shear Stress Analysis of Spur Gear Using ANSYS and Theoretical. Int. J. Mod. Stud. Mech. Eng. **2016**, 2, 9–14. [**Google Scholar**] [**CrossRef**]

49. Lahtivirta, J.; Lehtovaara, A. Modelling of spur gear contact using a local adaptive finite element mesh. Tribologia **2016**, 34, 41–55. **[Google Scholar]**

50. Lisle, T.J.; Shaw, B.A.; Frazer, R.C. External spur gear root bending stress: A comparison of ISO 6336:2006, AGMA 2101-D04, ANSYS finite element analysis and strain gauge techniques. Mech. Mach. Theory **2017**, 111, 1–9. **[Google Scholar] [CrossRef]**

51. Puneeth, M.L.; Mallesh, G. Static contact behavior of asymmetric spur gear. Mater. Today Proc. **2021**, 47, 3095–3104. **[Google Scholar] [CrossRef]**

52. Munoz, G.A. Lecture 7: Mesh Quality & Advanced Topics; ANSYS Inc.: Canonsburg, PA, USA, 2015; p. 37. Available online: **https://www.academia.edu/16970000/MESH_QUALITY_AND_ADVENCED_TOPICS_ANSYS_WORKBENCH_16_0** (accessed on 1 February 2022).

53. Sliwinski, P. Determination of the Theoretical and Actual Working Volume of a Hydraulic Motor—Part II (The Method Based on the Characteristics of Effective Absorbency of the Motor). Energies **2021**, 14, 1648. **[Google Scholar] [CrossRef]**

54. Garcia-Vilchez, M.; Gamez-Montero, P.J.; Codina, E.; Castilla, R.; Raush, G.; Freire, J.; Rio, C. Computational fluid dynamics and particle image velocimetry assisted design tools for a new generation of trochoidal gear pumps. Adv. Mech. Eng. **2015**, 7, 1–14. **[Google Scholar] [CrossRef]**

55. Gamez-Montero, P.J.; Castilla, R.; Codina, E. Methodology based on best practice rules to design a new-born trochoidal gear pump. Proc. Inst. Mech. Eng. Part C J. Mech. Eng. Sci. **2018**, 232, 1057–1068. **[Google Scholar] [CrossRef]**

56. Skowrońska, J.; Kosucki, A.; Stawiński, Ł. Overview of Materials Used for the Basic Elements of Hydraulic Actuators and Sealing Systems and Their Surfaces Modification Methods. Materials **2021**, 14, 1422. **[Google Scholar] [CrossRef]**

57. Student, M.; Hvozdetskyi, V.; Stupnytskyi, T.; Student, O.; Maruschak, P.; Prentkovskis, O.; Skačkauskas, P. Mechanical Properties of Arc Coatings Sprayed with Cored Wires with Different Charge Compositions. Coatings **2022**, 12, 925. **[Google Scholar] [CrossRef]**

12

PROTOTYPE AND EXPERIMENT WITH A DIGITALLY CONTROLLED KNEE EXOSKELETON POWERED BY HYDRAULICS

1. INTRODUCTION: AN OVERVIEW

Exoskeletons are wearable devices that enhance the physical strength of the individuals who wear them. These gadgets are utilized in a diverse range of applications. The medical application involves rehabilitating the motion and strength of patients' limbs who have limb injuries and post-stroke gait impairment. Exoskeletons are utilized by industrial workers to assist them in lifting big objects and engaging in demanding tasks without the risk of weariness and injury. Exoskeletons can be advantageous for soldiers, firefighters, and rescue workers as well [4,5].

Various types of exoskeleton devices with varying capacities have been created by researchers and industry in the past. Refer to [6] for a comprehensive discussion of previous exoskeletons. Research conducted by the authors' group and a literature review indicate that the majority of powered exoskeleton devices now being studied and sold utilize electro-mechanical mechanisms.

Hydraulic drives provide significant advantages over electro-mechanical drives, making them a superior choice for powered exoskeletons. Hydraulic drives are power-dense, enabling a substantial decrease in the weight and footprint of the exoskeleton device. Hydraulic drives can be set for energy recuperation, enhancing drive efficiency and enabling a smaller power supply. Hydraulic drives also enable motion locking without the need for a power source. Damped motion can be easily achieved using hydraulic drives, enabling smooth and natural motion patterns.

However, several obstacles have hindered hydraulically powered exoskeletons from becoming the dominant force in the exoskeleton market. Novel hydro-mechanical designs must be created to enable the integration of the hydraulic drive into the exoskeleton device in a compact manner. Secondly, small hydraulic components are required to construct a compact and lightweight hydraulic drive. Thirdly, rapid and effective control strategies suitable for hydraulic drives are required to operate the exoskeleton. The outdated perception of hydraulics as a cumbersome and outdated technology has deterred many researchers from considering hydraulic drives for exoskeleton actuation.

The Berkeley Lower Extremity Exoskeleton (BLEEX) was one of the first exoskeletons powered by hydraulics, created at the University of California, Berkeley. This exoskeleton enhanced the strength of the hip, knee, and ankle joints. Advancements resulted in the creation of exoskeleton devices ExoHiker and ExoClimber, which were said to be more pleasant for the users. Sarcos revealed its full-body exoskeleton XOS for military use about the same time. BLEEX utilized linear hydraulic actuators and mechanisms for motion transformation, whereas XOS employed rotational actuators directly at the limb joints. Recent research on hydraulically powered exoskeletons has concentrated on control strategies [14,15,16,17], designing small hydraulic power systems [18], and exploring different working fluids [19]. One major drawback of modern exoskeleton systems is their reliance on traditional servovalve-based resistance control, which is recognized for its low efficiency. Kaminaga et al. suggested using electro-hydraulic systems to control joints in humanoid robots and knee exoskeletons as a solution to the limitations of standard hydraulic systems. The design and control were specifically focused on obtaining improved back-drivability [22]. The knee joint power assist exoskeleton they designed utilized rotary actuators (vane motors) at the knee joint and included extra passive joints to enable secondary knee movements. Using hydraulic motors can be challenging due to internal leakages, which can affect efficiency and controllability, as well as low compactness. Recently, Lee et al. [26] and Jiang et al. [27] have investigated electro-hydraulic systems that use hydraulic cylinders, rather than hydraulic motors, to operate exoskeleton devices. Lee et al. included a multi-axial structure at the knee joint in their latest knee exoskeleton design to reduce misalignments. Sun et al. [29] introduced a lightweight electrohydrostatic actuator (LEHA) for knee exoskeletons by removing certain extra parts of conventional EHA and creating customized components to decrease the weight and size of their exoskeleton system.

Digital hydraulics has recently arisen as a novel technology that offers various advantages over traditional hydraulics in terms of high force and power density, reliability, durability, and cost-effectiveness. Digital hydraulics is well-suited for powering exoskeleton devices due to its benefits in compactness, resilience, lightweight power supply, and precise motion control. The leg exoskeleton created by Cao et al. is an early example of digital hydraulic concepts in exoskeleton devices, utilizing switching control.

The authors' research group has been working to utilize the benefits of digital hydraulics in creating lower limb exoskeleton systems. In 2017, the initial concept of a digital hydraulically powered knee exoskeleton was introduced. The exoskeleton included a digital hydraulic cylinder that operated the knee joint through a four-bar linkage mechanism. One major drawback of this design was the requirement for multi-chamber cylinders to create the digital cylinder, which are recognized for their high cost. In 2021, the authors introduced a new concept for a knee exoskeleton that uses a special mechanism to allow digital control of the exoskeleton using two basic hydraulic cylinders instead of a complex multi-chamber cylinder [33]. In this chapter , the authors aim to achieve the exoskeleton design discussed earlier. The design is initially optimized based on its dimensions and mass. Subsequently, detailed designs of each component are created, emphasizing their ability to bear loads. A prototype of the exoskeleton device is then created using this design. Experimental tests are undertaken to examine the device's performance and the control technique suggested by the writers in another location. 34. The essay effectively demonstrates the successful implementation of digital hydraulically driven exoskeletons. This offers the research community and industry a high-power hydraulic option as an alternative to electromechanically propelled exoskeletons. This work introduces digital hydraulics technology, along with its various benefits, to the realm of hydraulically powered exoskeletons.

This page is segmented into eight pieces, with this being the introductory section. Section 2 outlines the design of the knee exoskeleton device. Section 3 provides an in-depth analysis of the design optimization study. Section 4 contains the details of prototype development. The authors briefly introduce the control strategy in Section 5. The experimental experiments and findings are detailed in Section 6. Section 7 discusses how the study in this paper relates to previous research. The important conclusions of this paper are presented in Section 8.

2. KNEE EXOSKELETON DESIGN

Figure 1 displays the knee exoskeleton design created by the authors. The design includes the thigh and shank sections, which can be attached to the corresponding limb parts of a human. The components are linked via a revolute joint located at the knee. Two mechanisms, each comprising a hydraulic cylinder and a connecting rod, are used for motion actuation. Each mechanism has a junction between the cylinder rod and connecting rod that includes a guide pin restricted to follow a curved path. The trail is visible as grooves on the guide plates. There are two guide plates located on each side of the machinery to assist the guide pins. The front guide plate is not visible in Figure 1. The guide curve's geometric properties, in conjunction with the length of the linkages in the mechanisms, determine the transmission ratios of the mechanisms as discussed in Section 2.2.

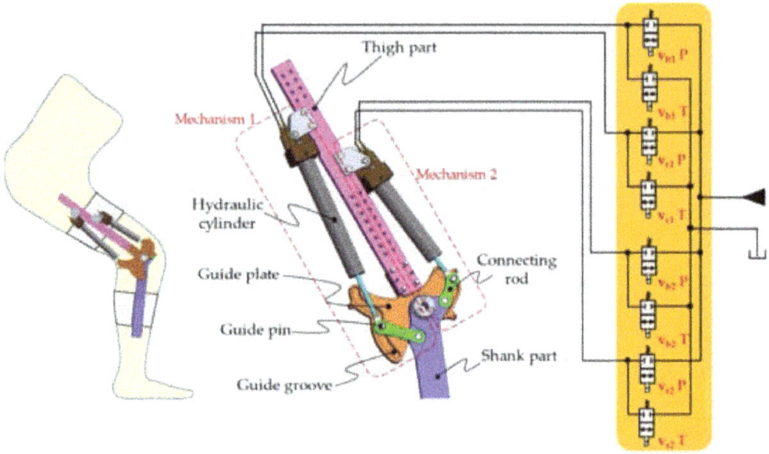

Figure 1. Knee exoskeleton attached to the human leg (left); details of the exoskeleton design and the digital hydraulic drive (right).

2.1. Hydraulic Drive

As shown in **Figure 1**, the digital hydraulic system driving the exoskeleton device consists of a pressure source, tank, digital valves (highlighted in yellow), and two hydraulic cylinders. The digital 2/2-way valves connect each cylinder chamber to the pressure source and tank.

There are eight 2/2-way valves in the hydraulic drive, and each valve can have two possible positions (on and off). Thus, the drive has 16 unique valve configurations, as illustrated in **Figure 2**.

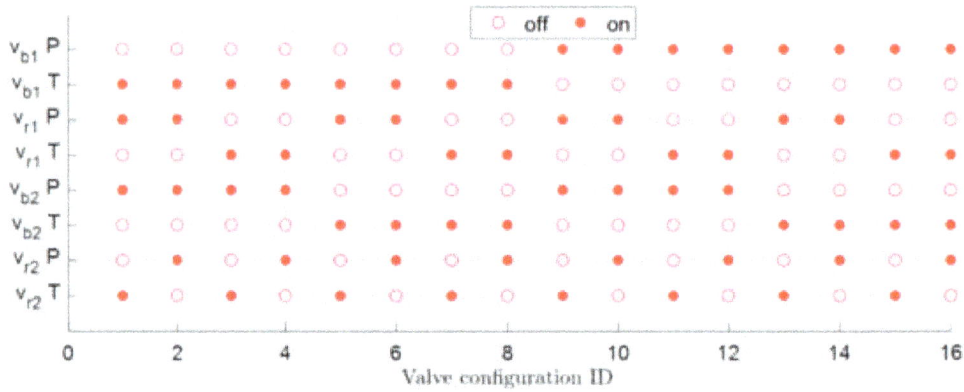

Figure 2. Valve positions for 16 unique valve configurations.

2.2. Transmission Ratio

The transmission ratio of a mechanism (TR) is defined as the ratio of the linear motion of the pistons in the cylinders and the angular motion of the knee joint:

$$T_{R,1,2} = \frac{dz_{1,2}}{d\psi}, \tag{1}$$

where z is the piston position in the cylinder and ψ is the knee angle, as indicated in **Figure 3.**

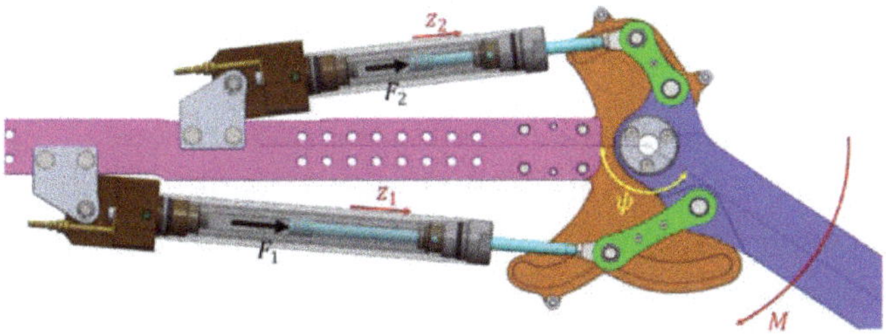

Figure 3. Illustration of the piston positions (z1,2), knee angle (ψ), cylinder forces (F1,F2(1,2), and knee torque (M).

The transmission ratio establishes the following relationship between the forces at the hydraulic cylinders (F) and the torque delivered at the knee joint (M) **[33]**:

$$M = T_{R,1}F_1 + T_{R,2}F_2. \tag{2}$$

To determine this knee torque requirement, the limb motion dynamics data from HuMoD database **[35]** are utilized. The database developed by researchers at TU Darmstadt, Germany **[36]** contains the biomechanical measurements (motion of the limb joints during typical gait cycles) and anthropomorphic parameters (length and mass of the limb parts) of the test subjects. This information is utilized in a multi-body numerical model of the human leg developed by the authors (a brief description is present in **Appendix A**) to determine the knee torque requirement in a typical gait cycle.

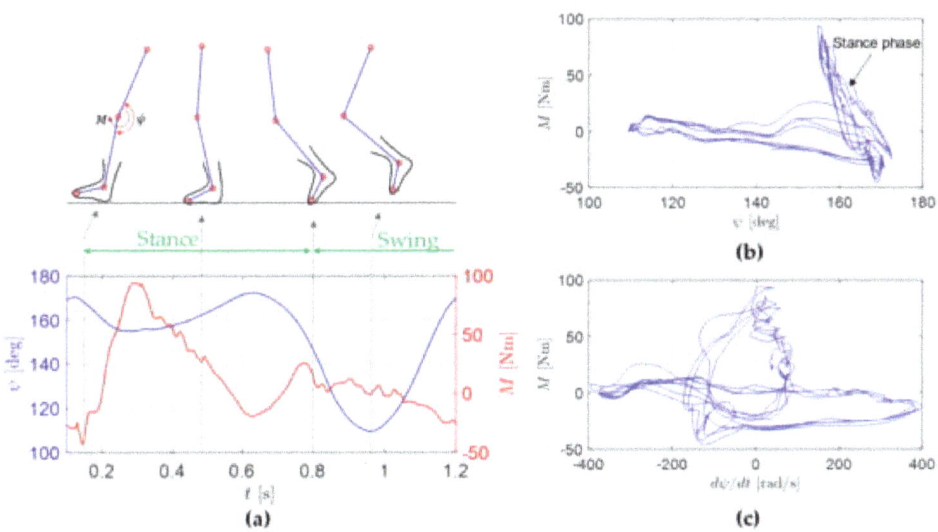

Figure 4. (a) Knee angle and torque over one gait cycle; (b) knee torque vs. knee angle over multiple gait cycles; (c) knee torque vs. knee angular speed over multiple gait cycles.

Figure 4a shows the knee angle and knee torque variation over a gait cycle. In the stance phase (where the foot touches the ground and the leg supports the load of the body), the knee torque requirement is high (up to 100 Nm). In contrast, in the swing phase (where the foot is off the ground), the knee torque remains low.

It is further notable from **Figure 4**b that the knee torque is high for knee angle ψ>155° and remains low for lower knee angles. This allows the mechanisms to be designed such that their transmission ratios are low for ψ<155° and high for ψ>155°° (**Figure 5**). Low transmission ratio for the majority of the knee angular span allows a smaller piston stroke of the hydraulic cylinders (as per Equation (1)), whereas high transmission ratio at peak torque requirement ensures a low hydraulic force requirement (as per Equation (2)), which enables using small hydraulic cylinders (in terms of bore size) working at medium pressure levels (≤≤200 bar).

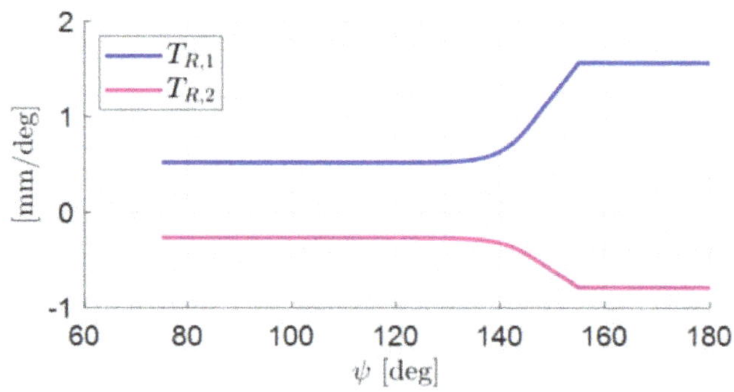

Figure 5. Transmission ratios of the two mechanisms.

The relationship between the transmission ratios of the two mechanisms is chosen as:

$$T_{R,2} = -0.5T_{R,1}, \tag{3}$$

with the negative sign indicating the opposite motion of the pistons for a given knee angular motion. Consequently, the knee torque expression from Equation (2) simplifies to:

$$M = T_{R,1}F_1 - 0.5T_{R,1}F_2. \tag{4}$$

Here, the cylinder force is:

$$F = p_bA_b - p_rA_r. \tag{5}$$

If the area ratio of the cylinders is chosen as:

$$A_b = 4A_r, \tag{6}$$

Equation (4) simplifies to:

$$M = T_{R,1}A_b(p_{b1} - 0.25p_{r1} - 0.5p_{b2} + 0.125p_{r2}). \tag{7}$$

From this expression, the cylinder bore diameter requirement can be determined. In particular, for the source pressure of 200 bar and the transmission ratio characteristic from **Figure 5**, the maximum torque (100 Nm) can be achieved with cylinders of bore diameter 8 mm.

Next, defining the effective force as,

$$F_{eff} = \frac{M}{T_{R,1}}, \tag{8}$$

the following expression is obtained:

$$F_{eff} = A_b(p_{b1} - 0.25p_{r1} - 0.5p_{b2} + 0.125p_{r2}). \tag{9}$$

For each of the 16 valve configurations shown in **Figure 2**, each of the pressure terms in Equation (9) is either equal to the source pressure or the tank pressure. Thus, each valve configuration corresponds to a unique effective force. **Figure 6** shows uniform stepping of the effective force, which is a direct result of the conditions imposed on the transmission ratios (Equation (3)) and the area ratios (Equation (6)).

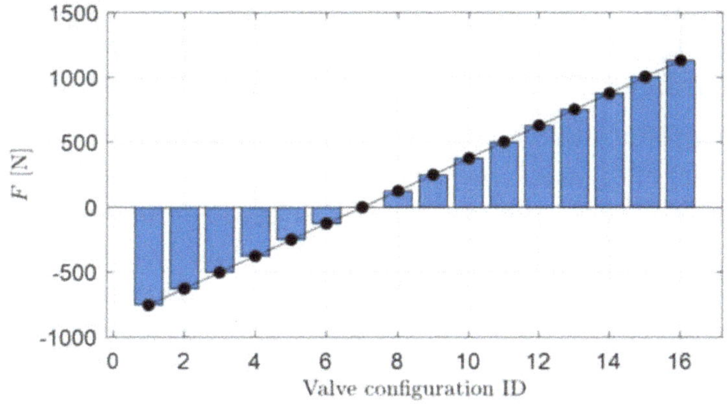

Figure 6. Effective force delivered by the hydraulic drive for different valve configurations.

3. DESIGN OPTIMIZATION STUDY

A major requirement of the exoskeleton device is that it should be compact and lightweight while providing its necessary function. To determine such a compact and lightweight design of the knee exoskeleton proposed by the authors, a design optimization study is conducted.

3.1. Objective Functions

For the user's comfort, it is important that the lateral width of the exoskeleton device (indicated as *w* in **Figure 7**) is as small as possible. In the exoskeleton design, this dimension is decided by the maximum distance between the guide curves of the two mechanisms. Thus, the first optimization objective is:

$$\text{minimize } w; w = \max|y_{B1}| + \max|y_{B2}|. \tag{10}$$

where $B1$ and $B2$ represent the guide curves of the two mechanisms.

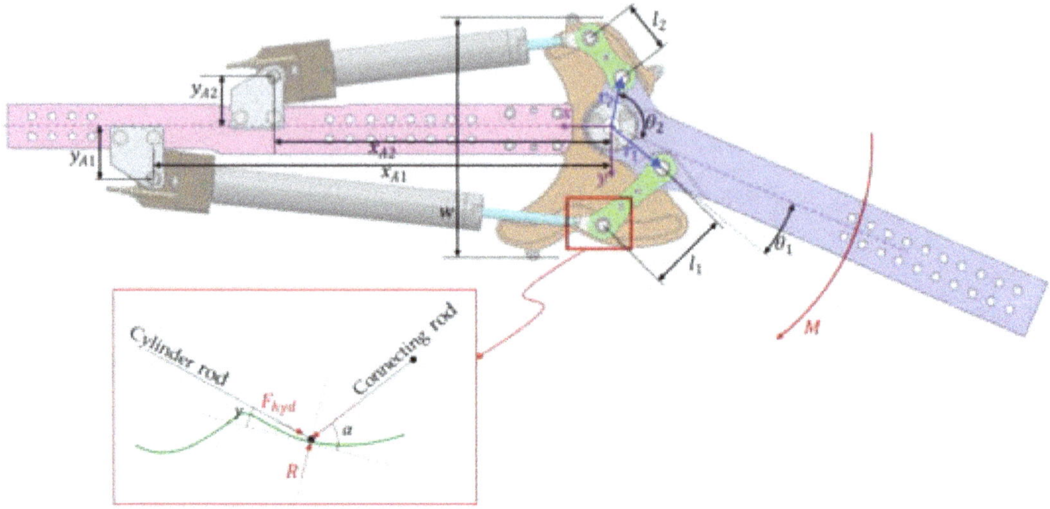

Figure 7. Illustration of critical geometrical dimensions used in the optimization problem. Inset shows the angles formed by the cylinder rod and connecting rod with the tangent of the guide curve at the guide pin.

The second optimization objective is related to loading at the guide curves. As shown in the inset of **Figure 7**, the guide curve needs to support the loads from the hydraulic cylinder and the connecting rod. Supporting higher loads will require thicker plates made of stronger material, which will lead to an increase in the weight of these plates. Thus, minimization of the load on the guide plate forms the second objective.

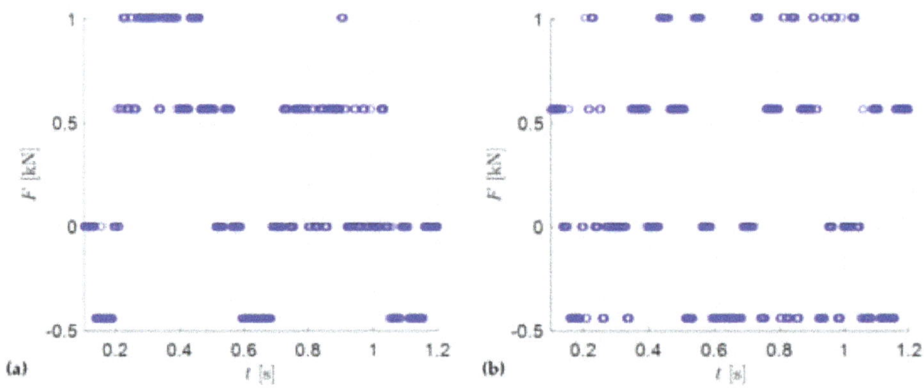

Figure 8. Force delivered by the hydraulic cylinders to support the required knee torque at different knee angles: (a) cylinder for mechanism 1; (b) cylinder for mechanism 2.

To determine the load on the guide plates and the contact force R, the gait cycle is divided into discrete time steps. For each time step in a gait cycle, from the knowledge of the knee torque requirement and the mechanism transmission ratio, the effective force requirement is determined (Equation (8)). The nearest discrete force level deliverable from the drive (shown in **Figure 6**) is chosen and the corresponding force from each cylinder ($Fhyd$) is determined. **Figure 8** shows the forces for each cylinder, where the four levels of forces observed are the result of four pressurization cases: (a) $p_b = 0$, $p_r = p_S$; (b) $p_b = pr = 0$; (c) $p_b = p_r = p_S$; (d) $p_b = p_S$, $p_r = 0$..

Next, the contact force at the guide plate is evaluated as:

$$R = F_{hyd}(\sin\gamma + \cos\gamma\tan\alpha), \tag{11}$$

where γ and α are the angles that the cylinder and connecting rod form with the tangent to the guide curve (as illustrated in the inset of **Figure 7**).

Thus, the second objective function is:

$$\text{minimize } R_{max}; R_{max} = \max(R_1, R_2). \tag{12}$$

3.2. Optimization Variables

The design variables for the optimization study are the dimensions $r1$, $r2$, $l1$, $l2$, $xA1, yA1,\ A2, yA2, \theta1,$ and $\theta2$, as indicated in **Figure 7. Table 1** shows the bounds imposed on these variables in the optimization process. The lower bounds of the crank lengths ($r1$ and $r2$) are set to 20 mm to allow enough space at the knee joint for installation of an angle sensor (shown in **Section 4.4**). The lower bounds of the connecting rods ($l1$ and 2) are set to 20 mm to allow enough space for realization of the revolute joints at their ends. The upper bounds of these linkages are set to 60 mm since linkages longer than this will lead to designs with lateral width (w) much higher than the typical width of the knee.

Table 1. Lower (*lb*) and upper (*ub*) bounds of the design variables in the optimization problem.

	$r_1(mm)$	$r_2\,(mm)$	$l_1\,(mm)$	$l_2(mm)$	$x_{A1}(mm)$	$y_{A1}(mm)$	$x_{A2}\,(mm)$	$y_{A2}(mm)$	$\theta_1(°)$	$\theta_2(°)°$
lb	2020	20	20	20	180	30	130	30	0	30
ub	60	60	60	60	350	40	300	40	80	190

The distance of each cylinder base mounting from the knee joint (A_1 and xA_2) is influenced by the stroke of that cylinder. The stroke of each cylinder is determined from the transmission ratio of the respective mechanism and the angular span of the knee joint. For the knee angle range $\psi \in [75°, 180°]$, the stroke values of the two cylinders are determined to be 90 mm and 45 mm. The higher values in the bounds of $xA1$ and $xA2$ account for the extra distance due to the lengths of the cylinder base and cylinder rods in extension. The y-coordinates of the cylinder base mounting ($yA1$ and $A2$) are limited to [30, 40] mm range. The lower bound ensures that a realistic mounting of the cylinders on the shank part is achievable, whereas the higher bound ensures the compactness of the device.

3.3. Optimization Constraints

The following constraints are imposed on the optimization problem:

- The design must satisfy the transmission ratio requirements described in **Section 2.2.**
- The design must be kinematically feasible for the full range of knee motion.
- During the full range of knee motion, interference between different parts of the exoskeleton device must be avoided.
- It can be observed that the groove for mechanism 1 has a relatively sharp peak near the middle. If the curvature of the groove at this peak becomes too high, the knee motion may experience jerky behavior when the guide pin passes through this region. To avoid this, the curvature of the groove must be smaller than a certain value (170 m^{-1} is considered as an appropriate limit).

3.4. Optimization Procedure and Results

The multi-objective optimization problem described in the preceding subsections is solved using the NSGA II algorithm [**37**] in MATLAB environment. The initial design space is populated with 1000 designs and the optimization algorithm is executed for 100 generations.

Figure 9 shows the results obtained from the optimization study in terms of the approximate pareto-front between the two objectives (minimization of the device width and minimization of the contact force at the guide plates). The existence of the pareto-front indicates that there is a clear trade-off between these objectives and no single design performs the best with respect to both the objectives.

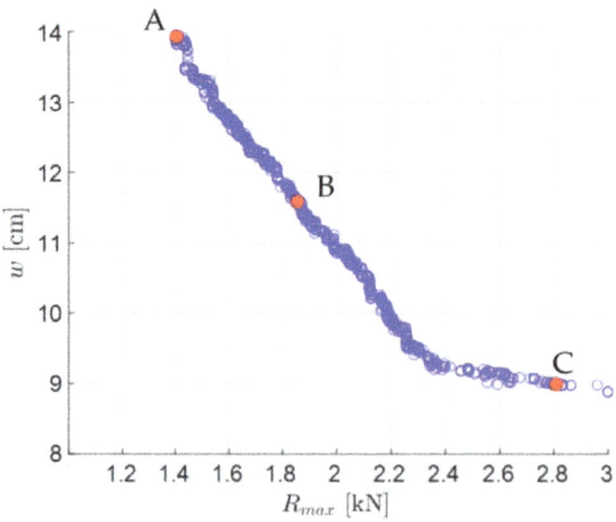

Figure 9. Approximate pareto-front obtained from the optimization study.

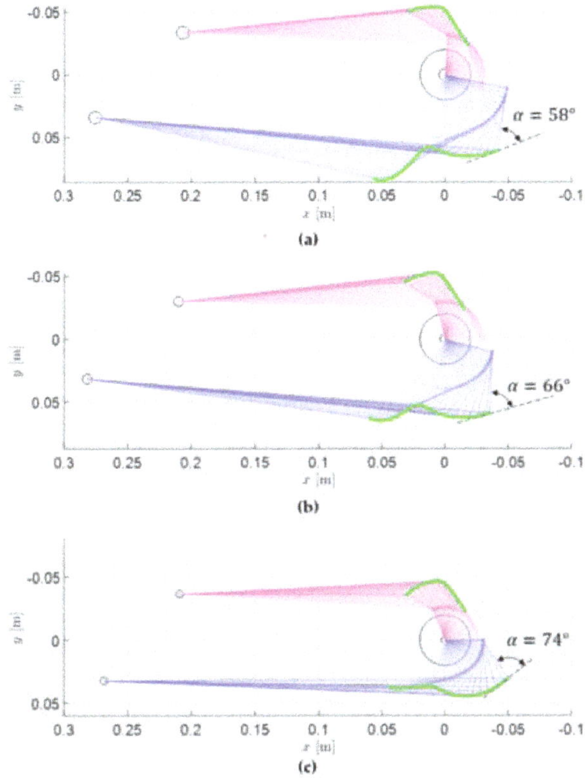

Figure 10. Three designs from the pareto-front: (a) design A; (b) design B; (c) design C. Mechanisms 1 and 2 are shown in blue and magenta, respectively. The guide groove profiles are shown in green.

Three designs chosen from the pareto-front are shown in **Figure 10**. Design A has a smaller contact force (1.4 kN) but is much larger in size (14 cm width). On the other hand, design C has a smaller size (9 cm width) but exhibits high contact force (2.8 kN). Design B is a good compromise between the two extreme designs ($Rmax$=1.9 kN , w=11.6 cm) and, hence, is chosen in this work. The illustrations of the exoskeleton device in **Figure 1, Figure 3**, and **Figure 7** show this chosen design. The values of the design variables for this design are reported in **Table 2.**

Table 2. Values of the design variables for the chosen design (design B).

$r1(mm)$	$r2$ (mm)	$l1$ (mm)	$l2(mm)$	$xA1(mm)$	$yA1(mm)$	$xA2$ (mm)	$yA2(mm)$	$\theta1$(°)	$\theta2$(°)
38.9	29.6	48.7	30.2	281.7	31.3	210.0	30.0	55.4	174.2

It is notable that the inverse relationship between the device size and contact force in **Figure 9** is the result of the fact that, as the size of the device is reduced, the angle between the connecting rod and guide groove (α in **Figure 7**) needs to increase, resulting in the increase in the contact force at the guide groove. This observation is illustrated for the three designs in **Figure 10**, where the value of α for the highest knee angle configuration is reported.

4. PROTOTYPE DEVELOPMENT

The exoskeleton design chosen from the optimization study in the previous section is realized in the form of a prototype. In the following subsections, key details related to the development/procurement of each component of the prototype and its integration in the prototype are presented.

4.1. Hydraulic Cylinders

The miniature hydraulic cylinders are obtained from HAWE Micro Fluid GmbH [**38**]. As shown in **Figure 11**, the cylinder design is based on the twin-tube principle, which allows location of both bore and rod side chamber ports at the cylinder base. The annular region between the outer and inner tubes allows fluid flow between the rod side chamber and the cylinder base.

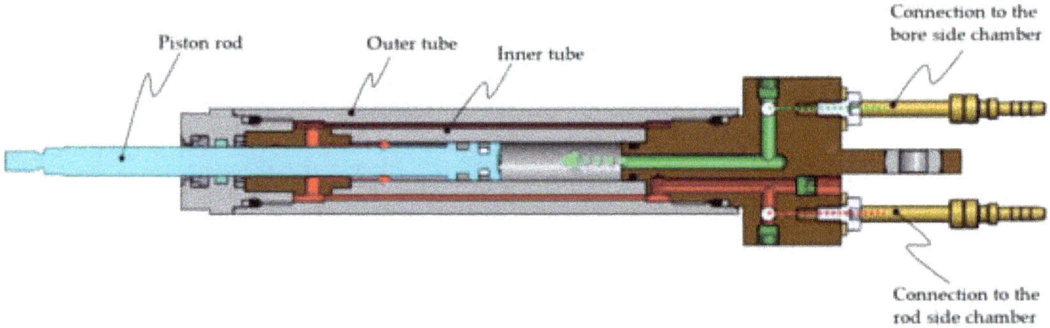

Figure 11. Sectioned view of the hydraulic cylinder from HAWE, illustrating the twin-tube design and the internal flow channels.

The strokes of the cylinders are 90 mm and 45 mm, which fulfils the requirements of the exoskeleton design. The bore and rod diameter of the cylinders are 8 mm and 6 mm, respectively, resulting in the area ratio of 2.29:1.

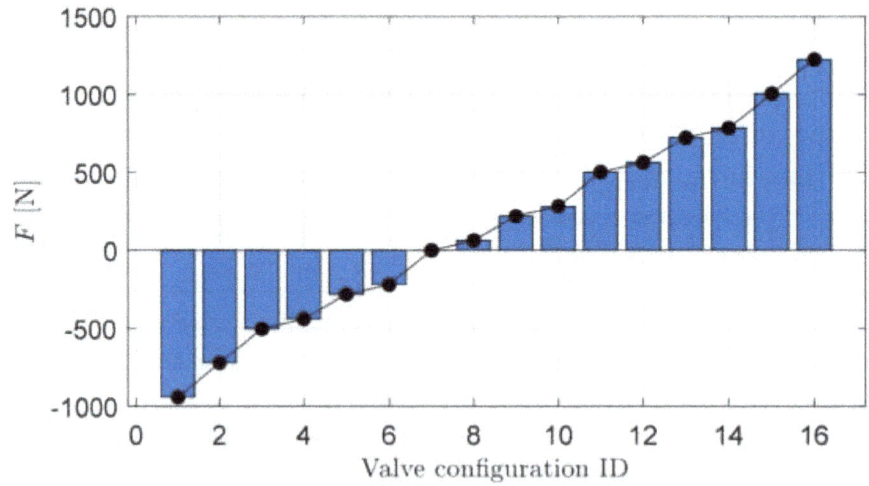

Figure 12. Effective force for different valve configurations using HAWE cylinders.

As this cylinder is an off-the-shelf component from HAWE, the desired area ratio of 4:1 could not be fulfilled. This results in non-uniformity of the force steps deliverable by the hydraulic drive. This non-uniformity is visible in **Figure 12**, which shows the forces for different valve configurations. The interpretation of each valve configuration in terms of the state of pressurization/depressurization of each of the cylinder chambers is illustrated in **Table 3**. From **Figure 12**, the force step non-uniformity is moderate and, thus, is tolerable for the sake of the first prototype.

Table 3. Cylinder chamber state for different valve configurations (u). $Vb1/2$ and $Vr1/2$ are the bore and rod side chambers of cylinder 1/2. Chamber state of 1 and 0 indicate pressurized and depressurized, respectively.

u	1	2	3	4	5	6	7	8	9	10	11	12	13	14	15	16
$Vb1$	0	0	0	0	0	0	0	1	0	1	1	1	1	1	1	1
$Vr1$	1	1	0	1	0	1	0	1	0	1	0	1	0	1	0	0
$Vb2$	1	1	1	0	1	0	0	1	0	1	1	0	1	0	0	0
$Vr2$	0	1	0	0	1	1	0	0	1	1	0	0	1	1	0	1

4.2. Linkage Parts and Guide Plates

In the mechanical design of the exoskeleton device, the thigh part, the shank part, and the hydraulic cylinders are aligned on the same plane (referred to as the central plane in **Figure 13**). This prevents the rise of any lateral moment during exoskeleton operation.

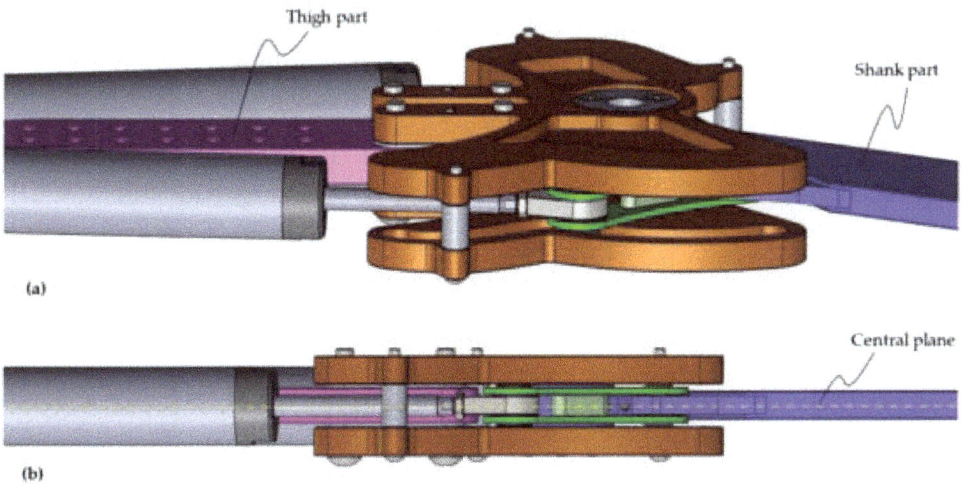

Figure 13. Illustration of the central plane and coplanarity of different linkages: (a) tilted view of the design; (b) bottom view of the design indicating the coplanarity of the thigh (magenta), cylinder (grey), and shank (blue) parts.

To link these coplanar cylinder rod and shank parts, the connecting rod part is realized via two individual parts that are interconnected via spacer elements

(**Figure 14**). Similarly, there are two guide plates that are interconnected via spacer elements (**Figure 13**). On the upper side (mechanism 2 side) of the device, there are two such connections. In contrast, the lower side (mechanism 1 side) has only one interconnection due to the space constraint arising from the presence of the shank part on this side.

Figure 14. Realization of the connecting rod linkage in the prototype.

To determine the appropriate dimensions of the key loaded parts (i.e., connecting rod elements, guide plates, and shank) and their material type, stress analyses are conducted using Finite Element Analysis (FEA) tool in Solidworks Simulation environment [**39**]. Brief details of this study are presented in the following subsections.

4.2.1. Connecting Rod Linkage

From the knowledge of the force at the hydraulic cylinder (**Figure 8**) and the orientation of the cylinder and the connecting rod (inset of **Figure 7**), the force experienced by the connecting rod linkage is determined as:

$$F_{cr} = F_{hyd}\frac{\cos\gamma}{\cos\alpha} \tag{13}$$

Figure 15 shows the forces acting on the connecting rod linkages. Higher force level at higher knee angles is a direct consequence of transmission ratio amplification as well as the high knee torque demand at these knee angles.

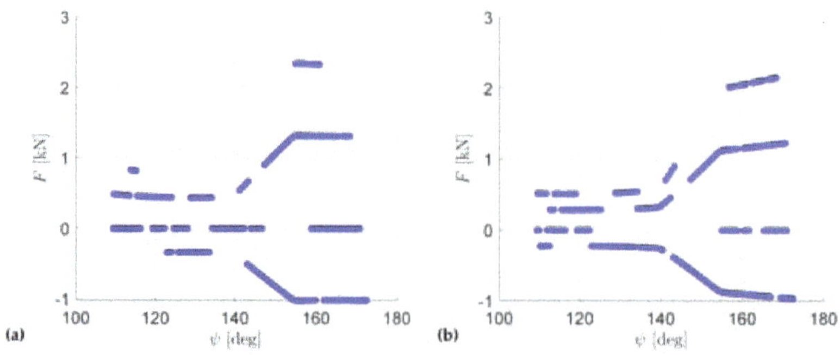

Figure 15. Forces (positive being compressive, negative being tensile) acting on the connecting rod linkages: (a) mechanism 1; (b) mechanism 2.

From **Figure 15** and the fact that the connecting rod linkage is realized via two parts, the maximum load to be supported by each part is 1.2 kN force in compression and 0.5 kN force in tension. **Figure 16** shows the stress experienced by these parts with 2 mm thickness. As the stress remains below 200 MPa, using steel E360 material (σYS=360 MPa) provides a decent factor of safety (1.8) for prototype purposes.

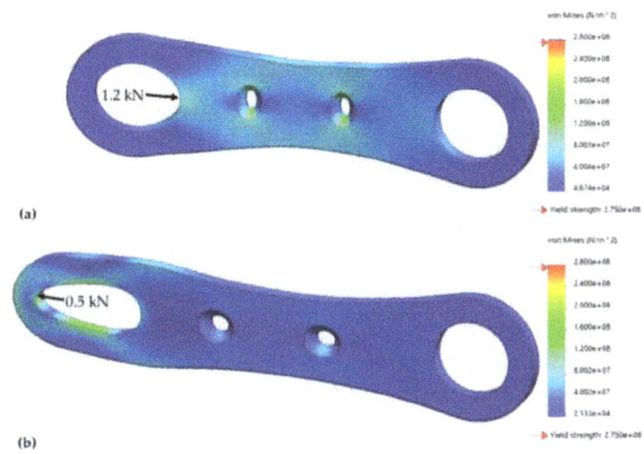

Figure 16. Stress analysis of the connecting rod elements in (a) compression and (b) tension.

4.2.2. Guide Plates

The guide plates need to support the load exerted by the guide pins of the two mechanisms. This contact force evaluation was explained in **Section 3.1** (Equation (11)). **Figure 17** shows the variation in these forces with the knee angle. The angle of 161°161° is the most severe position, where each of the two plates must support 1 kN load at the groove for mechanism 1 and 0.9 kN load at the groove for mechanism 2. These forces are acting outwards on the grooves from the knee joint perspective. The inwards forces are low (0.5 kN on mechanism 1 and 0.4 kN on mechanism 2) and compressive, and, thus, the material between the grooves and the knee joint can easily support them. Hence, the stress analysis is focused mainly on the outward loads.

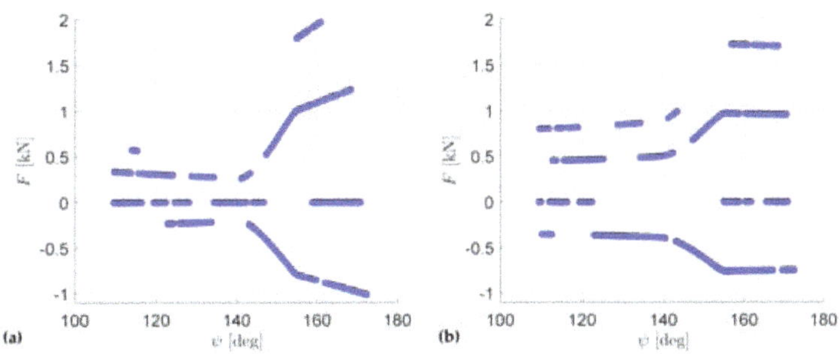

Figure 17. Forces (positive being outwards, negative being inwards) acting on the grooves of the guide plates: (a) grooves for mechanism 1; (b) grooves for mechanism 2.

Figure 18 shows that the maximum stress experienced by the guide plate is ~150 MPa. Thus, using steel E360 material (σYS= 360 MPa) provides a decent factor of safety (2.4). Furthermore, the maximum deformation of the guide groove remains below 40 μm (**Figure 19**), which is tolerable by the bearing at the guide pin.

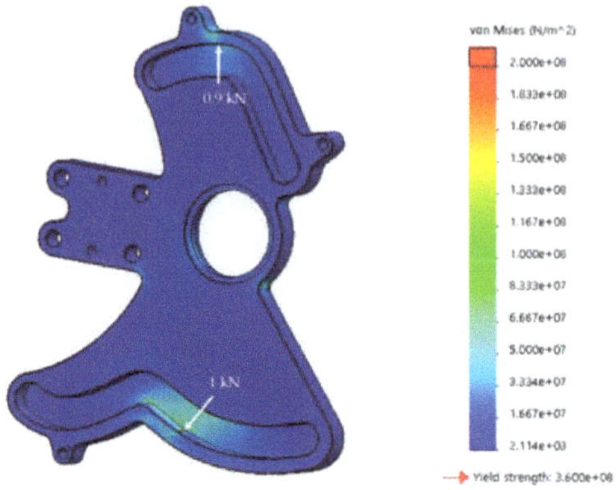

Figure 18. Stress analysis of the guide plate with the contact force exerted at the location of the guide pins for the knee angle of 161°161°.

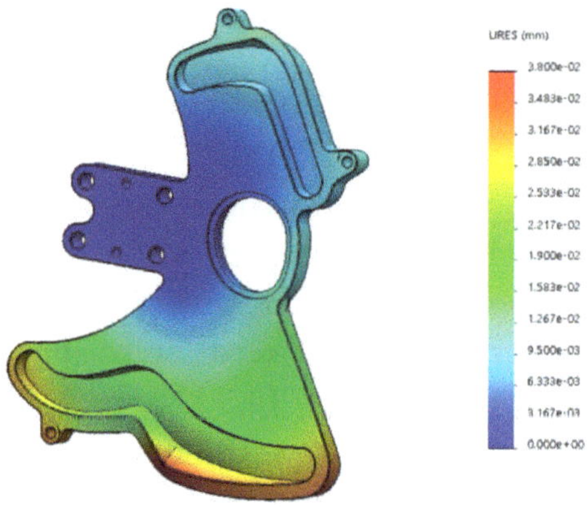

Figure 19. Deformation of the guide plate under the forces shown in Figure 18.

4.2.3. Shank Part

The portion of the shank part near the knee joint is critical from the strength perspective since there are three linkage connections (connections with the thigh part and two connecting rods) next to each other. Via an iterative procedure (involving design modification and stress analysis), the design of the shank part is developed such that it can handle the stresses involved.

The shank part is most stressed at the highest knee torque demand scenario when connecting rod 1 pushes the shank with a force of 2.3 kN and connecting rod 2 pulls with the force of 0.9 kN (as per **Figure 15**). Another scenario where the shank is significantly stressed is when the torque demand is moderate, but, to match the demand, both cylinders apply a pushing force. From Equation (4), it can be understood how the difference in the two high forces scaled by the transmission ratios can match a desired low torque.

Figure 20 shows the chosen design of the shank where the stresses remain below 75 MPa for both extreme scenarios, and, thus, using aluminum AW 5083 material (σYS=193 MPa) allows a good factor of safety (2.6).

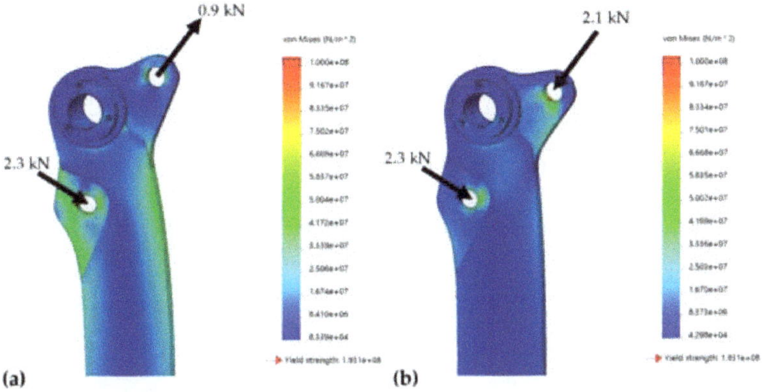

Figure 20. Stress analysis of the shank part: (a) high torque transfer case; (b) high internal force case.

4.2.4. Bearings

At all the joints, plain bearings from Igus [**40**] are used. These bearings are made of special material Iglidur and are known for being extremely wear-resistant, robust, and self-lubricating [**41**]. With the compressive strength of 100 MPa, the bearings of appropriate dimensions manufactured from this material are able to sustain the loads experienced in the normal operation of the device.

Finally, at the guide pin, to allow the pin to move in the guide curve with little friction, needle bearings are used. **Figure 21** shows the sectioned view of the design to reveal the bearings used.

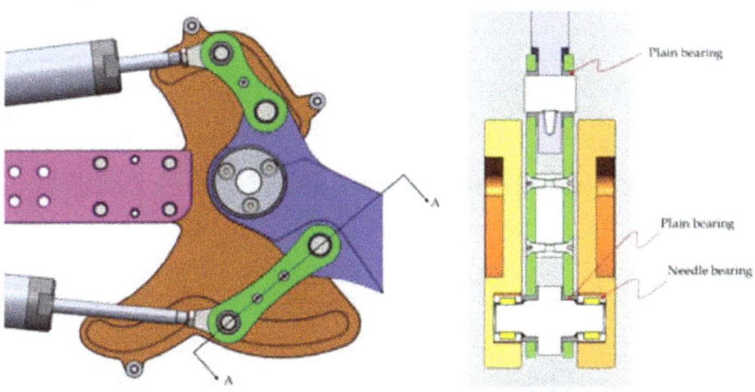

Figure 21. Illustration of the bearings (plain bearings from Igus and needle bearings) used in the design.

4.3. Valves

The miniature 2/2-way valves were obtained from researchers at Tampere University, Finland [**42**]. The eight valves are assembled in a compact fashion in a uniquely designed valve manifold, as shown in **Figure 22**. The flow capacity of each valve is 1.2 L/min at the pressure drop of 35 bar. The flows in the hydraulic drive during actuation of the gait cycle remain below 0.5 L/min except for one hydraulic line (connecting to the bore chamber of cylinder 1) in the swing phase, as shown in **Figure 23**, which goes up to 1.75 L/min. Thus, it is arguable that the pressure drop in this valve is too high for efficient operation and a larger valve would be preferable. However, this valve offers a key advantage of compactness, and such compact valves cannot be found elsewhere in the market.

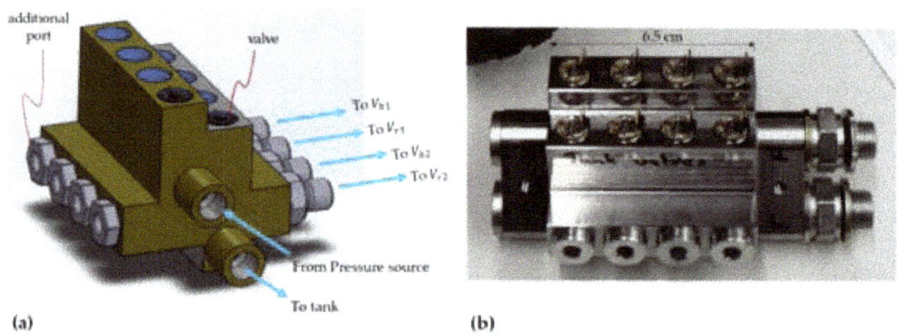

Figure 22. Valve manifold housing eight 2/2-way valves: (a) CAD; (b) procured valve block.

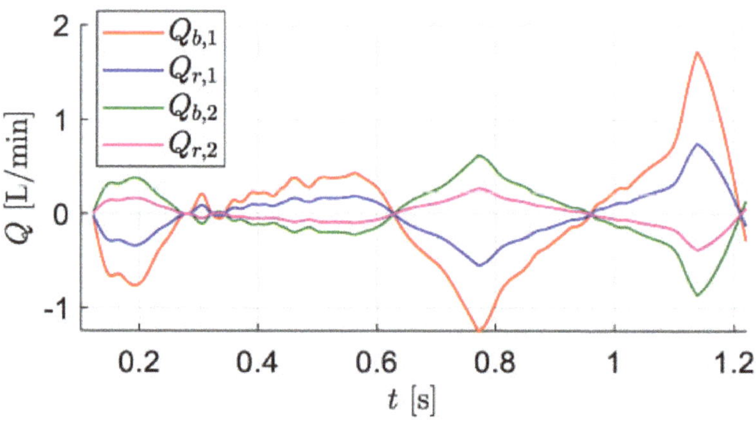

Figure 23. Flow in and out of the cylinder chambers over a gait cycle.

4.4. Sensors

For the control strategy described later in **Section 5**, the only information needed from the exoskeleton device is the knee angle. Thus, an angle sensor is integrated into the exoskeleton device. The hall effect sensor (model ETx25K) from Megatron is chosen due to its compact size [**43**]. The sensor's accuracy of ±1°±1° is adequate for the controller's operation.

Figure 24 shows the integration of this sensor in the device. A magnet holder is fabricated out of a plastic material that has a circular groove for the sensor magnet. The magnet is inserted into this groove and then the holder with the magnet is press-fitted into the joint cover (which rotates with the shank). Next, a sensor holder is fabricated and mounted to the guide plate. Finally, the sensor is mounted on the sensor holder. This arrangement ensures a precise distance between the magnet and the sensor, which is important for sensor's proper functionality.

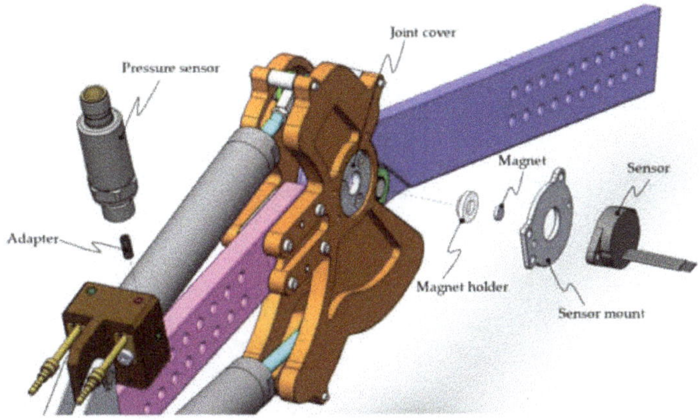

Figure 24. Sensor integration into the exoskeleton design.

During the testing phase, the pressure information from the cylinder chambers can be important. Thus, pressure sensors (AP019 from autosen [**44**]) are mounted on the cylinder base, as shown in **Figure 24.**

5. EXOSKELETON CONTROL STRATEGY

Control of an exoskeleton device depends on the functional requirements of the device. In this work, the authors consider the use case of patients suffering from paraplegia where they are unable to control their leg. Thus, the exoskeleton controller makes all the decisions in a gait cycle without any input from the user. The controller is tasked with following the characteristics of a typical gait cycle (i.e., track the knee angle observed in a typical gait cycle) and providing the necessary torque at the knee.

Two different strategies have been proposed by the authors for different phases in the gait cycle, a brief description of which is presented in the following subsections (a detailed description can be found in [**34**]).

5.1. Passive Control in the Stance Phase

From the typical knee motion characteristics in a gait cycle (**Figure 4**), it can be observed that the stance phase involves an inverse linear relationship between knee angle and knee torque. That is, as the knee angle decreases, the knee torque increases and vice versa. Thus, the knee motion in the stance phase can be realized via an elastic element in the knee joint. For the hydraulically driven knee exoskeleton proposed by the authors, the compressibility of the hydraulic oil can be exploited to achieve this elastic behavior.

As shown in **Figure 25**, during knee flexion, cylinder chambers $Vb1$ and $Vr2$ reduce in size. Thus, in the stance phase, both valves connecting to each of these chambers are turned off. As a result, when the knee flexes at the beginning of the stance phase, the hydraulic oil in these chambers becomes compressed. Consequently, the pressure in the chambers rises, resulting in an increase in the torque delivered by the drive. The opposite occurs in the knee extension at the latter part of the stance phase.

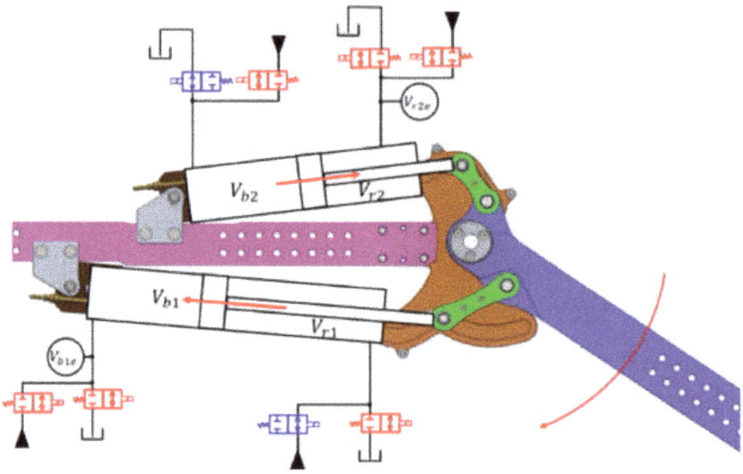

Figure 25. The valve configurations to realize the elastic behavior in the stance phase.

To ensure that the pressure in the cylinder chambers remains below their rated pressure (200 bar) in this elastic operation, extra volumes ($Vb1e$ and $Vr2e$) are added to the isolated cylinder chambers.

The size of these volumes can be determined via simple analysis, starting from the pressure build up equation for a closed volume (consisting of the cylinder chamber of area A and length z, and the extra volume Ve):

$$\frac{dp}{dt} = -\frac{K}{V_e + Az} A \frac{dz}{dt} . \tag{14}$$

Here, K is the bulk modulus of the fluid. Integrating the equation from the beginning of the elastic phase (instant a) to the middle of the elastic phase (instant b) and simplifying, the following expression for Ve is obtained:

$$V_e = \frac{A\left(z_a e^{-\frac{p_b - p_a}{K}} - z_b\right)}{1 - e^{-\frac{p_b - p_a}{K}}} . \tag{15}$$

From the design of the mechanism, the piston positions (za, zb) in the cylinder at different knee angles are known. Further, the mechanism is designed so that the cylinder pressure of 200 bar can supply the necessary peak torque in the stance phase. Thus, from instant a to b, the pressure should rise from 0 bar to 200 bar, i.e., in Equation (15), $pb-pa$=200 =200 bar. For a hydraulic fluid with bulk modulus of 14,000 bar, the sizes of these volumes are determined to be V_{b1e} and V_{r2e} == 19 mL. However, the presence of air in the fluid may lower the size requirements of these volumes. In the prototype, the valve block has additional ports (**Figure 22**), where these volumes of variable sizes (realized via appropriate fittings) can be mounted.

5.2. Model Predictive Control in the Swing Phase

In the swing phase, the exoskeleton device needs to track a desired knee motion typical of a gait cycle. To achieve this, a simplified form of model predictive control is employed.

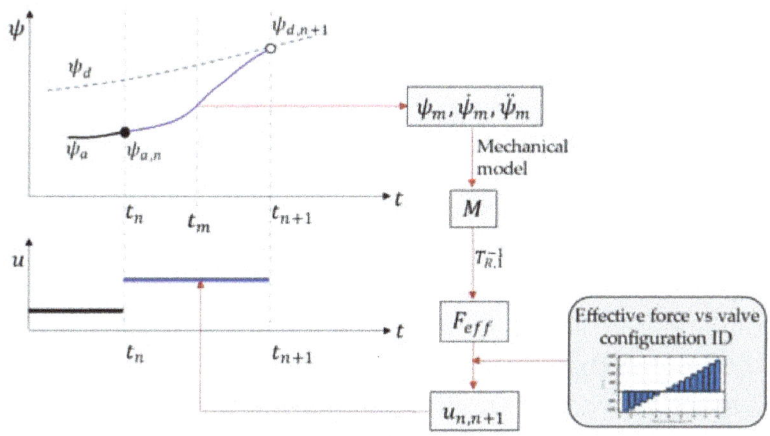

Figure 26. Illustration of the control strategy used in the swing phase of the gait cycle.

6. EXPERIMENTAL TESTS

Testing of novel exoskeleton devices is typically carried out in two phases. The first phase involves testing the operation of the exoskeleton without human in the loop. This phase is important for safety purposes. Once the motion of the exoskeleton device is found to be satisfactory and safe, then the second phase of testing is conducted with human wearing the exoskeleton device.

In this work, first-phase tests are conducted. The details of the experimental setup and test results are presented in the following subsections.

6.1. Experimental Setup

The exoskeleton prototype is mounted on a test rig, as shown in **Figure 27**. Two weights (1 kg each) are added on each side of the shank part to partially simulate the weight of the shank and foot of the wearer. The angle sensor attached to the knee joint measures the knee angle and transmits it to the data acquisition and control system. As described in **Section 5.2**, the measured knee angle is used in the motion control strategy. The pressure sensors are also present in the test setup for monitoring purposes. The details of the sensors used in the test setup are reported in **Table 4**.

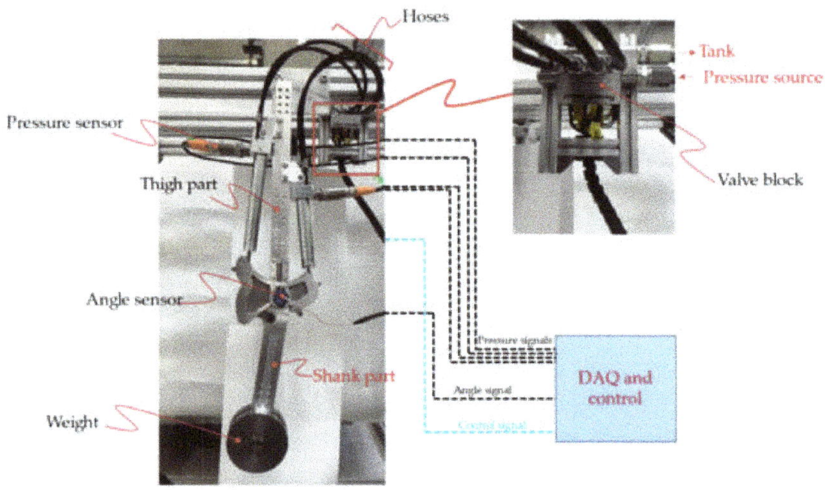

Figure 27. Experimental setup showing the exoskeleton device mounted on a test rig and connections to the data acquisition (DAQ) and control system. The inset indicates the valve block and connections to the pressure source and tank.

The hydraulic power supply used in this work is a traditional supply system available to the authors that is capable of supplying flow at a preset pressure level. For an exoskeleton device that would eventually be worn by a user, the hydraulic power supply will comprise a miniature pressure-controlled electrohydraulic pump that could fit in the backpack of the user. Focus on this aspect of hydraulic power supply will be a task for the future.

Table 4. Specification of the sensors used in the tests.

Sensor	Model	Accuracy	Measuring Range	Resolution
Angle sensor	Megatron ETx25K	±1%±1%	0°–360°0°–360°	0.07°0.07°
Pressure sensor	Autosen AP019	±0.2%±0.2%	0–2500–250 bar	-

Since the weights attached to the shank part are 2/5th of the typical weight of the shank and foot [**35**], the supply pressure is also set to 2/5th of the design pressure, i.e., 80 bar.

For data acquisition and control, a B&R X20 PLC system [**45**] is chosen. The system is flexible and can easily be adapted for any given I/O configuration, e.g., analog signals from the pressure sensors, bus connections, and driving the digital valves via "boost and hold" command [**42**].

The controller model described in **Section 5.2** is implemented in MATLAB/Simulink environment. The Simulink model is then imported into the programming tool "Automation Studio" from B&R using "Automation Studio Target for Simulink" software (which allows automatic code generation from the Simulink model). Finally, "Automation Studio" establishes a TCP/IP connection with the B&R target system.

During tests, the measurements are recorded with a sampling time of 0.5 milliseconds, and postprocessing is completed in the MATLAB environment.

6.2. Test Trajectory

The goal of the tests in this work is to validate the motion control strategy proposed by the authors for the swing phase (**Section 5.2**) and the capability of the exoskeleton device to follow the desired motion. As shown in **Figure 28**a, the knee angle trajectory in the swing phase can be approximated via a cosine curve. The equation describing this curve is:

$$\psi[^\circ] = 141^\circ + 31^\circ \cos(11(t - 0.1)). \tag{16}$$

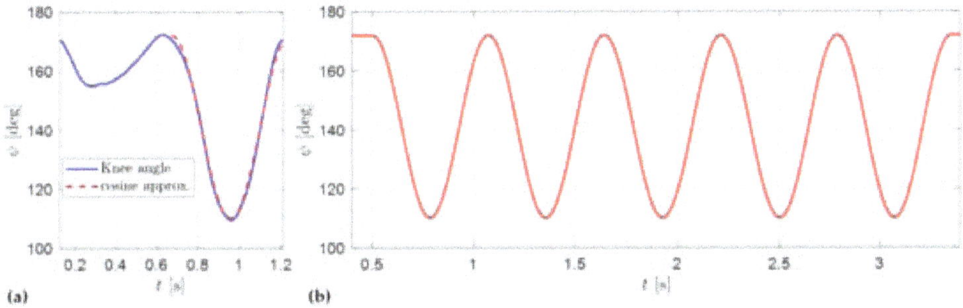

Figure 28. (a) Knee angle and a cosine curve approximation in the swing phase; (b) the repeating cosine curve used as the test trajectory.

This cosine curve is repeated to generate a test trajectory (**Figure 28**b), which is then used in the experiments.

6.3. Test Results

The experiments are conducted on the test setup using the control strategy described in **Section 5.2**. The only difference is the mechanical model, which, for the case of the test setup, is:

$$J\ddot{\psi} = M + mgl \sin \psi \,, \tag{17}$$

where J is the moment of inertia of the weights around the knee joint, M is the torque delivered by the hydraulic drive, m is the mass of the weights, g is the acceleration due to gravity, and l is the distance of the center of mass of the weights from the knee joint. Here, the inertia of the shank part is neglected since it is much lighter (~8 times) compared to the weights.

The exoskeleton device is asked to follow the test trajectory detailed in **Section 6.2**. The controller time step is set to 50 milliseconds.

Figure 29 shows the results from four sets of experiments. The conditions for the experiments are identical except for the initial knee angle and the pressure level of the source, which varies slightly between the experiments. The results indicate that the exoskeleton device is able to track the desired trajectory reasonably well. However, at some instances, it fails to reach the full knee extension as desired.

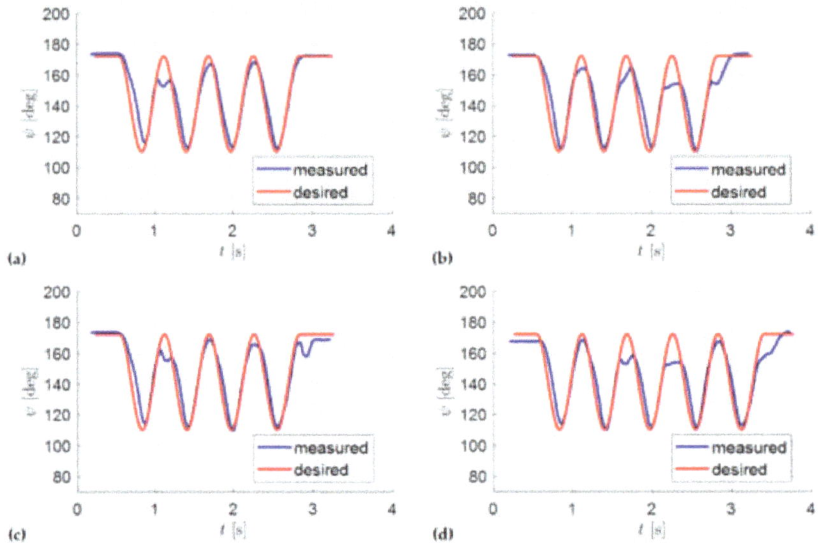

Figure 29. Results in the form of knee angle measured and desired (from the planned trajectory) for four sets of experiments: (a) Experiment A; (b) Experiment B; (c) Experiment C; (d) Experiment D.

For a quantitative analysis of the results, the error between the measured trajectory and desired trajectory is defined as:

$$e = \left|\psi_{meas} - \psi_d\right| . \tag{18}$$

The mean error (*emean*), its standard deviation (*estd*), and maximum error (*emax*) for the four sets of measurements are reported in **Table 5**. The error statistics are very similar across the experiments. Mean error of 5-6°5-6° is deemed to be tolerable by the user. However, the maximum error of 20-25°20-25° is of concern, which is caused by the inability of the device to reach full extension for some instances.

Table 5. Statistics for the error between the measured and desired trajectories.

Error Statistics	(a)	(b)	(c)	(d)
emean	4.6°4.6°	5.9°5.9°	5.1°5.1°	6.1°6.1°
estd	5.4°5.4°	5.8°5.8°	4.9°4.9°	5.0°5.0°
emax	24.5°24.5°	19.3°19.3°	21.7°21.7°	19.2°

A closer investigation into this discrepancy is illustrated via **Figure 30.** At t=1.596 s, the controller commands the valve configuration of u=8. As per **Table 3,** the state of the cylinder chambers should be: $\{Vb1,r1,Vb2,Vr2\}$={1,1,1,0}. However, as per **Figure 30**c, $Vb1$chamber starts to pressurize only at t=1.634 s. Thus, for 3838 ms, the state of the chambers is {0,1,1,0}0,1,1,0, which corresponds to u=1. Consequently, as per **Figure 12,** instead of negligible force (u=8), the drive delivers very high negative force, resulting in a sharp deceleration of the knee angle and a deviation from the desired trajectory.

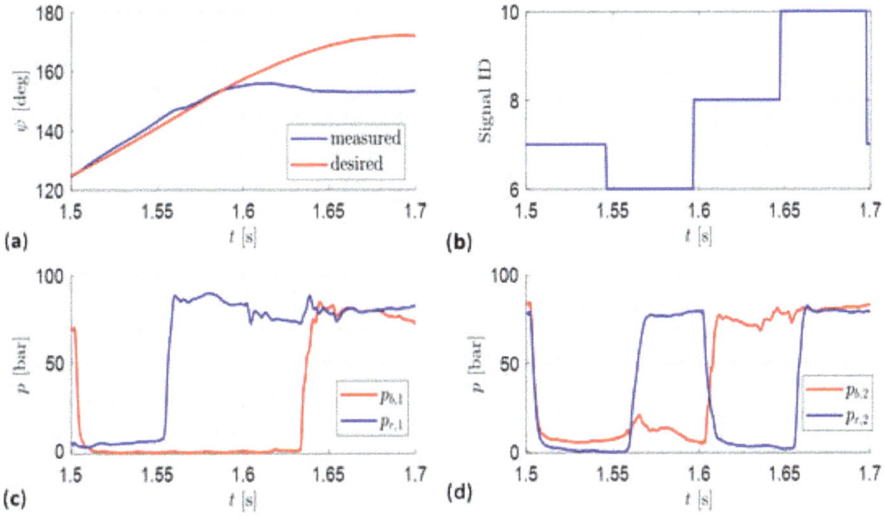

Figure 30. (a) Zoomed view of plot from Figure 29d; (b) valve configurations commanded by the controller; (c) pressure in cylinder 1 chambers; (d) pressure in cylinder 2 chambers.

The delayed pressurization of $Vb1$ chamber is explained via the following arguments. During knee extension, $Vb1$ chamber expands quickly. As gravity is helping the extension, the force needed from the drive is negative. From **Figure 12,** the valve configuration will remain below 7, i.e., $Vb1$ chamber remains connected to the tank. Thus, $Vb1$ is a fast-expanding chamber connected to tank, which needs high flow from the tank so that the chamber pressure can remain close to the tank pressure. However, the smallness of the valve (described in **Section 4.3**) restricts the amount of flow from the tank to the chamber. Consequently, the chamber pressure goes sub-atmospheric. Due the typical presence of entrained air in the fluid, its bulk modulus falls severely at such low pressures. Thus, when the chamber is connected to the pressure source, it takes more time to pressurize.

The disparity in trajectory tracking occurs specifically near the conclusion of knee extension when the controller commands a valve configuration that necessitates pressurized $Vb1$. When the controller directs different valve designs that do not need pressurization of this chamber, the discrepancy does not occur.

To solve this problem, a bigger valve can be utilized to maintain a high flow rate, preventing the pressure in chamber $Vb1$ from dropping below atmospheric pressure. Yet, the issue of finding compact arrangements with larger valves persists, necessitating the exploration of alternate solutions in the future.

7. DISCUSSION

The knee exoskeleton designed and tested in this study represents a notable progress in utilizing hydraulic technology for actuating exoskeleton devices. The digital hydraulically powered exoskeleton device offers more compactness, efficiency, and resilience compared to the traditional hydraulics technology utilized in previous exoskeletons. This design demonstrates compactness by the size of the valve block, which houses eight 2/2-way valves. Due to differences in the operations conducted by the device in this study and previous devices in the literature, a direct efficiency comparison is currently not feasible. The upcoming phase of complete gait cycle testing with user-in-loop will provide a direct efficiency comparison.

The exoskeleton device in this study is advantageous since it is specifically made for paraplegic patients and can provide 100% of the necessary torque at the knee joint. Most previous studies on exoskeleton development focus on either partial disability or power augmentation. The gadgets created can provide a knee torque ranging from 30% to 50%.

The exoskeleton technology developed in this work controls all decisions in the gait cycle for paraplegic patients who cannot control their legs. A simplified model predictive controller has been demonstrated to be adequate for this purpose, as shown in this study and the authors' previous research [34]. Devices designed for individuals with partial disability and power augmentation must consider the relationship between the force generated by the human and the device. Researchers working on these devices have utilized a range of intricate control mechanisms, including cascade force control, joint torque control, and repetitive learning control.

In the future, the scientists aim to expand the application of their exoskeleton technology to patients with partial knee control. An important challenge is determining the wearer's intended motion. Various methods for determining motion intent have been suggested by previous researchers; but, one that is suitable for the drive system employed in this study must be explored.

The last topic for consideration is the prototype's weight. The initial prototype created in this study has been noted to be excessively heavy for users' comfort, weighing 2.8 kg. Sun et al.'s LEHA has a weight of 2.5 kg, which includes the power supply. The high weight is mainly due to the twin-tube cylinder design, despite the optimized design. This design unnecessarily adds to the material usage, resulting in increased weight (0.62 kg and 0.54 kg for the cylinders). Yet, this is a clear drawback of utilizing ready-made components. Sun et al. achieved a low weight for their gadget with custom-designed components. In the future, the authors will investigate the possibility of manufacturing cylinders in-house to address the weight issue and achieve an area ratio of 4:1, which is optimal for uniform force stepping of the digital drive. The guide plate design's simplicity contributes to its high weight of 0.4 kg each. In the next iteration of the prototype, the guide plate design could be improved by creating a truss-frame-type structure, resulting in a substantial reduction in weight.

8. CONCISE SUMMARY

This article presents development and first-phase testing of a novel digital hydraulically driven knee exoskeleton prototype, the idea of which was conceived by the authors in their previous works. Via a multi-objective optimization study, a design is developed that is optimal with respect to its size and weight. The designs of the linkage parts are further refined while ensuring that each part can safely support the load exerted on it over a typical operational cycle. Subsequently, the prototype is manufactured and assembled with procured hydraulic cylinders, valves, and sensors. Finally, first-phase testing (without human) is conducted, where the prototype is mounted on a test stand and the controller is tasked with following a given trajectory that resembles the swing motion in a typical gait cycle. The results indicate that the prototype is able to track the desired motion, except with occasional discrepancies at full knee extension.

In the future, investigations to address the aforementioned discrepancy need to be conducted. These should be followed by testing the full gait cycle operation with a human in the loop.

Overall, the novel digital hydraulically driven knee exoskeleton device prototyped and tested in this work is a significant step towards bringing the advantages of digital hydraulics technology to the field of hydraulically driven exoskeleton devices. This can pave the way for widespread adoption of power-dense hydraulics technology in actuation of exoskeletons. The resulting compact lightweight exoskeleton devices will ultimately improve the comfort level of the wearers.

Furthermore, this work also showcases the gaps in the current hydraulics components' availability (e.g., miniature yet efficient valves and lightweight cylinders), which has historically hindered usage of hydraulics technology in exoskeleton actuation. This can be interpreted as an exciting opportunity for the industry and the research community for further research and development in this regard.

9. AT A GLANCE

Digital hydraulic drives are recognized for their high efficiency, power density, and durability. These benefits make them a compelling option for powering exoskeleton devices instead of electric drives. The study describes the creation of a prototype for a digital knee exoskeleton powered by hydraulics and its subsequent experimental evaluation. The device utilizes two small hydraulic cylinders and an innovative mechanism to convert linear motion into rotating motion. The gadget is operated via passive control during the stance phase and a simplified model predictive control method during the swing phase. The exoskeleton device's design is optimized for compactness and weight in this work. The design characteristics are adjusted to ensure the gadget can support operational loads. The concept is then transformed into a prototype using a combination of elements made in-house and components purchased externally. Finally, the design's performance and the control technique are examined by experimental experiments. The prototype exhibits several limitations concerning valve size and total weight, which will be rectified in future investigations.

Appendix A. Mechanical Model of Lower Limb Motion

The mechanical model of limb motion has been presented by the authors in previous works [**33,34**]. However, for the sake of completeness, the authors recapitulate the model in this appendix.

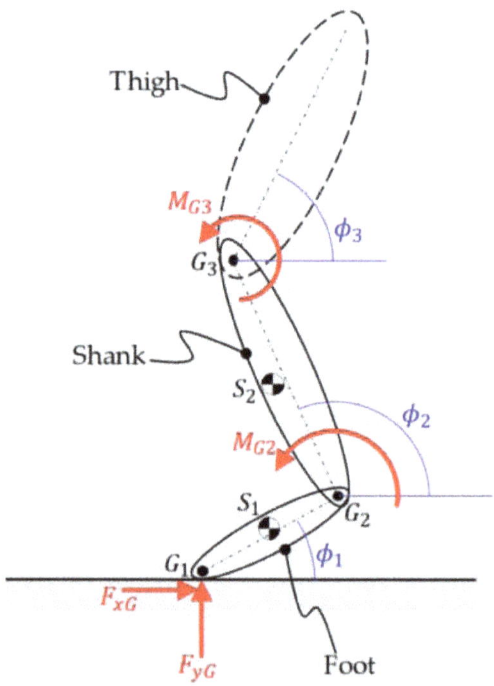

Figure A1. Planar view of limb with variables present in the mechanical model.

The lower limb motion is modelled via a planar system comprising foot, shank, and thigh. The motion of the upper body is until the thigh is considered to be known from the HuMoD database. Thus, the knee and ankle joint motions are the DOFs (degrees of freedom) considered in the model.

The kinetic and potential energy of the system is expressible as:

$$K_E = \frac{1}{2}\sum_{i=1}^{2}\left(m_i\dot{x}_{S_i}\cdot\dot{x}_{S_i} + J_i\dot{\phi}_i^2\right), \tag{A1}$$

$$P_E = g\sum_{i=1}^{2} m_i x_{S_i}\cdot e_y\,. \tag{A2}$$

where mi is the mass, $\boldsymbol{xSi}$ is the position of the center of mass, Ji is the moment of inertia, ϕi is the angular position of the ith limb part, and $\boldsymbol{ey}$ is the unit vector in vertical direction.

Next, the virtual work due to torques at the joints (MGi) and the ground reaction force ($\boldsymbol{FG}$) is:

$$\delta W = \sum_{i=2}^{3} M_{Gi}(\delta\phi_i - \delta\phi_{i-1}) + F_G^T\delta x. \tag{A3}$$

Substituting these expressions into the Lagrange equation of motion and simplifying, the following expression for the knee torque is obtained:

$$M = m_{eff,1}\ddot{\phi}_1 + m_{eff,2}\ddot{\phi}_2 - kB_FF_G - kV_{f0}. \tag{A4}$$

where $meff,1$ and $meff,2$ are the effective masses of foot and shank, $\boldsymbol{kBFFG}$ is the ground reaction term, and $\boldsymbol{kVf}0$ contains the terms related to the centripetal force, gravity, and the force from thigh. The complete expressions are lengthy, and, hence, are not shown here.

REFERENCES

1. Li, S.; Francisco, G.E.; Zhou, P. Post-stroke hemiplegic gait: New perspective and insights. Front. Physiol. **2018**, 9, 1021. **[Google Scholar] [CrossRef] [PubMed] [Green Version]**

2. Nilsson, A.; Vreede, K.S.; Häglund, V.; Kawamoto, H.; Sankai, Y.; Borg, J. Gait training early after stroke with a new exoskeleton—The hybrid assistive limb: A study of safety and feasibility. J. NeuroEng. Rehabil. **2014**, 11, 92. **[Google Scholar] [CrossRef] [PubMed] [Green Version]**

3. Voilqué, A.; Masood, J.; Fauroux, J.C.; Sabourin, L.; Guezet, O. Industrial Exoskeleton Technology: Classification, Structural Analysis, and Structural Complexity Indicator. In Proceedings of the 2019 Wearable Robotics Association Conference (WearRAcon), Scottsdale, AZ, USA, 25–27 March 2019; pp. 13–20. **[Google Scholar]**

4. Proud, J.K.; Lai, D.T.H.; Mudie, K.L.; Carstairs, G.L.; Billing, D.C.; Garofolini, A.; Begg, R.K. Exoskeleton application to military manual handling tasks. Hum. Factors **2022**, 64, 527–554. **[Google Scholar] [CrossRef] [PubMed]**

5. Osipov, A. Fire exoskeleton to facilitate the work of the fireman. E3S Web. Conf. **2019**, 126, 00015. **[Google Scholar] [CrossRef]**

6. De la Tejera, J.A.; Bustamante-Bello, R.; Ramirez-Mendoza, R.A.; Izquierdo-Reyes, J. Systematic review of exoskeletons towards a general categorization model proposal. Appl. Sci. **2021**, 11, 76. **[Google Scholar] [CrossRef]**

7. Scheidl, R. Digital Fluid Power for Exoskeleton Actuation—Guidelines, Opportunities, Challenges. In Proceedings of the Ninth Workshop on Digital Fluid Power, Aalborg, Denmark, 7–8 September 2017. **[Google Scholar]**

8. Zoss, A.; Kazerooni, H.; Chu, A. On the Mechanical Design of the Berkeley Lower Extremity Exoskeleton (BLEEX). In Proceedings of the 2005 IEEE/RSJ International Conference on Intelligent Robots and Systems, Edmonton, AB, Canada, 2–6 August 2005; pp. 3465–3472. **[Google Scholar]**

9. Zoss, A.B.; Kazerooni, H.; Chu, A. Biomechanical design of the Berkeley Lower Extremity Exoskeleton (BLEEX). IEEE/ASME Trans. Mechatron. **2006**, 11, 128–138. **[Google Scholar] [CrossRef]**

10. Kazerooni, H.; Racine, J.-L.; Huang, L.; Steger, R. On the Control of the Berkeley Lower Extremity Exoskeleton (BLEEX). In Proceedings of the 2005 IEEE International Conference on Robotics and Automation, Barcelona, Spain, 18–22 April 2005; pp. 4353–4360. **[Google Scholar]**

11. Xie, H.; Li, X.; Li, W.; Li, X. The Proceeding of the Research on Human Exoskeleton. In Advances in Intelligent Systems Research; Atlantis Press: Amsterdam, The Netherlands, 2014; pp. 754–758. **[Google Scholar]**

12. Karlin, S. Raiding Iron Man's Closet [Geek Life]. IEEE Spectr. **2011**, 48, 25. **[Google Scholar] [CrossRef] [Green Version]**

13. Huo, W.; Mohammed, S.; Moreno, J.C.; Amirat, Y. Lower limb wearable robots for assistance and rehabilitation: A state of the art. IEEE Syst. J. **2016**, 10, 1068–1081. **[Google Scholar] [CrossRef]**

14. Chen, S.; Chen, Z.; Yao, B.; Zhu, X.; Zhu, S.; Wang, Q.; Song, Y. Adaptive robust cascade force control of 1-DOF hydraulic exoskeleton for human performance augmentation. IEEE/ASME Trans. Mechatron. **2017**, 22, 589–600. **[Google Scholar] [CrossRef]**

15. Kim, H.; Shin, Y.J.; Kim, J. Design and locomotion control of a hydraulic lower extremity exoskeleton for mobility augmentation. Mechatronics **2017**, 46, 32–45. **[Google Scholar] [CrossRef]**

16. Yang, Y.; Dong, X.; Liu, X.; Huang, D. Robust repetitive learning-based trajectory tracking control for a leg exoskeleton driven by hybrid hydraulic system. IEEE Access **2020**, 8, 27705–27714. **[Google Scholar] [CrossRef]**

17. Chen, S.; Han, T.; Dong, F.; Lu, L.; Liu, H.; Tian, X.; Han, J. Precision interaction force control of an underactuated hydraulic stance leg exoskeleton considering the constraint from the wearer. Machines **2021**, 9, 96. **[Google Scholar] [CrossRef]**

18. Ouyang, X.; Ding, S.; Fan, B.; Li, P.Y.; Yang, H. Development of a novel compact hydraulic power unit for the exoskeleton robot. Mechatronics **2016**, 38, 68–75. **[Google Scholar] [CrossRef]**

19. Kosaki, T.; Li, S. A water-hydraulic upper-limb assistive exoskeleton system with displacement estimation. J. Robot. Mechatron. **2020**, 30, 149–156. **[Google Scholar] [CrossRef]**

20. Kaminaga, H.; Ono, J.; Nakashima, Y.; Nakamura, Y. Development of Backdrivable Hydraulic Joint Mechanism for Knee Joint of Humanoid Robots. In Proceedings of the 2009 IEEE International Conference on Robotics and Automation, Kobe, Japan, 12–17 May 2009; pp. 1577–1582. **[Google Scholar]**

21. Kaminaga, H.; Amari, T.; Niwa, Y.; Nakamura, Y. Electro-Hydrostatic Actuators with Series Dissipative Property and Their Application to Power Assist Devices. In Proceedings of the 2010 3rd IEEE RAS & EMBS International Conference on Biomedical Robotics and Biomechatronics, Tokyo, Japan, 26–29 September 2010; pp. 76–81. **[Google Scholar]**

22. Kaminaga, H.; Tanaka, H.; Nakamura, Y. Mechanism and Control of Knee Power Augmenting Device with Backdrivable Electro-Hydrostatic Actuator. In Proceedings of the 13th World Congress in Mechanism and Machine Science, Guanajuato, Mexico, 19–23 June 2011. **[Google Scholar]**

23. Kaminaga, H.; Amari, T.; Niwa, Y.; Nakamura, Y. Development of Knee Power Assist Using Backdrivable Electro-Hydrostatic Actuator. In Proceedings of the 2010 IEEE/RSJ International Conference on Intelligent Robots and Systems, Taipei, Taiwan, 18–22 October 2010; pp. 5517–5524. **[Google Scholar]**

24. Tanaka, H.; Kaminaga, H.; Nakamura, Y. Pressure feedback control based on singular perturbation method of an electro-hydrostatic actuator for an exoskeletal power-assist system. J. Robot. Mechatron. **2012**, 24, 354–362. **[Google Scholar] [CrossRef]**

25. Kurosawa, H.; Walker, P.S.; Abe, S.; Garg, A.; Hunter, T. Geometry and motion of the knee for implant and orthotic design. J. Biomech. **1985**, 18, 487–499. **[Google Scholar] [CrossRef]**

26. Lee, D.; Song, B.; Park, S.Y.; Baek, Y.S. Development and control of an electro-hydraulic actuator system for an exoskeleton robot. Appl. Sci. **2019**, 9, 4295. **[Google Scholar] [CrossRef] [Green Version]**

27. Jiang, J.; Wang, Y.; Cao, H.; Zhu, J.; Zhang, X. A novel pump-valve coordinated controlled hydraulic system for the lower extremity exoskeleton. Trans. Inst. Meas. Control **2020**, 42, 2872–2884. **[Google Scholar] [CrossRef]**

28. Lee, T.; Lee, D.; Song, B.; Baek, Y.S. Design and control of a polycentric knee exoskeleton using an electro-hydraulic actuator. Sensors **2020**, 20, 211. **[Google Scholar] [CrossRef] [Green Version]**

29. Sun, M.; Ouyang, X.; Mattila, J.; Chen, Z.; Yang, H.; Liu, H. Lightweight electrohydrostatic actuator drive solution for exoskeleton robots. IEEE/ASME Trans. Mechatron. **2022**, 1–12. **[Google Scholar] [CrossRef]**

30. Scheidl, R.; Linjama, M.; Schmidt, S. Is the future of fluid power digital? Proc. Inst. Mech. Eng. Part I J. Syst. Control. Eng. **2012**, 226, 721–723. **[Google Scholar] [CrossRef]**

31. Cao, H.; Ling, Z.; Zhu, J.; Wang, Y.; Wang, W. Design Frame of a Leg Exoskeleton for Load-Carrying Augmentation. In Proceedings of the 2009 IEEE International Conference on Robotics and Biomimetics (ROBIO), Guangxi, China, 13–19 December 2009; pp. 426–431. **[Google Scholar]**

32. Holl, E.; Scheidl, R.; Eshkabilov, S. Simulation Study of a Digital Hydraulic Drive for a Knee Joint Exoskeleton. In Proceedings of the ASME/BATH 2017 Symposium on Fluid Power and Motion Control, Sarasota, FL, USA, 4 December 2017. **[Google Scholar]**

33. Rituraj, R.; Scheidl, R.; Ladner, P.; Lauber, M. A Novel Design Concept of Digital Hydraulic Drive for Knee Exoskeleton. In Proceedings of the ASME/BATH 2021 Symposium on Fluid Power and Motion Control, Virtual, 13 December 2021. **[Google Scholar]**

34. Rituraj, R.; Scheidl, R. Advancements in the Control Strategy for Digital Hydraulically Driven Knee Exoskeleton. In Proceedings of the ASME/BATH 2022 Symposium on Fluid Power and Motion Control, Bath, UK, 19 September 2022. **[Google Scholar]**

35. HuMoD. Available online: **https://www.sim.informatik.tu-darmstadt.de/res/ds/humod/** (accessed on 30 March 2022).

36. Wojtusch, J.; Von Stryk, O. HuMoD—A Versatile and Open Database for the Investigation, Modeling and Simulation of Human Motion Dynamics on Actuation Level. In Proceedings of the 2015 IEEE-RAS 15th International Conference on Humanoid Robots (Humanoids), Seoul, Republic of Korea, 3–5 November 2015; pp. 74–79. **[Google Scholar]**

37. Deb, K.; Pratap, A.; Agarwal, S.; Meyarivan, T. A fast and elitist multiobjective genetic algorithm: NSGA-II. IEEE Trans. Evol. Comput. **2002**, 6, 182–197. **[Google Scholar] [CrossRef] [Green Version]**

38. HAWE Micro Fluid GmbH. Available online: **www.hawe.com** (accessed on 22 February 2022).

39. SOLIDWORKS Simulation. Available online: **https://www.solidworks.com/domain/simulation** (accessed on 1 June 2022).

40. Igus. Available online: **https://www.igus.co.uk/** (accessed on 3 July 2022).

41. Iglidur. Available online: **https://www.igus.co.uk/info/plain-bearings-x-material-data** (accessed on 3 July 2022).

42. Linjama, M.; Paloniitty, M.; Tiainen, L.; Huhtala, K. Mechatronic design of digital hydraulic micro valve package. Procedia Eng. **2015**, 106, 97–107. **[Google Scholar] [CrossRef]**

43. Megatron Kit Encoder Etx25K. Available online: **https://www.megatron.de/en/products/angle-sensors/encoder-etx25k-with-3d-hall-as-kit-versions.html** (accessed on 30 July 2022).

44. Autosen AP019 Pressure Sensor. Available online: **https://autosen.com/en/Process-Sensors/Pressure-sensors/Electronic-pressure-sensor-G1-4M-AP019** (accessed on 7 August 2022).

45. B&R X20 System. Available online: **https://www.br-automation.com/en/products/plc-systems/x20-system/** (accessed on 7 August 2022).

46. Yang, J.; He, Y.; Shi, P.; Yu, H. A review on human intent understanding and compliance control strategies for lower limb exoskeletons. Proc. Inst. Mech. Eng. Part I J. Syst. Control. Eng. **2022**, 236, 1067–1086. **[Google Scholar] [CrossRef]**

13

ENHANCING ENERGY EFFICIENCY IN HYDRAULIC HYBRID TRANSMISSIONS USING DIGITAL HYDRAULICS TECHNOLOGY

1. INTRODUCTION: AN OVERVIEW

Fluid power is a broad field of engineering technology that utilizes pressurized fluids to create, regulate, and transfer power. Hydraulic oil is commonly utilized because of its low freezing point, high boiling temperature, significant bulk modulus, and self-lubricating properties. Fluid power encompasses a distinct study field and academic pursuits that cater to many industries such as agriculture, construction, transportation, aerospace, marine, manufacturing, and other sectors that necessitate high power-to-weight ratios. Fluid power is known for its capacity to achieve high power densities, flexibility in transmission, ease of operation, low cost, controllability, and manageability. Fluid power technology is advantageous for carrying out important tasks that demand high power density and reliability, such as heavy-duty applications like construction trucks, material handling equipment, and military activities.

The fluid power sector is a crucial component of the United States economy. Thousands of companies in the United States use hydraulic power systems, as stated by the National Fluid Power Association (NFPA). These corporations are well-known as the leading industrial firms in the United States, employing over 845,000 workers with an annual payroll above $60 billion. The fluid power industry has various application areas. The four primary application areas of hydraulics and pneumatics are mobile hydraulics, industrial hydraulics, pneumatics, and aerospace.

Hydraulic hybrid technology, a component of mobile hydraulics, is recognized as a pioneering system in the automotive industry. Scholars extensively study this research issue to enhance the total efficiency of power transmission from a prime mover to the wheels. This technology enables significant fuel efficiency, energy savings, and several environmental benefits, making it a popular choice over traditional drivetrain designs [9]. Hydraulic hybrid drivetrains have numerous advantages compared to traditional manual/automatic gearboxes and continuously variable transmissions (CVTs). Hydraulic hybrid drivetrains consist of two power sources: a prime mover (internal combustion engine or electric motor) and a positive variable displacement pump/motor connected to a high-pressure storage device (accumulator). The variable displacement pump enables different speeds by controlling the flow rate, hence avoiding gearshift interruptions in traditional drivetrains (manual/automatic) when adjusting speed. Continuous Variable Transmissions (CVTs) do not encounter significant gearshift interruptions when adjusting speed. CVTs have restricted belt, sometimes referred to as chain, strengths that result in torsional moments, leading to excessive frictional forces and torsional moments. In addition to this benefit, incorporating a hydraulic accumulator in hydraulic hybrid drivetrains enhances power management [12]. The accumulator allows for the storage and release of energy, leading to increased overall efficiency and energy savings, as well as reduced carbon emissions and improved fuel economy.

Hydraulic hybrid drivetrains offer numerous advantages over electric hybrid transmissions, including their dominance over manual/automatic drivetrains and CVTs. Electric hybrid systems, like other systems, have a transmission source and an energy storage source (batteries). However, the batteries in electric systems can only utilize approximately 20% of the available energy during braking. The accumulators used in hydraulic hybrid transmissions can store and transmit a maximum of 80% of the braking energy, depending on the duty cycle. Performing this action while the vehicle is accelerating or decelerating might enhance fuel efficiency and lower pollutants [18,19]. It allows for the quick transmission of a significant amount of energy, resulting in increased power densities and power-to-weight ratios.

The benefits of hydraulic hybrid technology make it particularly suitable for demanding tasks such as heavy equipment production, road appliances, heavy load material handling equipment, and other high-power and energy-intensive equipment. Commercial vehicles, such heavy load vehicles engaged in high dynamic-drive-cycle activities, are now utilizing this gearbox technology. They transport heavy loads, making frequent pauses which necessitate increased braking and energy absorption. In addition to heavy-duty vehicles, certain utility vehicles meant to transport modest loads also utilize hydraulic hybrid technology. An example is the development of a sport utility vehicle in a research facility to showcase the hydraulic hybrid transmission concept in a passenger car. The vehicle was created to function in four distinct modes: hydrostatic driving, hybrid driving, a combination of hydrostatic and hybrid driving, and brake mode [22]. The chapter is structured as follows: Section 2 provides an overview of the hydraulic hybrid transmissions that currently exist. The text explains the various hydraulic hybrid powertrain configurations: series, parallel, and series-parallel, and details the operation of each. Section 3 showcases some significant applications of hydraulic hybrid drivetrains. Section 4 of this paper presents the problem statement, highlighting the shortcomings of the current hydraulic hybrid drivetrains. Furthermore, it emphasizes and presents the technique used to address these shortcomings. Section 5 provides a summary of digital hydraulics technology in fluid power. It includes the design of a cutting-edge digital pump, its fundamental operation, digital motoring/pumping data, and the ability to attain high efficiencies at various pump displacements. Section 6 illustrates the enhancement in efficiency of hydraulic hybrid transmission by including digital hydraulics. This was achieved through simulating a physical series hydraulic hybrid system using the specified digital motoring/pumping data. The document includes the circuit schematic that was used, a detailed simulation model, and the outcomes that were obtained. Section 7 ends with a concise overview of the key concepts derived from the study.

2. EXISTING HYDRAULIC HYBRID TRANSMISSION CONFIGURATIONS

There are three primary topologies of hydraulic hybrid transmission drivetrains: parallel, series, and series-parallel configurations. These hydraulic hybrid transmission setups utilize a variable displacement pump/motor, a high-pressure reservoir for energy storage (usually a nitrogen accumulator), and a low-pressure reservoir (fluid tank) [23]. The variable displacement pump/motor functions as a motor by turning hydraulic energy into mechanical energy to propel the vehicle. It functions as a pump by turning mechanical energy into hydraulic energy, which is then stored in the accumulator for future use.

The subsequent sections provide a concise overview of current hydraulic hybrid drivetrains (parallel, series, and series-parallel) and analyze their overall functionality.

2.1. Parallel Hydraulic Hybrid Configuration

A parallel hydraulic hybrid has a simple architecture, design, and implementation compared to other configurations [**24**]. It involves two sub-systems: a conventional mechanical transmission directly connected to the output shaft via a differential gearbox and a hydraulic transmission [**25**]. The sub-systems are coupled in parallel using a mechanical power transmitter like a gearbox, a chain, or a belt. **Figure 1** shows a parallel hydraulic hybrid configuration utilizing a belt mechanical power transmitter. As shown in **Figure 1**, the drivetrain consists of mechanical transmission (7), a clutch (2), and the prime mover (1), typically an internal combustion engine. The mechanical transmission (7) is directly connected to the engine (1) via the clutch component (2) and to the differential gearbox (8) on the drive shaft (9) to provide the needed power to each wheel. The hydraulic drivetrain comprises a variable displacement pump/motor (4), an accumulator (5), a fluid tank (6), and other hydraulic components, e.g., valves and fittings. The hydraulic pump/motor (4) is connected in parallel to the mechanical transmission (7) via the belt linkage (3). The clutch (2) allows engaging and disengaging the engine shaft from the rest of the system based on the accumulator's stored energy, thus controlling the engine on/off signal [**26**]. For instance, decoupling the engine from the system occurs when enough energy is in the accumulator to boost the wheels during acceleration. However, engaging the engine shaft to the transmission shaft is required to drive the wheels at low accumulator storage levels.

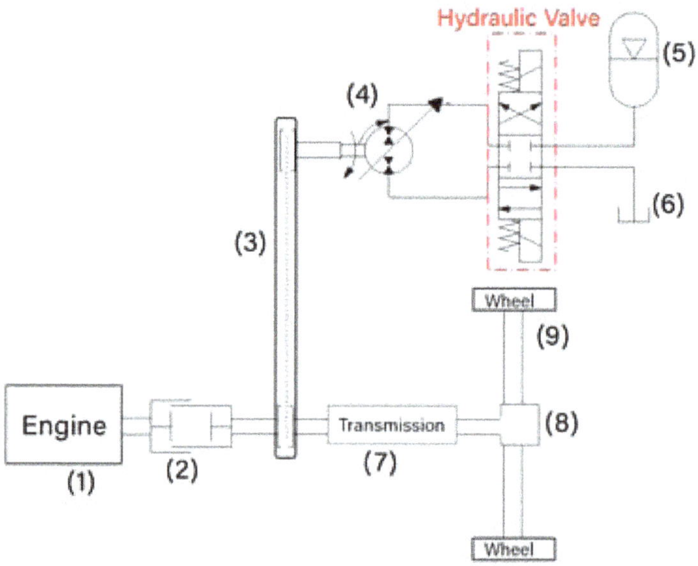

Item	Description
1	Engine
2	Mechanical Clutch
3	Mechanical Transmission Belt
4	Variable Displacement Motor/Pump
5	Gas Accumulator
6	Reservoir
7	Mechanical Transmission
8	Differential Gear Box
9	Output Shaft

Figure 1. Schematic for the parallel hydraulic hybrid configuration.

After introducing the basic concept beyond the parallel hybrid configuration, now, we briefly discuss its overall operation. The transmission duty cycle constitutes three main stages: acceleration, cruising, and braking. During acceleration, the accumulator acts as a hydraulic launch assist. It discharges its pressurized fluid to drive the hydraulic pump/motor (4) (operating as an energy source) and thus transforms the high-pressure energy into mechanical energy at the motor's shaft. The mechanical belt (3) transmits the generated mechanical power to the drive shaft (9), thus boosting the motion at the wheels. At the start of the cruising phase, the accumulator's pressure decreases below its pre-charged pressure, forcing the clutch to engage the engine shaft with the rest of the system to power the drive shaft (9) and drive the wheels. During the last stage, i.e., braking, the clutch (2) disengages the engine shaft from the drive shaft (9), the vehicle slows down. The rotary motion at the wheels is transmitted via the belt (3) to drive the hydraulic pump/motor (4) that will start acting as a pump to charge the accumulator. This cycle repeats every braking operation.

2.2. Series Hydraulic Hybrid Configuration

Unlike the parallel hydraulic hybrid configuration, the series hydraulic hybrid configuration does not utilize a mechanical transmission. Instead, it employs a variable displacement pump directly connected to the prime mover via a clutch [27]. As shown in **Figure 2**, in the series configuration, the mechanical transmission is replaced with a hydraulic variable displacement pump (3) connected in series to the hydraulic drivetrain consisting of a variable displacement pump/motor (6), accumulator (5), fluid tank (4), and other hydraulic components.

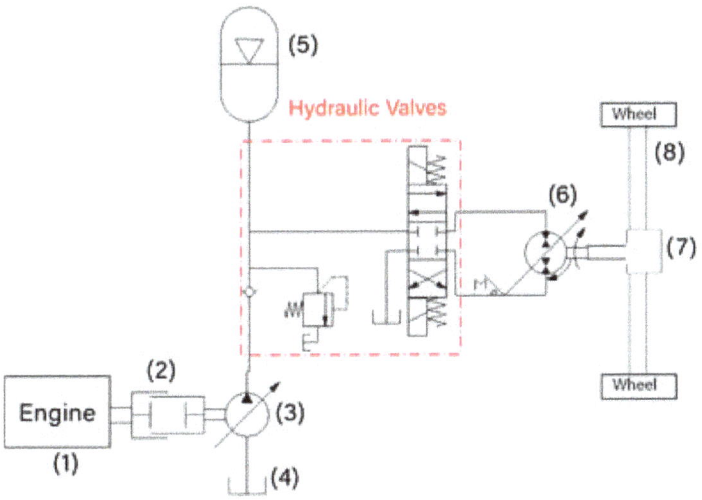

Item	Description
1	Engine
2	Mechanical Clutch
3	Variable Displacement Pump
4	Reservoir
5	Gas Accumulator
6	Variable Displacement Motor/Pump
7	Differential Gear Box
8	Output Shaft

Figure 2. Schematic for the series hydraulic hybrid configuration.

Similar to the parallel architecture, the operation comprises three main phases: acceleration, cruising, and braking. During the acceleration phase, the clutch (2) disengages the engine shaft from the pump shaft, and the pressurized hydraulic fluid stored in the accumulator (5) is delivered into the hydraulic pump/motor (6), allowing it to act as a motor. This operation converts the hydraulic energy in the accumulator into mechanical energy at the output shaft (8). After acceleration, cruising starts, and the fluid inside the accumulator drops to its minimum level. Its pressure decreases below the nitrogen's pre-charge pressure, forcing the engine (1) to operate and the clutch (2) to engage. Accordingly, the engine's energy is transmitted to the hydraulic pump (3), allowing the pump to deliver the fluid from the low-pressure reservoir (4) to the hydraulic pump/motor (6) to drive the wheels. The excess high-pressure fluid produced by the hydraulic pump is stored in the accumulator to be delivered when needed. During braking, the engine shuts down, the clutch disengages, and the hydraulic pump/motor (6) starts acting as a pump to push the fluid from the low-pressure reservoir into the accumulator. Thus, the kinetic energy from the wheels is converted into hydraulic energy stored in the accumulator. This stored energy will be used to accelerate the vehicle when needed.

2.3. Series-Parallel Hydraulic Hybrid Configuration

The series-parallel hydraulic hybrid system, also known as power-split, comprises the series and the parallel hydraulic hybrid layouts [**28**]. It integrates the two hydraulic hybrid configurations in one design [**29**]. Like the parallel configuration, the power-split architecture, presented in **Figure 3**, utilizes two sub-systems coupled via a mechanical transmitter, for example, a belt (3): a conventional mechanical drivetrain and a hydraulic drivetrain. The mechanical drivetrain employs a mechanical transmission (8) directly connected to the output shaft (10) via a differential gearbox (9) and the engine (1) via a mechanical clutch (2). The hydraulic drivetrain comprises two variable displacement pump/motors (4 & 5), accumulator (6), low-pressure reservoir (7), and other hydraulic components. Similar to the series architecture, the hydraulic transmission is directly connected to the output shaft (10).

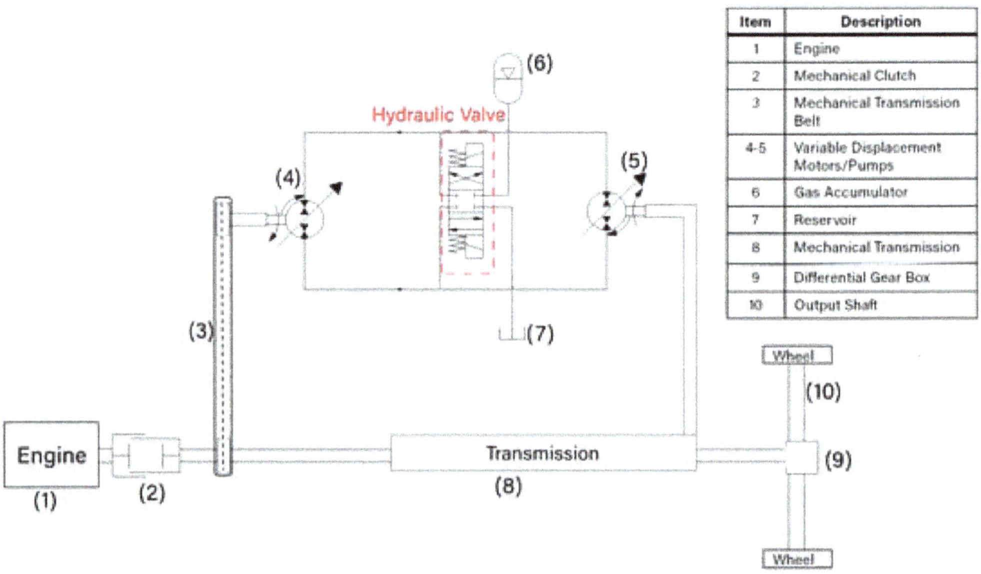

Figure 3. Schematic for the series-parallel hydraulic hybrid configuration.

This configuration enables operating the system in two different modes: pure hydraulic mode and hybrid mode. The pure hydraulic mode offers a simultaneous operation of the variable displacement pump/motors (4 & 5). However, the hybrid mode allows a seamless operation of the mechanical transmission (8) and the hydraulic drivetrain [**30**]. The seamless integration between the mechanical drivetrain and the hydraulic system allows transmitting the power from the engine to the wheels while maintaining optimal engine operation. This mechanism permits achieving high mechanical transmission performance in addition to the high-efficiency energy storage and regeneration, and thus operating the system at higher overall efficiency and lower fuel consumption. However, the design comes at the expense of increased system complexity, weight, and cost.

3. SUMMARY OF NOTABLE MOBILE APPLICATIONS USING HYDRAULIC HYBRID DRIVETRAINS

This section briefly reviews notable applications and proposed models for hydraulic hybrid drivetrains.

A comprehensive work presented by Stecki et al. [**31**] presented a review of the most notable computer simulation models of hydraulic hybrids available in the literature before 2005 and summarized the resulting simulation outcomes. The comprehensive report addresses the design, implementation, and simulation of more than eight prototype vehicles using CVT transmissions combined with hydrostatic regeneration. Most of the vehicles studied were on-road vehicles. The authors mentioned ADVISOR (National Renewable Energy Laboratory, Golden, CO, USA), the Advanced Vehicle Simulator developed by the National Renewable Energy Laboratory based in the US Department of Energy's (DOE's), as a simulation tool for developing hybrid drivetrain models. One of the vehicles discussed focused on a tactical vehicle, developed in MATLAB/Simulink which is created by MathWorks (an American privately held corporation based in California) and implemented in ADVISOR [**32**]. In this work, a Permo-Drive regenerative energy management system for an army tactical vehicle was modeled and simulated on MATLAB/Simulink. The model was then incorporated into the vehicle simulator (ADVISOR), a simulation tool for developing hybrid drivetrain models. The authors concluded their review by stating that extra fuel trails are required for examining the accuracy of the adopted fuel map and thus providing further validation of the model. Thy concluded that the accumulators might present a safety issue and recommended that their implementation require further accumulator technologies.

Besides the work of Stecki et al., the work of Stelson et al. [**10**] presents a simulation study of various hydraulic hybrid drivetrains comparing the three hydraulic hybrid configurations (series, parallel, and power-split). The prime mover is emulated as a combustion engine throughout their simulation studies. A dynamic programming approach is adopted, in which the fuel consumption was used as the optimization variable. The developed models reveal that the three configurations of the presented on-road vehicles (series, parallel, and power-split) improve fuel consumption for the two cycles: urban drive cycle and highway drive cycle. For the urban drive cycle, the three hydraulic hybrid configurations (series, parallel, and power-split) show an improvement of 14.7%, 53.2%, and 10.0%, respectively. Regarding the highway drive cycle, the series, parallel, and power-split architectures were enhanced by 8.9%, 11.9%, and 8.5%, respectively. Furthermore, throughout the simulation, the employed pump/motors were oversized. This forced the units to operate at partial displacements and thus led to a low overall hydraulic efficiency.

In 2004, Wu et al. **[33]** utilized the dynamic programming optimization methodology to optimize the accumulator's charging/discharging process for a Class VI parallel hydraulic hybrid delivery truck. The researchers used a 99.9 L accumulator operating at a maximum pressure of 36 MPa throughout this work. The proposed control algorithm involves two optimization parameters: the fuel consumption variable as the primary control parameter and the state of charge of the accumulator (SOC) as the secondary optimization parameter. The researchers estimated the fuel economy between 28% and 47%, depending on the efficiency of the hydraulic components. In 2009, Van Batavia **[34]** proposed a new control approach to monitor and regulate the hydraulic launch assist operation. The adopted strategy relies on operating the engine at its peak Torque-RPM performance. An approximate range of operation comprised between 1000 RPM to 5500 RPM and 125 N.m to 200 N.m for a Diesel engine rated for 110 kW capacity. The work claims that the proposed control strategy improved the fuel economy by 50% and reduced emissions by 40%. The proposed configuration used a variable displacement bent axis pump and a separate variable displacement bent axis motor. The bent axis units are known to have high efficiencies at 100% displacements of up to 98% **[35]**, yet this efficiency is only achievable at high displacements and is substantially lower at low displacements.

Buchwald et al. **[36]** introduced a Hydraulic—Internal combustion hybrid concept and four topologies for a passenger bus in Denmark. Besides the latter work, in 2008, Eaton Hydraulics **[37]** presented its hydraulic launch assist system for parallel drivetrain configurations. The objective was to improve power savings by supplementing engine power while launching the vehicle from a stopped position. The system was conceived for vehicles operating with heavy stop-and-go cycles, like refuse trucks, delivery vehicles, and buses. This system, developed in collaboration with the U.S. Environmental Protection Agency, depended on using a state-of-the-art pump/motor unit specifically designed for a parallel drivetrain configuration using an axial piston design.

Tollefson et al. **[38]** examined a new configuration of internal combustion engine/hydro-pneumatic accumulator energy storage for automobiles. The proposed architecture requires one single hydrostatic pump-motor (P/M) unit. Computer-based simulations were conducted over a wide range of parameter variations using a suitable control strategy. The simulation results revealed that the novel concept allowed significant fuel savings. The proposed hydraulic accumulator passenger car could achieve up to 65 mpg during city driving conditions, unlike the present automobiles that can attain only 25–30 mpg under similar conditions.

Similarly, Kepner [**39**] presents a Hydraulic Power Assist (HPA) demonstration on a full-size sport utility vehicle. It exhibits a description of the vehicle, the resulting fuel economy data, and performance and noise outcomes. The vehicle is constructed as a collaboration between Ford Motor Company, United States Environmental Protection Agency (EPA), and Advanced Technology Division, in the context of a Cooperative Research and Development Agreement (CRAD). It utilizes a valve block provided by Ifield Technologies, a pump/motor connected in parallel to the driveshaft with the conventional powertrain, and carbon fiber accumulators acting as HPA. Results show that the utilization of the HPA improved the Environmental Protection Agency (EPA) and the Heavier Acceleration city cycles by 23.6% and 35.5%, respectively. Bozic [**40**] presented a hybrid configuration comprising a combustion engine as the prime mover, hydrostatic transmission, and a coexisting vehicle electric generation. According to the literature consulted, it is evident that the implementation of hybrid powertrains significantly contributes to energy savings, in some cases, as high as 50% [**33,34**]. Yet, the system's efficiency may be compromised by the efficiency of the heart of the hybrid transmission; that is, the pump, the motor, or the pump/motor units. These units can operate at an efficiency as high as 98%, provided the units are operating at 100% displacement, but this may be the case of only a fixed segment of a duty cycle and may not be true at all operating conditions.

4. INADEQUACIES OF THE EXISTING HYDRAULIC HYBRID DRIVETRAINS

Despite the prominence of fluid power transmission, the existing state-of-the-art fluid power drivetrains are ineffective [**3**]. A study published by the United States Department of Energy shows that, among all industries, fluid power drives have low efficiencies between 9% and 60%, depending on the application, with an average efficiency not exceeding 22% [**8**]. The insufficiency in the fluid power transmissions' efficiencies is notably associated with the performance of the employed hydraulic units. The variable displacement pump/motor is an essential hydraulic unit employed in all hydraulic applications; it is the heart of fluid power systems [**41**]. Its efficiency significantly affects the performance and overall efficiency of fluid power transmission [**42,43**].

The conventional variable displacement pump/motors, i.e., existing state-of-the-art hydraulic pumps, are efficient only when operating at full displacements, whereas at partial displacements, their performance decreases until operating at low efficiencies down to 30% [**42,44,45**]. The insufficiency of the conventional pumps at partial or low displacements affects the system's overall efficiency due to the high pressure that may remain in the displacement chamber, in addition to leakage and other mechanical losses [**46**]. Given the limited efficiency of the conventional variable displacement pump/motors, the overall efficiency of the hydraulic hybrid drivetrains is relatively low, around 64–81% [**47**]. This critical problem has motivated many researchers in multiple fluid power applications to propose, design, and implement new techniques and methodologies to improve the performance of hydraulic hybrid transmissions by enhancing the maximum efficiency of variable displacement vane and piston pumps [**48,49,50,51,52,53,54,55**]. For instance, the work in [**49**] addresses the pre-defined problem by proposing a dynamic seven-stage model that predicts the thickness of the fluid film between the valve plate and cylinder block in axial piston pumps to get accurate information on lubrication. In [**51**], relying on a coating methodology, the authors improved the overall efficiency map of axial piston pumps by minimizing the friction losses by applying nanocoatings to the pump's slippers. Besides these works, the work in [**54**] provides research results for a previously executed research study. The resulting outcomes can be used to enhance the efficiency of many technological systems in power engineering and transport involving hydraulic pumps and motors. Similarly, the work in [**55**] focuses on improving the numerical algorithm used for solving the optimal control problem for the three-dimensional heat equation, which can improve many technological systems like hydraulic units.

All the proposed techniques in the pre-listed works still result in low overall efficiencies when operating at partial volumetric displacements. There is no noticeable improvement in the pumps' lowest efficiency at low volumetric displacements [**56**]. Accordingly, the work presented in [**42**] proposed integrating digital technologies into fluid power systems through designing digital variable displacement pump/motors that allow maintaining high efficiencies over a wide range of displacements.

Building on the results from [**41,42,57,58**], this work aims at enhancing the performance and efficiency of a hydraulic hybrid transmission by simulating the implementation of an experimentally-validated digital pump/motor into a hydraulic hybrid transmission. As will be proved through a physical-based simulation, unlike the ordinary variable displacement pump/motors, the adopted digital pump/motors pursue to overcome the efficiency limitation in hydraulic hybrid drivetrains by increasing the maximum overall efficiency of the transmission.

5. DIGITAL HYDRAULICS

Digital Hydraulics is a pioneering sub-field of fluid power, which relies on advanced on/off control technologies to enhance hydraulic systems' performance, reinforce advancements in fluid power applications, and minimize energy consumption [**59**]. This technology can replace conventional fluid power components, such as directional control valves, with multiple switching on/off valves in a parallel configuration to improve efficiency [**41**]. Such valves' configurations promote a superior design in this application because they use a poppet-style actuation that does not allow leakage across the valve, commonly seen in spool valves. In turn, this configuration reduces the system's losses, enables more control, increases the performance and efficiency, and reduces the cost of maintaining components [**60,61,62**]. This technology has also been utilized in various other configurations to improve the efficiency of conventional systems, including replacing traditional check valves on fixed displacement pumps. It also enables the opportunity to limit and divert flow in conventional fixed displacement pumps allowing variable displacements [**44**].

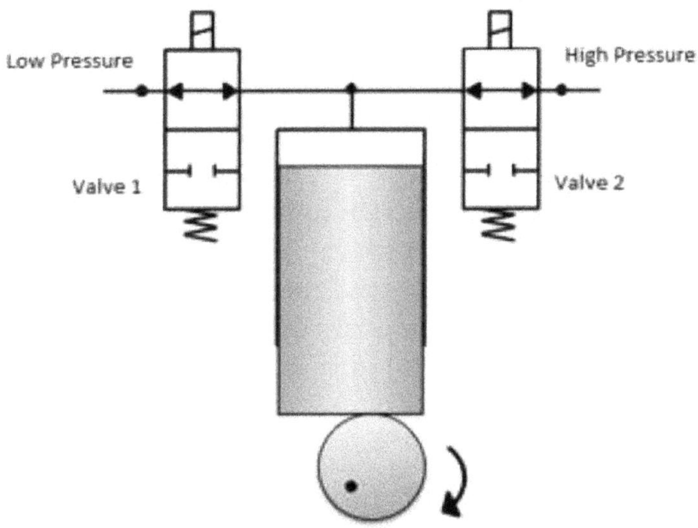

Figure 4. Single-piston digital pump/motor configuration [64].

Some conventional piston pump/motors, e.g., reciprocating piston pumps, utilize mechanical check valves on the inlet and outlet ports of the pump to separate the system's high- and low-pressure sides. During pumping, this configuration allows fluid to enter through the low-pressure check valve when the pump is on its intake stroke. When the pump exhausts its cylinder, pressure closes the intake check valve and opens the exhaust check valve, displacing fluid to the high-pressure side of the system. This configuration limits the pump to fixed displacements since the pressure differential in the chamber is the only method of controlling the actuation of the intake and exhaust valves. Thus, the work in [**42**] proposes an electrically actuated digital inline 3-piston pump that replaces the conventional valves with digital switching on/off valves. The proposed digital pump/motor is a 28cc/rev unit comprising three displacement chambers, each with two high-speed switching on/off valves. **Figure 4** shows the pump/motor's configuration for a single displacement chamber. The figure shows that Valve 1 controls the flow at the low-pressure side, and Valve 2 controls the flow at the high-pressure side. The proposed configuration allows opening or closing the valves at any time throughout the pumping/motoring cycle. This configuration adds a degree of controllability and flexibility to the system, enabling various digital pumping/motoring techniques, referred to as flow diverting and limiting operating strategies [**44,63**].

Besides the electrically actuated digital pump, our team designed and implemented a mechanically actuated digital pump, which can also achieve high efficiencies over a wide range of displacements. The following sub-sections exhibit an overview of the internal mechanical structure of the mechanically actuated digital pump, intro-duce the electrically actuated pump's most efficient operating strategy, and present the experimental testing and validation on the electrically actuated pump.

5.1. Internal Mechanical Structure of the Mechanically Actuated Digital Pump

This subsection gives a brief introduction to the mechanical structure of the mechanically actuated digital pump, which can achieve efficiencies higher than the ones reported in this paper. The internal mechanical structure of the intended digital pump, shown in **Figure 5**, comprises three main sub-systems: (1) gear-cam unit, (2) cam-follower unit, and the (3) valve manifold.

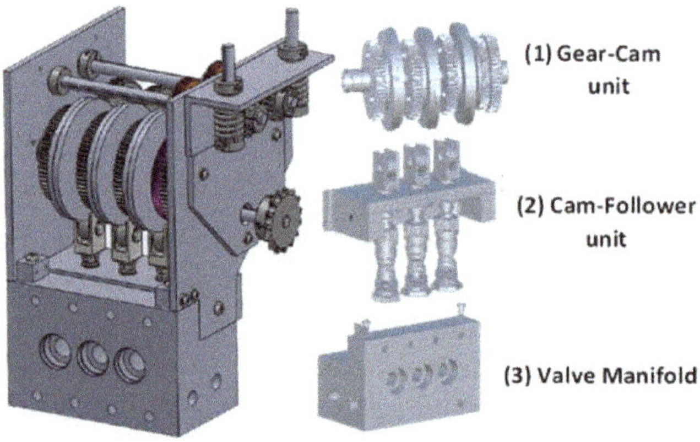

Figure 5. CAD drawing showing the basic units of the mechanically actuated digital pump.

The gear-cam unit utilizes three main half-masking cams, where each cam consists of two sub-cams that are phased relative to one another to allow achieving different cam profiles. This unit is responsible for the phasing mechanism of the camshafts, where it allows varying the cams' orientation relative to the shaft during rotation. The cam-follower unit consists of the three followers, their mount bracket, and the three valves to be actuated via the cam-follower mechanism. This unit allows converting the rotational motion at the cam to a linear motion at the follower and thus controlling the valves' opening and closing. Therefore, varying the cam profile using the phasing mechanism permits varying the opening and closing of the valves and thus enables achieving variable pump displacement. A detailed description of the mechanical structure and the pumping mechanisms is presented in [**65**].

5.2. Operating Strategies of the State-of-the-Art Electrically Actuated Digital Pump

The electrically actuated digital variable displacement pump/motor [**42**] relies on the control flow entering and leaving the displacement chambers. The pump/motor displacement variation can be achieved by controlling the actuation of two on/off valves on the intake and exhaust ports of the pump. The control and timing of the valves' actuation enable various pumping/motoring operating strategies. Four different operating strategies have been explored and tested [**42**] utilizing the state-of-the-art digital pump. The four different operating modes are Partial Flow-Diverting (PFD), Partial Flow-Limiting, Sequential Flow-Diverting (SFD), and Sequential Flow-Limiting (SFL). The most efficient operating strategy discovered is the SFL [**63**]. SFL is a technique that permits the states of each cylinder to be enabled or disabled, sending either 100% or 0% displacement to the output valve, respectively. This method can be used for both pumping and motoring cycles.

An example of a 100% displacement pumping cycle is shown in **Figure 6.** This cycle achieves similar functionality as the mechanical check valve. During pumping, the intake valve is opened for the entire stroke (A) and closed when the exhaust stroke begins (B). The exhaust valve is then opened and held until the exhaust stroke is completed (C). This transfers 100% of the fluid entering the piston chamber on the intake stroke, disregarding the dead volume in the cylinder.

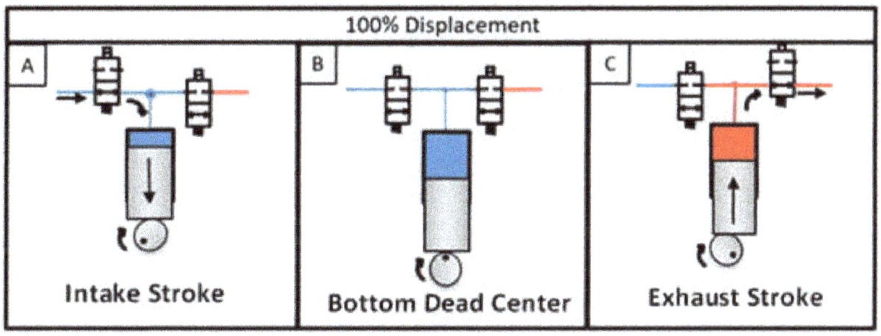

Figure 6. Enabled chamber operating at 100% displacement during SFL mode.

An example of 0% displacement for pumping is shown in **Figure 7**. During this cycle, all intake and exhaust valves remain closed for the entire process, allowing the fluid to be compressed and decompressed. As the cylinder begins its intake stroke, the piston starts to decompress the fluid in the dead volume of the cylinder (A) until the cylinder reaches the Bottom Dead Center (B), which allows depressurizing the fluid and creating a partial vacuum in the chamber. Once the piston begins the exhaust stroke (C), the fluid is compressed back to its original state. Since the intake and exhaust valves remained closed during this cycle, 0% displacement is delivered by the cylinder. Thus, this cycle allows the chamber's dead volume fluid to be depressurized and pressurized, and no fluid is delivered to the system.

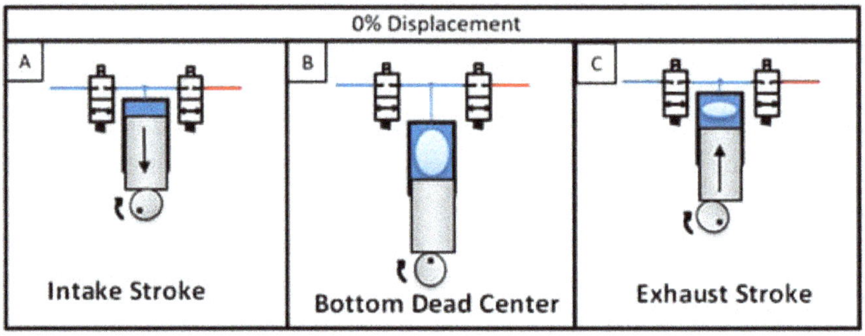

Figure 7. Disabled chamber operating at 0% displacement during SFL mode.

The partial vacuum created during the intake stroke is called chamber voiding. The chamber voiding phenomenon was experimentally inspected by **[66]**. It was experimentally proved that the gas created during the chamber voiding phase, i.e., (B), would dissolve entirely as the fluid is re-compressed.

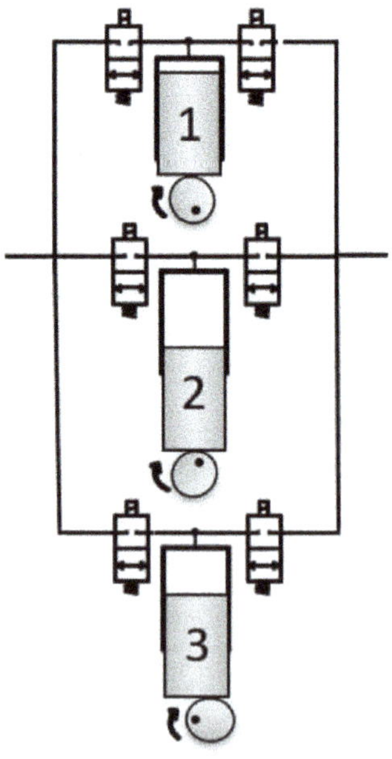

Figure 8. Three-piston digital inline piston pump configuration.

The central concept of this strategy relies on varying the state of the chambers se-quentially to change the 'ump's displacement. To illustrate, a three-piston pump configuration is shown below in **Figure 8**. To achieve 66% pumping displacement, two chambers (1 & 2) would operate at 100% displacement, while the third chamber (3) would run at 0% displacement.

Sequencing is averaged over various pump revolutions to achieve a wide range of displacements between 0, 33%, 66%, and 100% [**42**]. As a case in point, to achieve 11.11% displacement, the first cycle of the pump would turn on one cylinder (1) and turn off two cylinders (2 & 3). The following two pump revolutions would sequence all cylinders (1, 2 & 3) off. This operation achieves 1/9th, or 11%, displacement over the three cycles. A ~1% displacement resolution can be achieved by averaging the sequences over 32 shaft revolutions.

5.3. Experimental Validation of the Electrically Actuated Digital Pump/Motor

The experimental test stand used for testing the SFL operating strategy is shown in **Figure 9**. The test stand involves the intended 28 cc/rev digital pump, two accumulators (one at the suction side and one at the discharge side), manifold housing with six on/off valves (two for each piston chamber), and other hardware used to complete the hydraulic circuit, data acquisition, and control. The two accumulators were added to the system to reduce pressure and flow ripples and avoid cavitation problems during the suction phase. For instance, if the hydraulic system experienced sudden stops during suction, rapid changes in the system pressure will generate pressure ripples causing shock waves. In this case, the system's pressure will go above the accumulator's pre-charged pressure. The accumulator at the discharge port will act as a shock cushioning. It will allow the excessive flow to enter, compressing the gas cushion (nitrogen bladder) and thus cushion and minimize the shock. The second accumulator (one located at the suction port) is required because the pump is bi-directional, i.e., the pump can discharge fluid from either port. A detailed description of the hardware used is presented in [**66**].

The experimental testing of the SFL pumping/motoring strategy was conducted under steady-state conditions at a temperature T = 50 °C and different pump displacements, 25%, 50%, 75%, and 100% displacement. Moreover, the SFL pumping and motoring cycles were examined at different system pressures ranging from 4 MPa to 18 MPa and shaft speeds up to 700 RPM. Operating the digital pump at shaft speeds beyond 700 RPM requires high-speed switching valves, affecting the valves' variability and causing non-linearity and errors in the valves' timing (opening and closing time). A timing valve correction algorithm was developed in [**67**] and utilized throughout the experimental testing to overcome this problem and account for the valve timing error. The developed algorithm obtains the delay in the valve timing using the pressure readings at the low- and high-pressure ports, thus regulating the valves' opening and closing time. Although the adopted algorithm enabled high-speed switching, the valves' opening and closing speeds are not high enough. Thus, the maximum operating shaft speed was limited to 700 RPM throughout the research study.

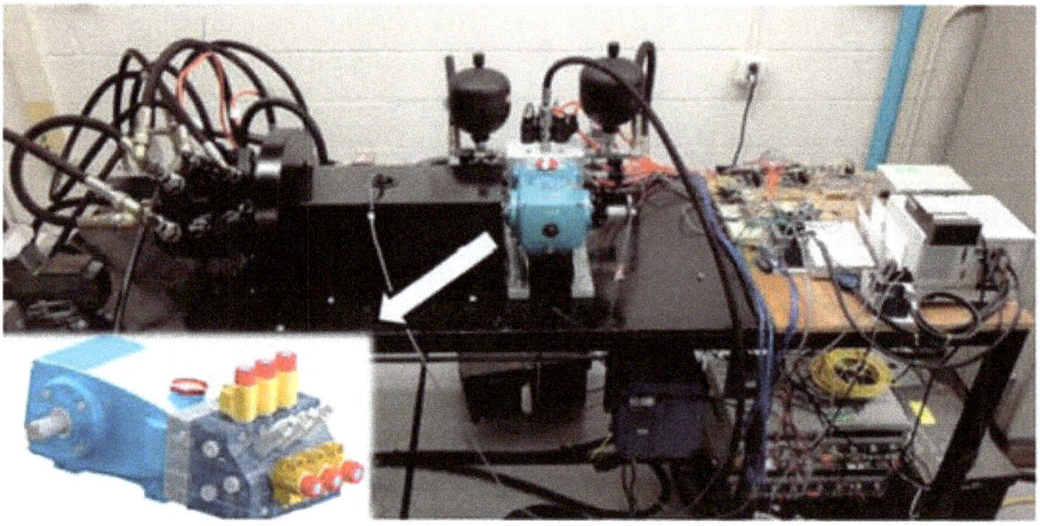

Figure 9. The experimental test stand for testing different operating strategies of the electrically actuated digital 3-piston pump [42].

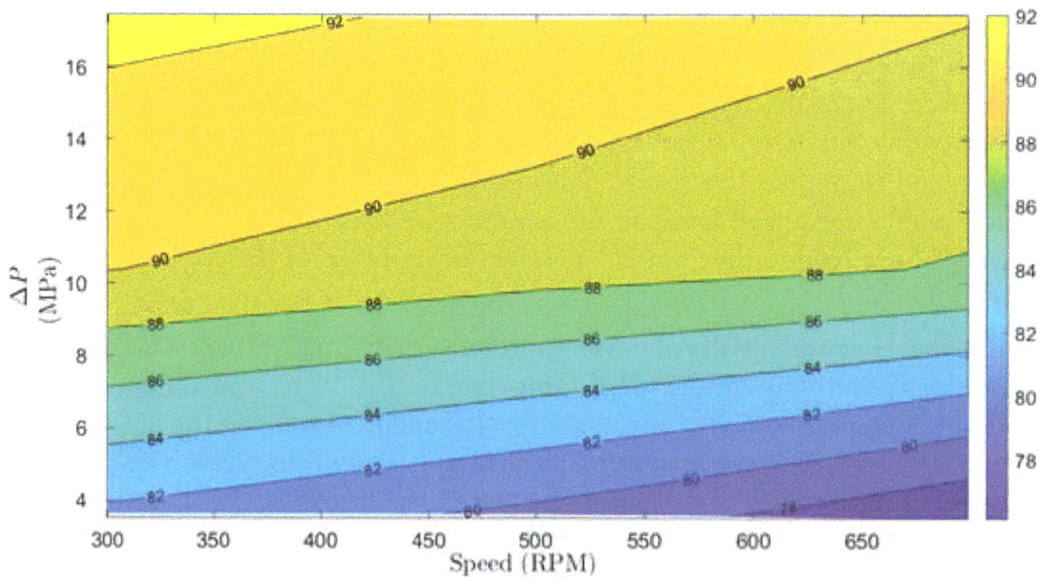

Figure 10. Efficiency map of the electrically actuated digital pump/motor operated in the SFL pumping strategy at 100% displacement.

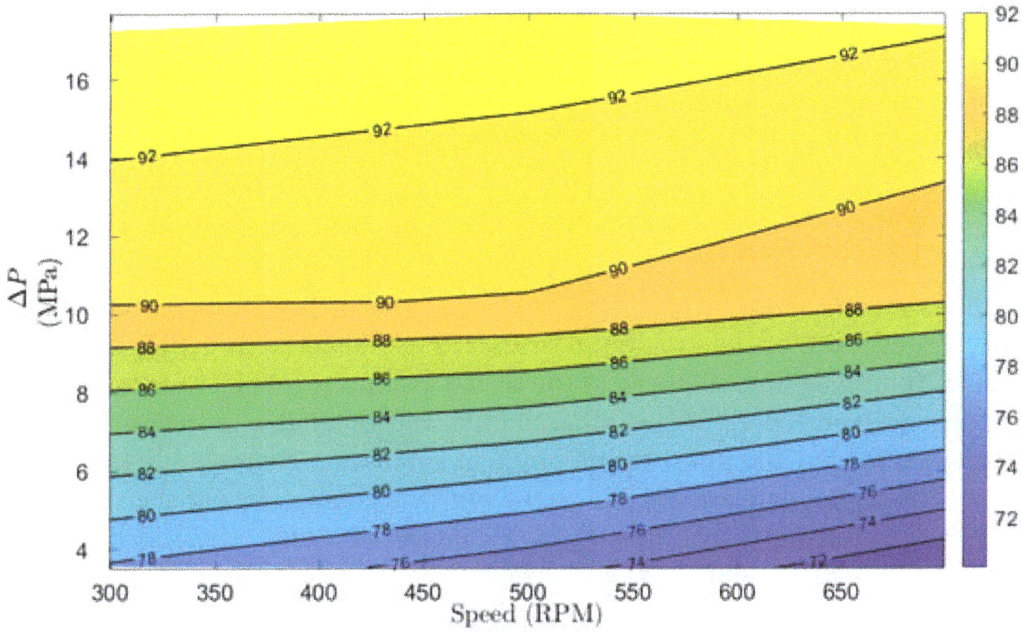

Figure 11. Efficiency map of the electrically actuated digital pump/motor operated in the SFL motoring strategy at 100% displacement.

After executing the experimental testing, the required data (torque, pressure, flow, etc.) were acquired at various pump displacements (25%, 50%, 75%, and 100% displacement), system pressures (4 MPa, 10 MPa, and 18 MPa), and shaft speeds (300 RPM, 500 RPM, and 700 RPM). Then, the efficiencies were calculated according to the obtained data.

The experimental outcomes resulting from operating the SFL strategy at 100% displacement during the pumping and motoring cycles are shown in **Figure 10** and **Figure 11**. **Figure 10** illustrates the calculated efficiency curves across various system pressures and shaft speeds at 100% displacement. As shown, at high shaft speeds and low pressures, the pump can achieve efficiencies of at least 78%, while at low shaft speeds and high pressures above 16 MPa, the pump can achieve high efficiencies up to 92%. Furthermore, the system pressure significantly impacts the overall pump's efficiency, where the pump's efficiency significantly increases as pressure increases.

Similarly, the efficiency map corresponding to operating the pump with the SFL strategy at 100% displacement during the motoring mode is shown in **Figure 11**. As shown in the figure, the motoring mode enables efficiencies of at least 72% and up to 92%. Unlike the results of the pumping mode, in motoring mode, the efficiency curves are almost parallel to the x-axis, which means that the shaft's speed has a negligible impact on the pump's efficiency. For instance, we can still achieve efficiencies from 72% to 92% at any RPM, depending on the system's pressure.

6. SIMULATION MODEL: HYDRAULIC HYBRID DRIVETRAINS UTILIZING DIGITAL PUMP/MOTORS

This section presents a physics-based series hydraulic hybrid drivetrain simulation conducted on MATLAB Simscape, a graphical programming software created by MathWorks [**68**]. The simulated MATLAB Simscape model mimics regular terrain hydraulic vehicles' series hydraulic hybrid transmission. The goal of the model is to investigate the improvement of the hydraulic hybrid transmission performance after replacing the conventional pump/motors with the digital pump/motor units. The intended state-of-the-art digital units are emulated by applying the digital motoring/pumping data acquired at the tested displacements (25%, 50%, 75%, and 100% displacement), pressures (4 Mpa, 10 MPa, and 18 MPa), and shaft speeds (300 RPM, 500 RPM, and 700 RPM) to the physical simulation components (pump and motor). The digital data are applied to the physical components of the simulation model by converting it into lookup tables imported to the physical pump and motor, employed throughout the simulation.

The following sub-sections present the designed circuit schematic, display the simulation model, design, and functionality during the three transmission stages (acceleration, cruising, and braking), and discuss the resulting outcomes.

6.1. Circuit Schematics

The adopted series hydraulic transmission, illustrated in **Figure 12**, comprises a variable displacement pressure compensated pump (5) driven by an electric motor (2), a mechanical clutch (3) to engage and disengage the motor's shaft from the rest of the system, a pressurized gas accumulator (1) for energy storage-boost purposes, a low-pressure reservoir (8), and a variable displacement pump/motor (11) for power transmission. The pump/motor (11) operates as a motor during acceleration and cruising, i.e., motoring mode, and works as a pump during braking, i.e., pumping mode. A check valve (6) is added to block the backflow (flow from and into the accumulator) from entering the pump's discharge port. Two 4/3 way (9 & 10) and one 3/2 way (7) directional flow control valves allow a seamless operation during the motoring and pumping modes. The 4/3 way valves (9 & 10) are closed-center valves with spring return having two directly actuated solenoid coils, while the 3/2 way valve (7) is a spring return valve with one actuated solenoid coil. As demonstrated through the simulation model, valve (9) is employed during the motoring mode, i.e., acceleration and cruising, where the user controls the valve based on the direction of motion, i.e., forward or backward drive. However, valve (10) is used during the pumping mode, i.e., regenerative braking, and it is controlled based on the direction of rotation received by the unit's shaft (11). Valve (7) allows switching between valves (9) and (10) based on the system's operation mode. During the motoring mode, i.e., acceleration and cruising, valve (7) is set at its un-actuated position, allowing pumping the fluid into the motor (11) through valve (9). However, during the pumping mode, i.e., at braking, the mechanical braking signal will be converted into an electrical signal at valve (7), forcing it to switch to the other position and thus allowing the motor (11) to pump the flow from the tank into the accumulator through the valve (10).

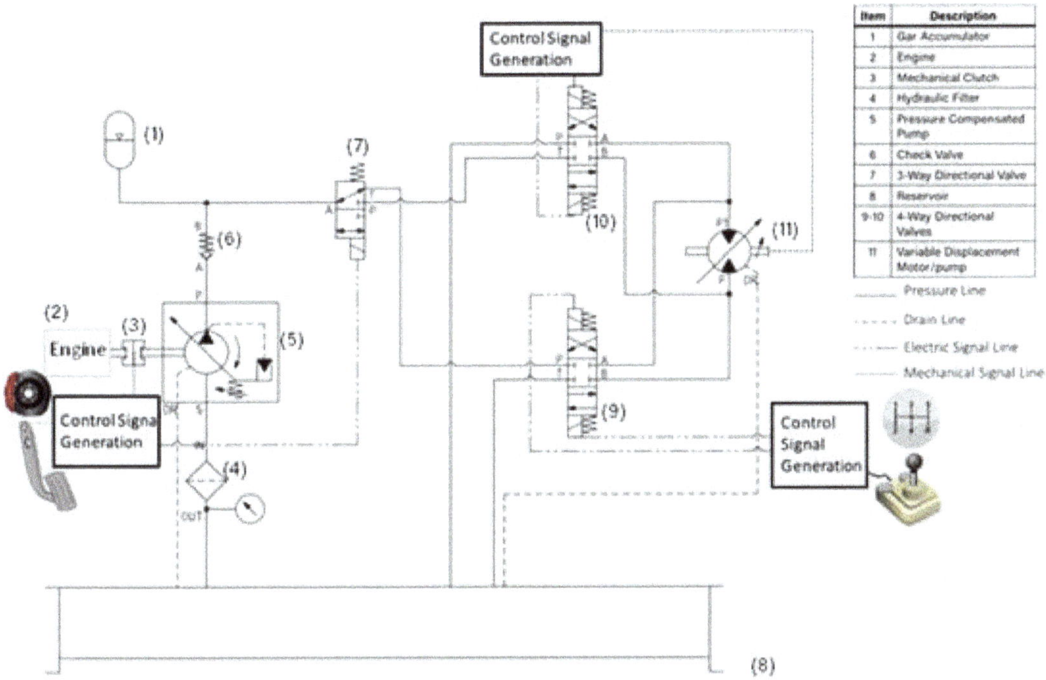

Figure 12. A circuit illustrating the simulated series hydraulic hybrid transmission.

6.2. Simulation Model, Design, and Operation

Figure 13 exhibits the physical model for the series hydraulic hybrid transmission illustrated in **Figure 12**. The adopted system is modeled and simulated using Simscape Fluids and Drivetrains libraries on MATLAB-Simulink. The simulation model is constructed by integrating the required numerical block diagrams with the physical paradigms. All the hydraulic physical components, i.e., pump, motor, valves, and other hydraulic units, are compiled using interfaceable physical connections. The figure shows that the model involves a pressure compensated pump and a variable displacement motor that operates as a motor when pumping high-pressure fluid into one of its ports and as a pump when receiving a mechanical signal at its output shaft. Furthermore, the model involves physical hydraulic sensors to sense the flow and the pressure at specific points in the system. For instance, pressure-flow sensors are installed at the pump's discharge port, accumulator's port, and motor's inlet and outlet ports. The accumulator's flow sensor is installed into the accumulator. Then, a positive flow rate at the accumulator's port designates flow into the accumulator; during charging. However, a negative flow denotes flow out of the accumulator; during discharging. Besides the pressure-flow sensors, mechanical sensors, e.g., rotational motion and torque sensors, are added at the pump's and motor's shaft to measure the resulting shaft's speed and torque. Acquiring this data (torque and speed) allowed us to compute the input and output mechanical powers. The power data was then exported into the MATLAB workspace for calculating the average input and output powers. The average input and output powers were individually calculated over the entire cycle using the mean function in MATLAB.

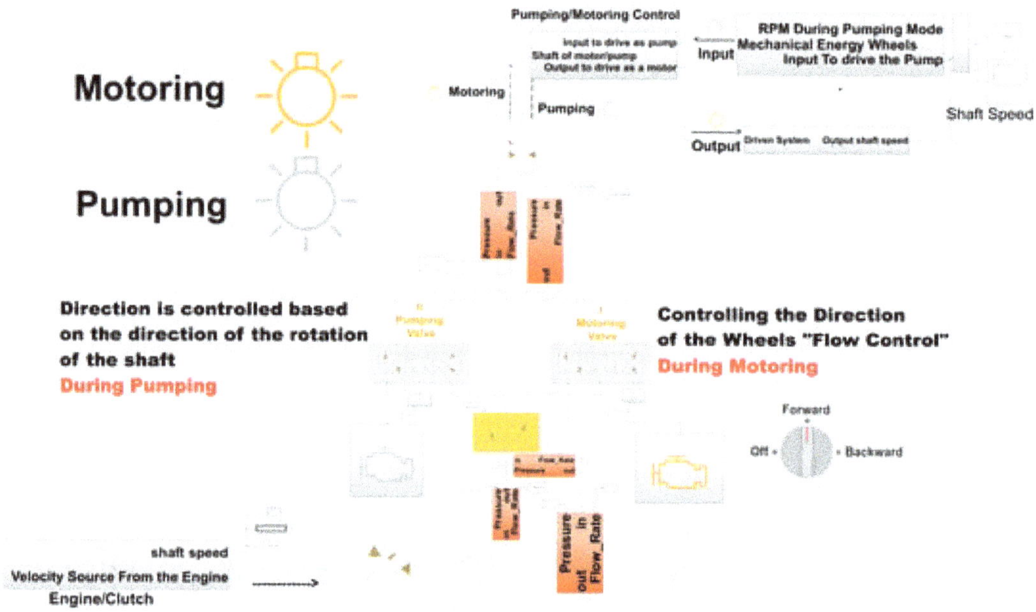

Figure 13. A MATLAB Simscape simulation model for the series hydraulic hybrid transmission.

The digital pumping/motoring data presented in **Section 5.2** are converted into lookup tables and then imported to the physical pumping/motoring units (the utilized pressure-compensated pump and the variable displacement motor) to incorporate the pre-introduced digital pump/motor. The pumping data is imported into the physical motor (11) during braking as a lookup table. Also, it is imported into the physical pump (5) during cruising, knowing that the pump will be off during acceleration. Similarly, the motoring data is applied to the physical motor as a lookup table during acceleration and cruising. Thus, the physical pump and motor (pressure-compensated pump and the variable displacement motor) are sized based on the imported lookup tables, which allows setting the maximum and minimum pressures in the system in addition to the required flow rate. Then, the accumulator and other hydraulic components, e.g., valves, fittings, hoses, etc., are sized and selected based on the acquired system's pressure and flow rate.

Based on the digital pumping/motoring data, the hydraulic units (5) and (11) will operate between 300 RPM and 700 RPM with a maximum displacement of 28 cc/rev. Furthermore, the hydraulic units will work under a volumetric efficiency up to 94% and overall efficiency of at least 78% and up to 90% during pumping. However, unit (11) will operate during motoring under volumetric and total efficiencies up to 99% and 92%, respectively. Furthermore, the simulation model accounts for the system's internal losses, e.g., losses in hydraulic pipelines, fittings, and valves. These internal losses are simulated by adding hydraulic resistance, capacitance, and inductance to the distributed parameter model using the simscape block models for hydraulic pipelines, fittings, and hoses.

During the series hydraulic hybrid transmission, the gas accumulator serves as a hydraulic lunch assist device to provide auxiliary flow during high demand requirements (within the acceleration stage), forcing the gas inside the accumulator to expand and compress under adiabatic conditions. Based on the intended application, the accumulator is sized using the following equations [**69**]:

$$P_1 V_1^k = P_2 V_2^k \quad (1)$$

$$V_1 = \frac{V_x (P_3/P_1)^{1/k}}{1 - (P_3/P_2)^{1/k}} \quad (2)$$

where $P1$, $P2$, and $P3$ in Pa denote the accumulator's pre-charged pressure, system maximum operating pressure, and system minimum operating pressure, respectively, set based on the experimentally validated data of the digital pump. $V1$ and $V2$ in L designate the nitrogen's volume at $P1$ and $P2$, respectively, Vx in L denotes the required fluid volume, and k=1.4 is the nitrogen constant during adiabatic operation. To prevent the accumulator's poppet (poppet that opens and closes, allowing fluid to enter and leave) from fully closing during the expansion of the nitrogen bladder, the accumulator's pre-charged pressure $P1$ has to be set between 60% to 80% of the minimum system pressure [**70,71**]. Thus, for $P2$ = 1.9e + 7 Pa, $P3$ = 5.3e + 6 Pa and $P1$ = 0.75$P3$, we get $V1$ = 36.3 L and Vx = 18 L.

The adopted physical model comprises two emulation options, options one and two shown in **Figure 14**, for simulating the intended system. Option one allows for a user interface simulation. The user can interact with the transmission model by switching between the motoring and the pumping modes, thus mimicking the driving and braking situations. Given this option, the user generates the duty cycle by choosing when to drive and brake. On the other hand, option two enables a complete auto-simulation over the three transmission stages (acceleration, cruising, and braking). The simulation will be done automatically based on the provided duty cycle for this option.

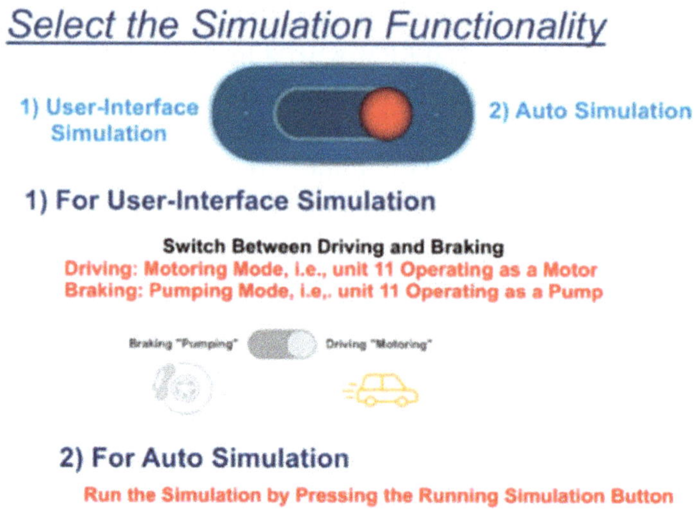

Figure 14. User interface illustrating the functionality of the simulation model.

Relying on option two, several drive cycles were adopted to test the simulation model's capability over the three transmission stages. The total time frame of the tested drive cycles ranged from 200 s/cycle to 300 s/cycle, where the acceleration and braking phases required long times up to 60 s and 40 s, respectively. The acceleration and braking timing is due to the accumulator's charging and discharging time, which was selected based on the experimentally acquired pump data that was limited with a maximum and minimum pressure around 19.1 MPa and 5.3 MPa, respectively.

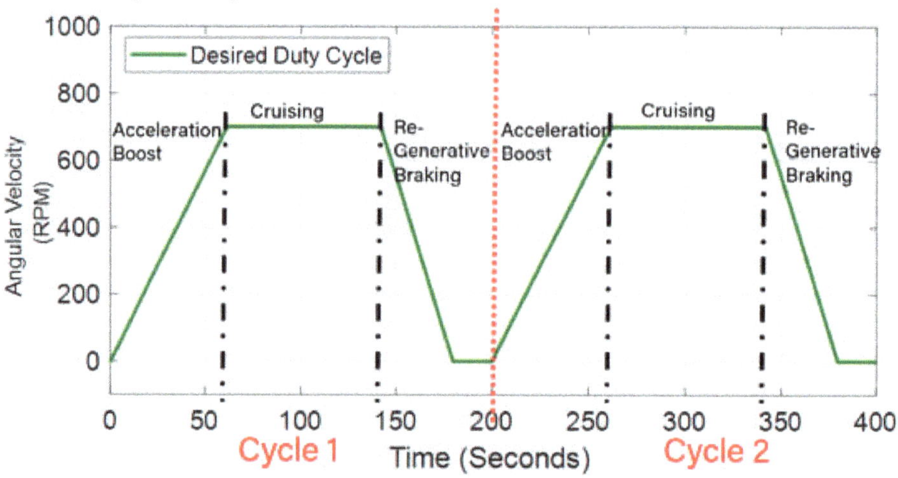

Figure 15. The desired duty cycle.

One of the tested duty cycles is shown in **Figure 15**. The figure shows that the physical system is simulated for 400 s for two consecutive cycles. Each cycle comprises the three stages of the series hydraulic hybrid transmission (acceleration, cruising, and braking). During the first cycle, the acceleration phase occurs between

0 and 60 s, the cruising starts at time = 60 s and continues until time = 140 s, and the re-regenerative braking takes place within 140 and 180 s. The second cycle starts 20 s after the first cycle, which accounts for the complete stopping of the vehicle. Similarly, the three transmission stages (acceleration, cruising, and braking) occur for 60, 80, and 40 s, respectively, during the second cycle.

To fully simulate the hydraulic hybrid series transmission system for achieving the desired duty cycle in **Figure 15**, it is required to apply the desired operating angular velocity at the system's input port and specify its load. The system's operating RPM is simulated by applying an input signal builder using a physical velocity source. The transmission's total mass was included within the physical paradigms of the utilized hydraulic units and all components in the Simscape environment. The transmission's load is simulated using four Simscape Tire blocks, which simulate the tires by their radius and ideal inertia. In [**72**], Valeski approximates the average tire's weight and diameter to be around 10 kg and 40 cm, respectively. Thus, the calculated moment of inertia on each tire was around 0.37 $kg{\cdot}m^2$. Given the calculated moment of inertia on each tire, a moderate physical torque source with an ideal moment of inertia of 1.5 $kg{\cdot}m^2$ was applied at the motor's shaft. Now, we explain the operation of the simulation model during each stage.

6.2.1. Acceleration

During acceleration, the accumulator is fully-charged with pressurized fluid, and its pre-charged pressure is greater than the system's operating pressure. At this phase, the pump is off, and thus the accumulator acts as an auxiliary power supply. The accumulator discharges its stored energy into the motor (11), operating in its motoring mode, to accelerate the vehicle. The valves' opening/closing operation is controlled based on the accumulator's pre-charged pressure. For instance, during this phase, because the pre-charged pressure of the nitrogen inside the accumulator is greater than the system's pressure, no signal is generated at the valve (7). Then, valve (7) is kept at its un-actuated position to pass the flow through the valve (9) and block it at valve (10).

6.2.2. Cruising

At the end of the acceleration phase, the fluid inside the accumulator starts decreasing, and the pre-charged pressure becomes below the system's pressure. Namely, the accumulator discharged almost all its stored energy, and thus the system requires an additional hydraulic power supply to drive the wheels. At this stage (cruising), the drop in the pre-charged pressure triggers an electric signal at the engine and forces the clutch to engage the engine to the rest of the system. The pump starts pumping the fluid from the low-pressure reservoir (hydraulic tank) into the hydraulic motor (11) that is still operating as a motor to execute the motion at the wheels. Like the acceleration phase, the fluid is delivered to the motor through the valve (9), where valve (10) is still closed.

6.2.3. Braking

During braking, the motor (11) switches to the pumping mode, and the clutch decouples the engine from the rest of the system, forcing the pump (5) to shut down. The pre-charged pressure of the nitrogen gas inside the accumulator drops below the system's pressure. At this stage, an electric signal is triggered at valve (7), forcing it to switch to the second position and thus pumping the fluid through valve 10 and blocking it at valve 9. The motor utilizes the braking kinetic energy at its output shaft to pump fluid from the tank into the accumulator converting the mechanical energy into hydraulic energy stored in the accumulator to be delivered during acceleration.

7. RESULTING OUTCOMES

After introducing the simulation model and discussing its fundamental operation, throughout the following sections: **Section 7.1** and **Section 7.2**, we present the simulation outcomes resulting from applying the digital motoring/pumping data into units (5) and (11) at moderate load-speed conditions (1.5 kg·m^2 and 700 RPM).

7.1. Pressure-Flow Outcomes

Figure 16 and **Figure 17** illustrate the variations in the system's pressure and flow throughout the proposed duty cycle. They show the pressure-flow outcomes at different ports in the system during the three transmission stages (acceleration, cruising, and braking). As shown in the figures, during the acceleration phase (0 to 60 s in cycle one and 200 to 260 s in cycle two), the pump's shaft angular velocity is kept at zero, then triggered after 40 s. The pump started operating after 40 s because the accumulator had enough stored energy at the start of the acceleration phase (from 0 to 40 s). Thus, the clutch disengages the engine's shaft from the pump's shaft, causing a zero RPM at the pump's input shaft. The accumulator acts as a hydraulic launch assist and discharges its stored energy into the motor. Thus, the flow rate at the accumulator's port flows in the negative direction, i.e., from the accumulator into the system, as shown in **Figure 16.** A pressure drop of about 10 MPa is created between the motor's inlet and outlet ports, as shown in **Figure 17,** allowing it to convert the hydraulic energy into mechanical energy enabling the wheels to accelerate until the output shaft speed reaches 600 RPM. As the accumulator delivers the auxiliary power, the fluid's pressure at the accumulator's port, shown in **Figure 16,** decreases because the accumulator's poppet is still open to deliver fluid to the system. After the accumulator discharges about 80% to 85% of its stored energy, i.e., at time = 40 s, the pump starts operating via the clutch-engine mechanism maintaining the system's output angular velocity at about 600 RPM, where its discharge pressure, captured in **Figure 16**, increases up to 6–7.5 MPa. After 20 s, at time = 60 s, the vehicle undergoes the cruising stage. The pump operates at its maximum speed (700 RPM) with an operating pressure of around 7.5 to 8 MPa. The motor's output shaft operates at almost moderate speed (500 RPM). During this stage, the pressure at the accumulator's port, shown in **Figure 16**, is the same as the system's pressure because the accumulator's poppet is closed. Thus, the accumulator is neither charging nor discharging. After 80 s, the vehicle undergoes the braking phase, in which the clutch starts disengaging the engine shaft from the pump shaft resulting in a decrease in its RPM. The motor (11) switches to its pumping mode during this phase, pumping the fluid from the tank into the accumulator.

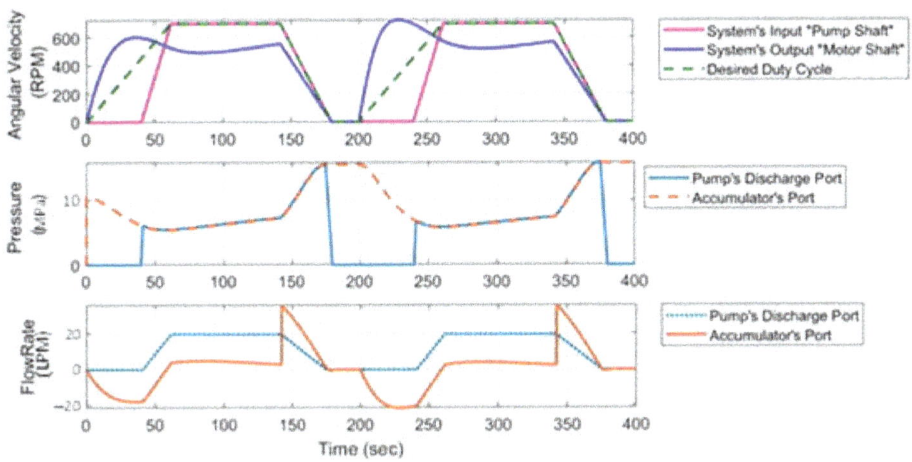

Figure 16. The resulting pressure and flow simulation outcomes at the pump's discharge and accumulator's ports throughout the desired duty cycle.

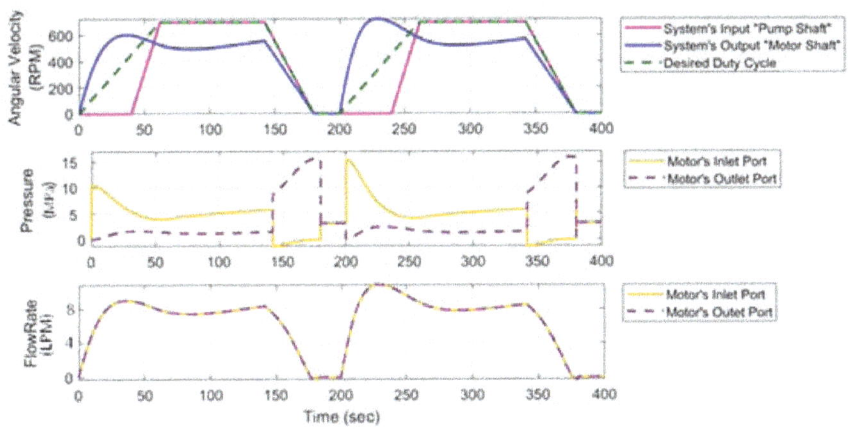

Figure 17. The resulting pressure and flow simulation outcomes at the Motor's inlet and outlet ports throughout the desired duty cycle.

The pressure at the motor's outlet port, shown in **Figure 17**, increases up to 17.5 MPa causing a high pressure at the accumulator's port and the pump's discharge port, shown in **Figure 16**. The accumulator starts charging, where its flow rate, shown in **Figure 16**, flows in the positive direction, i.e., from the system into the accumulator.

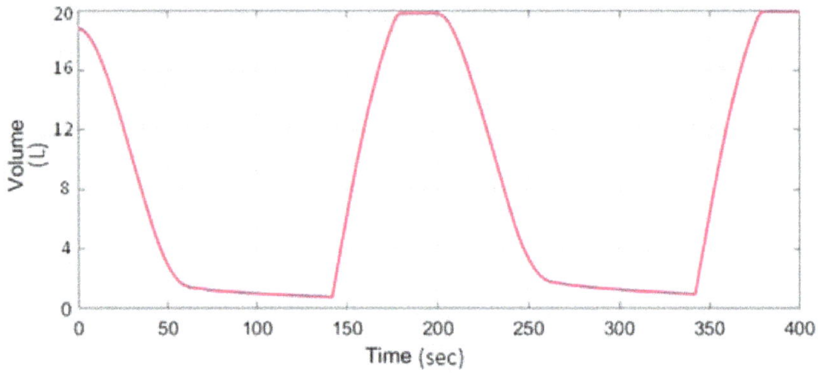

Figure 18. Simulation outcomes for the fluid volume inside the accumulator.

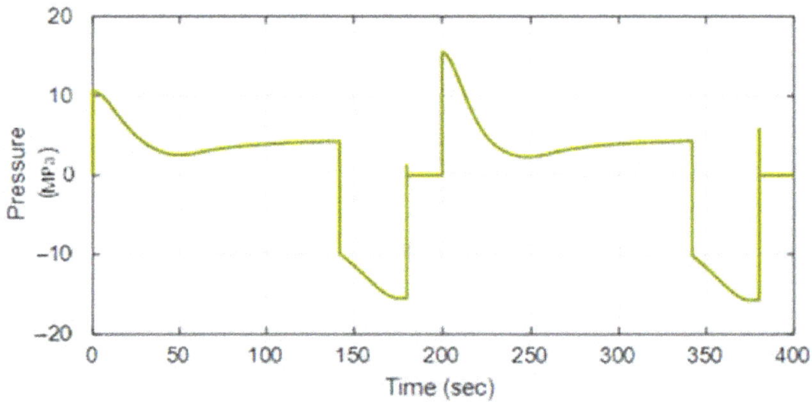

Figure 19. Simulation results for the pressure drop across the motor, i.e., the pressure difference between the inlet and outlet motor's pressure.

Figure 18 and **Figure 19** show the variations in the accumulator's fluid volume and the motor's pressure drop, respectively, throughout the desired duty cycle for the three transmission stages (acceleration, cruising, and braking). As discussed previously, during the acceleration phase, the accumulator discharges its stored energy into the motor's inlet port to accelerate the motion at the wheels. Thus, the motor operates in its motoring mode. The accumulator's discharging process causes a decrease in fluid volume, shown in **Figure 18**, and a positive pressure drop of about 10 MPa at the motor, shown in **Figure 19**. As shown in **Figure 19**, the 10 MPa pressure drop across the motor slightly decreases until 5–6.5 MPa. Then, because the accumulator is neither charging nor discharging during the cruising phase, its fluid volume, shown in **Figure 18**, is kept almost constant. During this phase, the pressure drop across the motor, shown in **Figure 19**, is maintained at around 7.5 MPa. During braking, the accumulator's fluid volume, shown in **Figure 18**, increases because the accumulator is charging. A negative pressure drop is created across the motor, as shown in **Figure 19**, where the motor is operating in its pumping mode.

7.2. Power-Efficiency Outcomes

Figure 20 shows the load-speed outcomes and the resulting mechanical power at the pump and motor shafts, i.e., the system's input and output shafts, respectively, throughout the three transmission stages (acceleration, cruising, and braking). During acceleration, the pump is not operating, and thus the accumulator is the only power supply in the system. The accumulator discharges its stored energy into the motor. The motor converts the hydraulic energy into mechanical energy resulting in an output mechanical power of around 1300 W in cycle one and about 2000 W in cycle two. The difference in the mechanical powers in cycles one and two is due to the accumulator's initial stored energy. During cruising, the pump starts operating at around 700 RPM, producing an input mechanical power of about 2100 W. This results in an output mechanical power of around 1000 W at the motor's shaft due to the 1.5 kg·m^2 inertia added at the wheels. During braking, the motor switches to its pumping mode. It utilizes the braking energy and converts it into hydraulic energy stored in the accumulator.

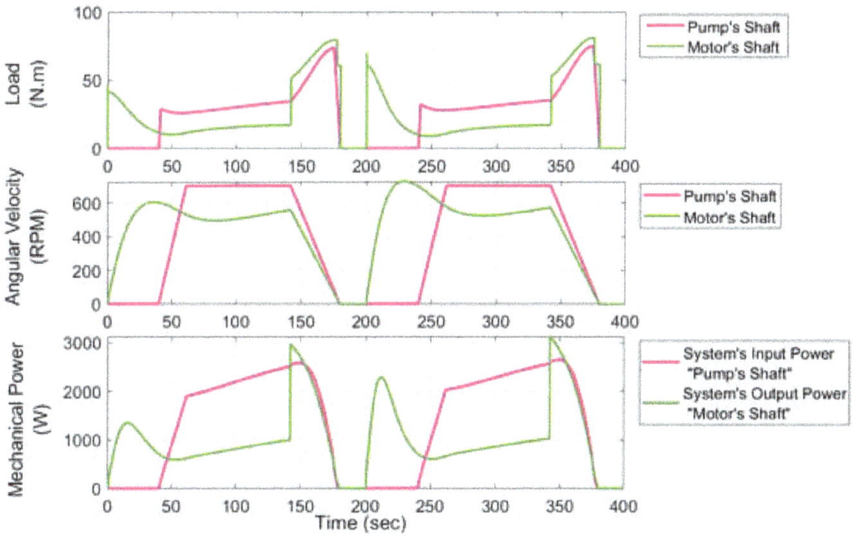

Figure 20. Load-Speed and mechanical power simulation results.

Simulating the system to achieve the desired duty cycle at moderate load-speed conditions (1.5 kg·m2 and 700 RPM) results in average input power of 876 W at the system's input shaft. This results in an average output power of 780 W at the system's output shaft, which outputs an overall efficiency of around 89%. This result reveals a significant improvement in the efficiency of the series hydraulic hybrid drivetrains. Due to the rated efficiency of the conventional hydraulic pump/motors, the total average efficiency of the series hydraulic hybrid transmission is around 64% to 81%, which is relatively low [47]. The use of the advanced multi-piston digital pump improved the system's performance by enhancing overall efficiency by a minimum of 8% and up to 25%.

8. CONCISE SUMMARY

Series hydraulic hybrid drivetrains typically exhibit an average efficiency ranging from 64% to 81%, mostly because they employ traditional variable displacement pumps that experience inefficiencies when operating at partial displacements. This study evaluates the enhancement in efficiency of a sequential hydraulic hybrid transmission through the integration of hydraulic hybrid technology with digital hydraulics. The suggestion is to substitute the traditional piston pump/motor used in series hydraulic hybrid drivetrains with the multi-piston digital pump created and utilized in previous studies. The suggested approach is evaluated by the implementation of a physical-based simulation of a hydraulic hybrid drivetrain series using MATLAB Simscape. The advanced multi-piston digital pump is replicated by inputting its digital driving and pumping data into the power supply units, specifically the pump and motor used in the simulation model. The simulation model was used to assess the performance of the hydraulic hybrid drivetrain series and to analyze the enhancement of the system's overall efficiency. The final simulation results showed an overall efficiency of 89%, surpassing the standard series hydraulic hybrid drivetrain's efficiency range of 64% to 81%. The use of advanced digital pump technology increases the overall efficiency of hydraulic hybrid drivetrains by a minimum of 8% and up to 25%. The 89% efficiency achieved can be further improved by utilizing a more sophisticated digital unit. The digital pump used in the simulation model was a prototype designed as a proof of concept, with a maximum shaft speed of 700 RPM. A refined model would operate effectively across a wider variety of inputs, leading to improved outcomes. Our future goal is to upgrade the proof-of-concept digital pump into a faster and more advanced unit for experimental testing of the planned transmission.

9. AT A GLANCE

Hydraulic hybrid drivetrains in automobiles offer a popular alternative to traditional drivetrain designs because of their significant energy efficiency, adaptable power transfer, and user-friendly operation. Hydraulic hybrid drivetrains have several environmental advantages in comparison to alternative power transmission solutions. They offer robust vehicles such as commercial transportation, construction equipment, wagon handling, drilling machines, and military trucks that have the capability to improve fuel efficiency and reduce carbon emissions.

Although hydraulic hybrid transmissions are common, modern hydraulic hybrid drivetrains have very low efficiency, ranging from 64% to 81%. The low efficiency is caused by using traditional variable displacement pumps and motors, which suffer from substantial power losses during the drive cycle and are unable to sustain good operating efficiency at lower volumetric displacements. The study introduces and confirms a novel approach to enhance the overall effectiveness of hydraulic hybrid drivetrains by substituting traditional pump/motor units with digital versions. The digital pump/motor can achieve higher overall efficiency than conventional pump/motors across a broad variety of operating circumstances. A prototype of a digital pump/motor was constructed and tested as a proof of concept. The experimental data were incorporated into a complex physics-based simulation model of a hydraulic hybrid transmission with many domains. The proposed methodology improves the overall efficiency of a hydraulic hybrid transmission series, leading to energy savings. By simulating the system under moderate load-speed circumstances, an overall efficiency of approximately 89% was attained. Our simulation results show that using the latest digital pump can increase the total efficiency of the series hydraulic hybrid powertrain by up to 25% compared to the average efficiency.

REFERENCES

1. McCloy, D.; Martin, H.R. Control of Fluid Power: Analysis and Design, 2nd ed.; Ehlh: Chichester, UK, 1980; Available online: **https://ui.adsabs.harvard.edu/abs/1980ehlh.book.....M/abstract** (accessed on 10 October 2021).
2. Durfee, W.; Sun, Z. Fluid Power System; A National Science Foundation Engineering Research Center: Minneapolis, MN, USA, 2009; pp. 1–48.
3. Vacca, A. Energy Efficiency and Controllability of Fluid Power Systems. Energies **2018**, 11, 1169. **[Google Scholar] [CrossRef] [Green Version]**
4. Konami, S.; Nishiumi, T. Hydraulic Control Systems; John Wiley & Sons: Singapore, 2016; pp. 187–299. **[Google Scholar] [CrossRef]**
5. Fabis-Domagala, J.; Domagala, M.; Momeni, H. A Concept of Risk Prioritization in FMEA Analysis for Fluid Power Systems. Energies **2021**, 14, 6482. **[Google Scholar] [CrossRef]**
6. What is Fluid Power? National Fluid Power Association. 2021. Available online: **https://www.nfpa.com/home/AboutNFPA/What-is-Fluid-Power.htm** (accessed on 1 December 2021).
7. U.S. Fluid Power Industry Brief. 2021. Available online: **https://www.nfpa.com/home/industry-stats/Industry-Brief.htm** (accessed on 1 December 2021).

8. Love, L.; Lanke, E.; Alles, P. Estimating the Impact (Energy, Emissions and Economics) of the US Fluid Power Industry. In Oak Ridge National Laboratory (Issue December). 2012. Available online: **http://www.osti.gov/servlets/purl/1061537/** (accessed on 1 December 2021).

9. Kwon, H.; Sprengel, M.; Ivantysynova, M. Thermal modeling of a hydraulic hybrid vehicle transmission based on thermodynamic analysis. Energy **2016**, 116, 650–660. [**Google Scholar**] [**CrossRef**]

10. Stelson, K.A.; Meyer, J.J.; Alleyne, A.G.; Hencey, B. Optimization of a passenger hydraulic hybrid vehicle to improve fuel economy. Proc. JFPS Int. Symp. Fluid Power **2008**, 2008, 143–148. [**Google Scholar**] [**CrossRef**] [**Green Version**]

11. Chen, J.-S. Energy Efficiency Comparison between Hydraulic Hybrid and Hybrid Electric Vehicles. Energies **2015**, 8, 4697–4723. [**Google Scholar**] [**CrossRef**] [**Green Version**]

12. Gary, J. Hydrid Transmission. Acad. Manag. Rev. **2006**, 31, 386–408. [**Google Scholar**]

13. Zhang, Z.; Chen, J.; Wu, B. The control strategy of optimal brake energy recovery for a parallel hydraulic hybrid vehicle. Proc. Inst. Mech. Eng. Part D J. Automob. Eng. **2012**, 226, 1445–1453. [**Google Scholar**] [**CrossRef**]

14. Rydberg, K.-E. Energy Efficient Hydraulic Hybrid Drives. In Proceedings of the 11th Scandinavian International Conference on Fluid Power, SICFP'09, Linköping, Sweden, 2–4 June 2009. [**Google Scholar**]

15. Tvrdić, V.; Podrug, S.; Šuljić, I.; Matić, B. Hydraulic hybrid vehicle configurations and comparison with hybrid electric vehicle. In Proceedings of the Comtemporary Issues in Economy & Technology—CIET 2018, Split, Croatia, 1–2 June 2018; pp. 548–556. [**Google Scholar**]

16. Florida, N.; López, C.; Pocomucha, V. CORE View metadata, citation and similar papers at core. ac. uk. J. Dedik. Pendidik. **2012**, 2, 35–43. [**Google Scholar**]

17. Shan, M. Modeling and Control Strategy for Series Hydraulic Hybrid Vehicles. Ph.D. Thesis, The University of Toledo, Toledo, OH, USA, December 2009. [**Google Scholar**]

18. Tavares, F.; Johri, R.; Salvi, A.; Baseley, S.; Filipi, Z. Hydraulic Hybrid Powertrain-In-the-Loop Integration for Analyzing Real-World Fuel Economy and Emissions Improvements. SAE Technical Paper. 2011. Available online: **https://doi.org/10.4271/2011-01-2275** (accessed on 20 October 2021).

19. Midgley, W.; Cebon, D. Comparison of regenerative braking technologies for heavy goods vehicles in urban environments. Proc. Inst. Mech. Eng. Part D J. Automob. Eng. **2012**, 226, 957–970. [**Google Scholar**] [**CrossRef**]

20. Li, C.-T.; Peng, H. Optimal configuration design for hydraulic split hybrid vehicles. In Proceedings of the 2010 American Control Conference, Baltimore, MD, USA, 30 June–2 July 2010; pp. 5812–5817. **[Google Scholar] [CrossRef]**
21. Kim, N.; Rousseau, A. A Comparative Study of Hydraulic Hybrid Systems for Class 6 Trucks. SAE Tech. Pap. **2013**, 2. **[Google Scholar] [CrossRef]**
22. Sprengel, M.; Bleazard, T.; Haria, H.; Ivantysynova, M. Implementation of a Novel Hydraulic Hybrid Powertrain in a Sports Utility Vehicle. IFAC-PapersOnLine **2015**, 48, 187–194. **[Google Scholar] [CrossRef]**
23. Bottiglione, F.; Mantriota, G.; Valle, M. Power-Split Hydrostatic Transmissions for Wind Energy Systems. Energies **2018**, 11, 3369. **[Google Scholar] [CrossRef] [Green Version]**
24. Jae, Y.; Filipi, Z.; Kim, Y.J.; Filipi, Z. Simulation Study of a Series Hydraulic Hybrid Prop System for a Light Truck. SAE Trans. **2021**, 116, 147–161. **[Google Scholar]**
25. Hui, S.; Junqing, J. Research on the system configuration and energy control strategy for parallel hydraulic hybrid loader. Autom. Constr. **2010**, 19, 213–220. **[Google Scholar] [CrossRef]**
26. Zhou, S.; Walker, P.; Tian, Y.; Zhang, N. Mode switching analysis and control for a parallel hydraulic hybrid vehicle. Veh. Syst. Dyn. **2020**, 59, 928–948. **[Google Scholar] [CrossRef]**
27. Sim, T.P.; Li, P.Y. Analysis and Control Design of a Hydro-Mechanical Hydraulic Hybrid Passenger Vehicle. In Proceedings of the Dynamic Systems and Control Conference, Hollywood, CA, USA, 12–14 October 2009; pp. 667–674. **[Google Scholar] [CrossRef] [Green Version]**
28. Cheong, K.L. Design and Analysis of Hydraulic Hybrid Passenger Vehicles. Ph.D. Thesis, University of Minnesota, Minneapolis, MN, USA, 2015. **[Google Scholar]**
29. Ramdan, M.I.; A Stelson, K. Optimal design of a power-split hybrid hydraulic bus. Proc. Inst. Mech. Eng. Part D J. Automob. Eng. **2016**, 230, 1699–1718. **[Google Scholar] [CrossRef]**
30. Sprengel, M.; Ivantysynova, M. Recent Developments in a Novel Blended Hydraulic Hybrid Transmission. SAE Tech. Pap. **2014**, 2014. **[Google Scholar] [CrossRef]**
31. Stecki, J.; Matheson, P. Advances In Automotive Hydraulic Hybrid Drives. Proc. JFPS Int. Symp. Fluid Power **2005**, 2005, 664–669. **[Google Scholar] [CrossRef] [Green Version]**

32. Matheson, P.; Stecki, J. Development and simulation of a hydraulic-hybrid powertrain for use in commercial heavy vehicles. SAE Tech. Pap. **2003**, 112, 114–123. **[Google Scholar] [CrossRef]**

33. Wu, B.; Lin, C.-C.; Filipi, Z.; Peng, H.; Assanis, D. Optimal Power Management for a Hydraulic Hybrid Delivery Truck. Veh. Syst. Dyn. **2004**, 42, 23–40. **[Google Scholar] [CrossRef]**

34. Van Batavia, B.L. Hydraulic Hybrid Vehicle Energy Management System; SAE Technical Paper; SAE International: Warrendale, PA, USA, 2009. **[Google Scholar] [CrossRef]**

35. Achten, P.A.J.; Schellekens, M.P.A.; Murrenhoff, H.; Deeken, M. Efficiency and Low Speed Behavior of the Floating Cup Pump. J. Commer. Veh. **2004**, 113, 366–376. **[Google Scholar] [CrossRef]**

36. Buchwald, P.; Christensen, G.; Larsen, H.; Sunn Pedersen, P. Improvement of citybus fuel economy using a hydraulic hybrid propulsion system—A theoretical and experimental study. SAE Tech. Pap. **1979**, 88, 1042–1056. **[Google Scholar] [CrossRef]**

37. Hydraulic Launch Assist. 2013. Available online: **https://www.eaton.com/SEAsia/ProductsSolutions/Hydraulics/ProductsServices/HydraulicLaunchAssist/index.htm** (accessed on 1 October 2021).

38. Tollefson, S.; Beachley, N.H.; Fronczak, F.J. Studies of an Accumulator Energy-Storage Automobile Design with a Single Pump/Motor Unit. SAE Trans. **1985. [Google Scholar] [CrossRef]**

39. Kepner, R.P. Hydraulic power assist—A demonstration of Hydraulic Hybrid Vehicle regenerative braking in a road vehicle application. SAE Tech. Pap. **2002**, 111, 826–833. **[Google Scholar] [CrossRef]**

40. Bozic, A. Introducing Hydraulic-Electric Synergy into Hybrid Transmission Using the Free-piston Engine Technology. In Proceedings of the Commercial Vehicle Engineering Congress and Exhibition and Powertrain & Fluid Systems Conference, Rosemont, IL, USA, 29 October–1 November 2007. **[Google Scholar] [CrossRef]**

41. Breidi, F.; Garrity, J.; Lumkes, J. Design and Testing of Novel Hydraulic Pump/Motors to Improve the Efficiency of Agricultural Equipment. Trans. ASABE **2017**, 60, 1809–1817. **[Google Scholar] [CrossRef]**

42. El-Breidi, F. Investigation of digital pump/motor control strategies. Ph.D. Thesis, Purdue University, West Lafayette, IN, USA, 2016. **[Google Scholar]**

43. Yang, H.-Y.; Pan, M. Engineering research in fluid power: A review. J. Zhejiang Univ. A **2015**, 16, 427–442. **[Google Scholar] [CrossRef] [Green Version]**

44. Merrill, K.J.; Breidi, F.Y.; Lumkes, J. Simulation based design and optimization of digital pump/motors. In Fluid Power Systems Technology; American Society of Mechanical Engineers: New York, NY, USA, 2013. [**Google Scholar**]

45. Li, M.; Foss, R.; Stelson, K.A.; Van De Ven, J.D.; Barth, E.J. Design, Dynamic Modeling, and Experimental Validation of A Novel Alternating Flow Variable Displacement Hydraulic Pump. IEEE/ASME Trans. Mechatron. **2019**, 24, 1294–1305. [**Google Scholar**] [**CrossRef**]

46. Wilhelm, S.R.; Van De Ven, J.D. Synthesis of a Variable Displacement Linkage for a Hydraulic Transformer. In Proceedings of the International Design Engineering Technical Conferences and Computers and Information in Engineering Conference, Washington, DC, USA, 28–31 August 2011. [**Google Scholar**] [**CrossRef**] [**Green Version**]

47. Chen, Y.L.; Liu, S.A.; Jiang, J.H.; Shang, T.; Zhang, Y.K.; Wei, W. Dynamic analysis of energy storage unit of the hydraulic hybrid vehicle. Int. J. Automot. Technol. **2013**, 14, 101–112. [**Google Scholar**] [**CrossRef**]

48. Ivantysynova, M. Innovations in pump design-what are future directions? Proc. JFPS Int. Symp. Fluid Power **2008**, 2008, 59–64. [**Google Scholar**] [**CrossRef**] [**Green Version**]

49. Zhang, C.; Huang, S.; Du, J.; Wang, X.; Wang, S.; Zhang, H. A new dynamic seven-stage model for thickness prediction of the film between valve plate and cylinder block in axial piston pumps. Adv. Mech. Eng. **2016**, 8. [**Google Scholar**] [**CrossRef**] [**Green Version**]

50. Shang, L.; Ivantysynova, M. A temperature adaptive piston design for swash plate type axial piston machines. Int. J. Fluid Power **2016**, 18, 38–48. [**Google Scholar**] [**CrossRef**]

51. Rizzo, G.; Massarotti, G.; Bonanno, A.; Paoluzzi, R.; Raimondo, M.; Blosi, M.; Veronesi, F.; Caldarelli, A.; Guarini, G. Axial piston pumps slippers with nanocoated surfaces to reduce friction. Int. J. Fluid Power **2015**, 16, 1–10. [**Google Scholar**] [**CrossRef**]

52. Inaguma, Y.; Hibi, A. Reduction of friction torque in vane pump by smoothing cam ring surface. Proc. Inst. Mech. Eng. Part C J. Mech. Eng. Sci. **2007**, 221, 527–534. [**Google Scholar**] [**CrossRef**]

53. Wieczorek, U.; Ivantysynova, M. Computer Aided Optimization of Bearing and Sealing Gaps in Hydrostatic Machines—The Simulation Tool Caspar. Int. J. Fluid Power **2002**, 3, 7–20. [**Google Scholar**] [**CrossRef**]

54. Sazonov, Y.A.; Mokhov, M.A.; Gryaznova, I.V.; Voronova, V.V.; Tumanyan, K.A.; Frankov, M.A.; Balaka, N.N. Development and Prototyping of Jet Systems for Advanced Turbomachinery with Mesh Rotor. Emerg. Sci. J. **2021**, 5, 775–801. **[Google Scholar] [CrossRef]**

55. Kostikov, Y.A.; Romanenkov, A.M. Approximation of the Multidimensional Optimal Control Problem for the Heat Equation (Applicable to Computational Fluid Dynamics (CFD)). Civ. Eng. J. **2020**, 6, 743–768. **[Google Scholar] [CrossRef]**

56. Manring, N.D. Valve-Plate Design for an Axial Piston Pump Operating at Low Displacements. J. Mech. Des. **2003**, 125, 200–205. **[Google Scholar] [CrossRef]**

57. Helmus, T.; Breidi, F.; Lumkes, J. Simulation of a variable displacement mechanically actuated digital pump unit. In Proceedings of the Eight Workshop on Digital Fluid Power, Tampere, Finland, 24–25 May 2016; pp. 95–106. **[Google Scholar]**

58. Chehade, A.; Breidi, F.; Pate, K.S.; Lumkes, J. Data-driven Adaptive Thresholding Model for Real-time Valve Delay Estimation in Digital Pump/Motors. Int. J. Fluid Power **2019**, 20, 271–294. **[Google Scholar] [CrossRef]**

59. Winkler, B. Recent Advances in Digital Hydraulic Components and Applications. In Proceedings of the The Ninth Workshop on Digital Fluid Power, Aalborg, Denmark, 7–8 September 2017; Available online: **https://www.et.aau.dk/digitalAssets/377/377525_id117-recent-advances-in-digital-hydraulic-components-and-applications.pdf** (accessed on 1 December 2021).

60. Brandstetter, R.; Deubel, T.; Scheidl, R.; Winkler, B.; Zeman, K. Digital hydraulics and "Industrie 4.0". Proc. Inst. Mech. Eng. Part I J. Syst. Control. Eng. **2016**, 231, 82–93. **[Google Scholar] [CrossRef]**

61. Peng, S.; Branson Iii, D.T.; Guglielmino, E.; Boaventura, T.; Caldwell, D.G. Performance Assessment of Digital Hydraulics in a Quadruped Robot Leg. 2012. Available online: **http://asmedigitalcollection.asme.org/ESDA/proceedings-pdf/ESDA2012/44861/227/4451304/227_1.pdf?casa_token=fWcoONtKGfUAAAAA:9c3YrOxPOrRp-PmWLcv8DUFtas4Qe6HnyCq02vGig-lSARq94FAcQnGOpL9W-XpcM-sdwGk** (accessed on 5 November 2021).

62. Pinto, L.P.G.; Belan, H.C.; Locateli, C.C.; Krus, P.; De Negri, V.J.; Lantto, B. New perspectives on digital hydraulics for aerospace applications. In Proceedings of the Aerospace Technology Congress, Solna, Stockholm, 11–12 October 2016. **[Google Scholar]**

63. Breidi, F.; Helmus, T.; Lumkes, J. The impact of peak-and-hold and reverse current solenoid driving strategies on the dynamic performance of commercial cartridge valves in a digital pump/motor. Int. J. Fluid Power **2015**, 17, 37–47. **[Google Scholar] [CrossRef]**

64. Breidi, F.; Chehade, A.; Lumkes, J. Monitoring Digital Technologies in Hydraulic Systems Using CUSUM Control Charts. In Fluid Power Systems Technology; American Society of Mechanical Engineers: New York, NY, USA, 2019. **[Google Scholar]**

65. Helmus, T. Investigation and Implementation of Mechanically Actuated Valves for Digital Hydraulic Units. Ph.D. Thesis, Purdue University, West Lafayette, IN, USA, 2017. **[Google Scholar]**

66. Holland, M.A. Design of Digital Pump/Motors and Experimental Validation of Operating Strategies Doctor of Philosophy. 2012. Available online: **http://www.purdue.edu/policies/pages/teach_res_outreach/c_22.html** (accessed on 1 December 2021).

67. Breidi, F.; Garrity, J.; Lumkes, J., Jr. Investigation of a real-time pressure based valve timing correction algorithm. In Fluid Power Systems Technology; American Society of Mechanical Engineers: New York, NY, USA, 2017. **[Google Scholar]**

68. Simscape—MATLAB & Simulink. 2021. Available online: **https://www.mathworks.com/products/simscape.html** (accessed on 6 October 2021).

69. Sizing Tool—Reasontek Corp. 2021. Available online: **https://www.reasontek.com/sizing-tool/** (accessed on 23 November 2021).

70. Cheema, J. Air and Oil Do Mix: Selecting the Right Accumulator Is Critical for Wind Farm Applications. Hydraulics and Pneumatic. 2013. Available online: **http://ezproxy.derby.ac.uk/login?url=http://search.ebscohost.com/login.aspx?direct=true&db=edsbl&AN=RN332050940&site=eds-live** (accessed on 28 November 2021).

71. Xu, Z.; Li, W.; Liu, X.; Chen, Z. Dynamic characteristics of coupling model of valve-controlled cylinder parallel accumulator. Mech. Ind. **2019**, 20, 306. **[Google Scholar] [CrossRef]**

72. Valeski, B. Average Tire Weight (With 10 Examples)—Survival Tech Shop. 2021. Available online: https://www.survivaltechshop.com/tire-weight/ (accessed on 1 December 2021).

INDEX

www.ingramcontent.com/pod-product-compliance
Ingram Content Group UK Ltd.
Pitfield, Milton Keynes, MK11 3LW, UK
UKHW052002051026
15624UKWH00003B/520

9 781787 153158